Galaxies at High Redshift

At the turn of the twenty-first century a golden era is occurring in observational cosmology. The new generation of large telescopes, combined with the capabilities of the Hubble Space Telescope and other space missions, allow astronomers to directly observe galaxy assembly over cosmic time. These developments demand a new generation of scientists trained in methods suited to the study of distant galaxies.

This timely volume contains the lectures delivered at the XI Canary Islands Winter School of Astrophysics, reviewing both scientific results and the main questions in the field. It covers the study of normal galaxies, distant galaxies, and studies based on far-infrared diagnostics; it reviews quasar absorption lines and the properties of nearby galaxies. Each chapter is written by a world expert in the field, making the book an essential reference for all astronomers working in the field of high-redshift galaxies.

Galaxies at High Redshift

XI Canary Islands Winter School of Astrophysics
Santa Cruz de Tenerife, Tenerife, Spain
November 15–26, 1999

Edited by: I. Pérez-Fournon, M. Balcells,
F. Moreno-Insertis and F. Sánchez

Instituto de Astrofísica de Canarias

PUBLISHED BY THE PRESS SYNDICATE OF THE UNIVERSITY OF CAMBRIDGE
The Pitt Building, Trumpington Street, Cambridge, United Kingdom

CAMBRIDGE UNIVERSITY PRESS
The Edinburgh Building, Cambridge CB2 2RU, UK
40 West 20th Street, New York, NY 10011-4211, USA
477 Williamstown Road, Port Melbourne, VIC 3207, Australia
Ruiz de Alarcón 13, 28014 Madrid, Spain
Dock House, The Waterfront, Cape Town 8001, South Africa

http://www.cambridge.org

© Cambridge University Press 2003

This book is in copyright. Subject to statutory exception
and to the provisions of relevant collective licensing agreements,
no reproduction of any part may take place without
the written permission of Cambridge University Press.

First published 2003

Printed in Great Britain at the University Press, Cambridge

Typeface CMR 10/11.75 pt   *System* LATEX 2_ε [TB]

A catalogue record of this book is available from the British Library

ISBN 0 521 82591 1 hardback

Contents

Participants	viii
Group Photograph	x
Preface	xiii
Acknowledgments	xv

Galaxy Formation and Evolution: Recent Progress

R. Ellis

1 Introduction	1
2 Galaxy Formation and Cosmology	2
3 Star Formation Histories	5
4 Morphological Data from HST	10
5 Constraining the Masses of Distant Galaxies	17
6 Origin of the Hubble Sequence	20
7 Conclusions	24
References	25

Galaxies at High Redshift

M. Dickinson

1 Introduction	29
2 How to Find High Redshift Galaxies	30
3 The Hubble Deep Fields	40
4 Properties of Distant Galaxies in the HDF and Elsewhere	44
5 Properties of Lyman Break Galaxies	48
6 Clustering at High Redshift	56
7 The Most Distant Galaxies	59
8 Future Directions	61
References	62

High-Redshift Galaxies: The Far-Infrared and Sub-millimeter View

A. Franceschini

1 Introduction	69
2 Dust in Galaxies	71
3 Evaluating the Dust Emission Spectra	77
4 Generalized Spectro-Photometric Models of Galaxies	79
5 Infrared and Sub-mm Line Spectra	83
6 IR Starburst and Ultra-Luminous Galaxies in the Local Universe	86
7 IR Galaxies in the Distant Universe: Pre-ISO/SCUBA Results	93
8 The Breakthrough: Discovery of the CIRB	94
9 Deep Sky Surveys with the Infrared Space Observatory (ISO)	100
10 Explorations of the Deep Universe by Large Millimeter Telescopes	105
11 Interpretations of Faint IR/mm Galaxy Counts	107
12 Nature of the Fast Evolving Source Population	111
13 Global Properties: The SFR Density and Contributions to the CIRB	119
14 Conclusions	124
References	125

Quasar Absorption Lines

J. Bechtold

   1 Introduction 131
   2 Analysis of Absorption Line Spectra 132
   3 Absorption by Material Associated with the QSO 138
   4 The Lyman Alpha Forest 140
   5 Damped Lyman Alpha Absorbers and other Metal-Line Absorbers 153
   6 Imaging of QSO Absorbers 165
   7 Future Prospects 167
   References 169

Stellar Population Synthesis Models at Low and High Redshift

G. Bruzual A.

   1 Introduction 185
   2 The Population Synthesis Problem 186
   3 The Isochrone Synthesis Algorithm 188
   4 Evolutionary Population Synthesis Models 189
   5 Stellar Ingredients 189
   6 Spectral Evolution at Fixed Metallicity 191
   7 Dependence of Galaxy Properties on Stellar Metallicity 193
   8 Calibration of the Models in the C-M Diagram 194
   9 Observed Color-Magnitude Diagrams and Integrated Spectra 197
   10 Comparison of Model and Observed Spectra 200
   11 Different Sources of Uncertainties in Population Synthesis Models . . . 200
   12 Summary and Conclusions 211
   References 219

Elliptical Galaxies

K. C. Freeman

   1 The Structure of Elliptical Galaxies 223
   2 The Rotation of Ellipticals 225
   3 The Fundamental Plane 228
   4 The Stellar Content of Elliptical Galaxies 231
   5 Dwarf Ellipticals 234
   6 The Formation of Ellipticals 236
   7 Globular Cluster Systems in Elliptical Galaxies 237
   8 The Inner Regions of Ellipticals 240
   References 243

Disk Galaxies

K. C. Freeman

   1 The Structure of Disk Galaxies 245
   2 Star Formation Law in Disks 248
   3 Star Formation History of Disks 249
   4 Dark Matter in Disk Galaxies 251
   5 The Tully-Fisher Law 253
   6 Bars 256
   7 Bulges 258

8 Pure Disk Galaxies	261
9 The Velocity Dispersion of the Gas Disks	262
10 S0 Galaxies	264
References	265

Dark Matter in Disk Galaxies

K. C. Freeman

1 Dark Halos in Spiral Galaxies	267
2 The Shapes of Dark Halos	269
3 Dark Matter in Dwarf Irregular Galaxies	270
4 Dark Matter in Our Galaxy	271
5 The Tully-Fisher Law	272
6 Dark Matter in High Velocity Clouds	272
7 Cusped Halos	272
8 Scaling Laws for Dark Halos	274
9 Summary	274
References	274

Participants

Acosta, José	Instituto de Astrofísica de Canarias (Spain)
Azzaro, Marco	Isaac Newton Group-ORM (Spain)
Balcells, Marc	Instituto de Astrofísica de Canarias (Spain)
Barden, Marco	Max-Planck-Institut für Extra. Physik (Germany)
Bechtold, Jill	University of Arizona (USA)
Bergmann, Marcel	University of Texas at Austin (USA)
Bon, Edi	Belgrade Astronomical Observatory (Yugoslavia)
Bongiovanni, Angel	CIDA (Venezuela)
Booth, Jane	University of Oxford (UK)
Bouche, Nicolas	University of Massachusetts (USA)
Bruzual, Gustavo	CIDA (Venezuela)
Cabrera, Fernando	Instituto de Astrofísica de Canarias (Spain)
Carrasco, E. Rodrigo	University of Sao Paulo (USP) (Brazil)
Cenarro, A. Javier	Universidad Complutense de Madrid (Spain)
Colbert, James	UCLA Astronomy (USA)
Conselice, Christopher J.	University of Wisconsin-Madison (USA)
Courty, Stephanie	Observatoire de Paris-Meudon (LAEC) (France)
Cristóbal, David	Instituto de Astrofísica de Canarias (Spain)
Crivellari, Lucio	Instituto de Astrofísica de Canarias (Spain)
Croft, Steve	University of Oxford (UK)
da Costa, Domingos	IAG - University of Sao Paulo (Brazil)
Delgado, Aránzazu	Universidad Complutense de Madrid (Spain)
Dickinson, Mark	Space Telescope Science Institute (USA)
Domínguez, Carlos	Instituto de Astrofísica de Canarias (Spain)
Drory, Niv	University of Munich (Germany)
Ellis, Richard S.	University of Cambridge (UK)
Falcón, Jesús	University of Nottingham (UK)
Franceschini, Alberto	Universitat di Padova (Italy)
Freeman, Kenneth C.	Mt. Stromlo Observatory (Australia)
Fogh, Lisbeth	Copenhagen University Observatory (Denmark)
García, Begoña	Instituto de Astrofísica de Canarias (Spain)
García, César E.	Universidad Complutense de Madrid (Spain)
Gilbank, David	University of Durham (UK)
González, Antonio C.	Kapteyn Astronomical Institute (The Netherlands)
González, Eduardo	Instituto de Astrofísica de Canarias (Spain)
Hughes, Mark	University of Sheffield (UK)
Jarvis, Matthew	University of Oxford (UK)
Jiménez, Elena	INTA (Spain)
Knudsen, Kirsten K.	Copenhagen University Observatory (Denmark)
Kurk, Jaron	University of Leiden (The Netherlands)
Loaring, Nicola	University of Oxford (UK)
Maier, Christian	Max-Planck Institut für Astronomie (Germany)
Marín, Antonio	Instituto de Astrofísica de Canarias (Spain)
Marri, Simone	Max-Planck-Institut für Astrophysik (Germany)
Martini, Paul	Ohio State University (USA)
Matute, Israel	Osservatorio Astronomico di Padova (Italy)
Mayen, Christophe	Observ. Midi-Pyrènès-Laboratoire d'Astroph. (France)
Milvang-Jensen, Bo	University of Nottingham (UK)

Monreal, Ana	Instituto de Astrofísica de Canarias (Spain)
Moreno–Insertis, Fernando	Instituto de Astrofísica de Canarias (Spain)
Noll, Stefan	Ruprecht-Karls-Universitaet Heidelberg (Germany)
Orozco, Verónica	INTA (Spain)
Papadopoulos, Padeli P.	University of Leiden (The Netherlands)
Pascual, Sergio	Universidad Complutense de Madrid (Spain)
Penny, Gail E.	University of Glasgow (UK)
Pérez, Pablo G.	Universidad Complutense de Madrid (Spain)
Pérez-Fournon, Ismael	Instituto de Astrofísica de Canarias (Spain)
Pignatelli, Ezio	SISSA (Italy)
Priddey, Robert	University of Cambridge (UK)
Quilis, Vicent	University of Durham (UK)
Rawlings, Steve	University of Oxford (UK)
Reuland, Michiel	University of Leiden (The Netherlands)
Richards, Eric	Arizona State University (USA)
Romano, Emilio	Kapteyn Astronomical Institute (The Netherlands)
Rosa, Daniel	INAOE (México)
Rubiño, José A.	Instituto de Astrofísica de Canarias (Spain)
Sharp, Robert	University of Cambridge (UK)
Solórzano, Carmen	University of Sheffield (UK)
Sokasian, Aaron	Harvard University (USA)
Stanic, Natasa	Belgrade Astronomical Observatory (Yugoslavia)
Sullivan, Mark	University of Cambridge (UK)
Tantalo, Rosaria	University of Padova (Italy)
Temporin, Sonia G.	University Innsbruck (Austria)
Toft, Sune	University of Copenhagen (Denmark)
Trujillo, Ignacio	Instituto de Astrofísica de Canarias (Spain)
von Kuhlmann, Bernd	Max-Planck Institut für Astronomie (Germany)
Weilbacher, Peter	Georg-August Universitaet Goettingen (Germany)
White, Simon	Max-Planck Institut für Astrophysik (Germany)
Wolf, Marsha J.	University of Texas at Austin (USA)

1	Marc Balcells	26	Kirsten K. Knudsen	52	David Cristóbal
2	Ezio Pignatelli	27	Begoña García Lorenzo	53	Lourdes González
3	Jaron Kurk	28	Michiel Reuland	54	Eduardo González
4	Niv Drory	29	Marcel Bergmann	55	José A. Rubio
5	Rosaria Tantalo	30	Padeli Papadopoulos	56	Ignacio Trujillo
6	Lisbeth Fogh Olsen	31	Bernd von Kuhlmann	57	Alberto Franceschini
7	Elena Jiménez Bailón	32	Israel Matute	58	Kenneth C. Freeman
8	Jesús Falcón Barroso	33	Marsha J. Wolf	59	Ismael Pérez-Fournon
9	Matthew Jarvis	34	Carlos Domínguez	60	Nieves Villoslada
10	Vicent Quilis	35	Christian Maier	61	Gustavo Bruzual
11	Sonia Giovanna Temporin	36	Christopher J. Conselice	62	Robert Sharp
12	Sergio Pascual	37	Mark Hughes	63	Robert Priddey
13	Daniel Rosa	38	Aránzazu Delgado	64	Edi Bon
14	Steve Rawlings	39	Carmen Solorzano Inarrea	65	Natasa Stanic
15	Simone Marri	40	Marco Barden	66	Aaron Sokasian
16	Pablo G. Pérez	41	Unknown	67	Gail E. Penny
17	Jane Booth	42	Jill Bechtold	68	Stefan Noll
18	Steve Croft	43	A. Javier Cenarro	69	José A. Acosta Pulido
19	Nicolas Bouche	44	Antonio Marín	70	A. César González García
20	Nicola Loaring	45	Mark Sullivan	71	Verónica Orozco de la Torre
21	Milvang-Jensen	46	Christophe Mayen	72	Stephanie Courty
22	Laura Colombol	47	Ana Monreal	73	Emilio Romano
23	Alister Graham	48	David Gilbank	74	Eleazar R. Carrasco Damele
24	Marco Azzaro	49	James Colbert	75	Peter Weilbacher
25	Ignacio de la Rosa	50	Sune Toft	76	Angel Bongiovanni
		51	Paul Martini		

Preface

The study of galaxy formation and evolution made a dramatic change in the last decade of the 20th Century. The high spatial resolution of the Hubble Space Telescope, combined with the large collecting power of new 8-10m class telescopes, have stretched tenfold the redshift domain over which the stellar light of normal galaxies can be analyzed. As a result, nearly 90% of the cosmic time is now accessible to direct observation. Observational cosmology, once dealing with the search for the basic cosmological parameters, puts increasing emphasis today on the study of the intrinsic properties of galaxies and their evolution with look-back time. We can study the wide range of diagnostics which have, over the years, built our understanding of galaxies in the Local Universe: galaxy morphologies, colors, luminosities, dynamical masses, metal content, gas content, star formation rates, clustering, merging activity. Theories of galaxy formation need not resort exclusively to examining the fossil record in nearby galaxies, but can be directly tested against observations of real galaxies at different cosmological epochs. Distant galaxies are now confirmed to be less massive and more irregular than present-day giant galaxies, in general confirmation with the predicions that galaxy assembly is a gradual process.

These developments, which are paving the way to what may become a golden era of observational cosmology, pose new challenges to researchers. Taking advantage of the new oportunities requires that we adjust research methods. Traditional observational projects are being replaced by large programs aimed toward the generation of extensive public databases, with much data becoming available for interpretative work. Resort to multi-wavelength information is becoming critical, both to obtain diagnostics on a wide range of physical phenomena and to cope with the redshift displacement and/or dust obscuration of the rest-frame energy distributions of samples observed at different cosmic epochs. Near-, far-infrared and sub-millimeter techniques are becoming increasingly important. Innovative methods need to be used to identify and select galaxy samples at the faintest limits. Strong selection effects operate on distant and near samples, and need to be well understood. Finally, the wealth of new information on distant galaxies requires an unbiased, precise knowledge of the properties of nearby galaxies so that evolution can be adequately gauged.

The XI Canary Islands Winter School of Astrophysics brought together eighty advanced graduate students, postdoctoral researchers and interested scientists for an up-to-date review of developments in the research of galaxies at high redshift, following lectures by world experts in the field. This book presents a compilation of the lectures given at the School. Both theoretical and observational lines of research are covered. Reviews are given of the science behind the redshift surveys that opened the domain up to z=1, and of the Hubble Deep Fields that mapped the Universe up to z$\sim$3, with an extensive discussion of the optical and near-infrared properties of Lyman break galaxies. The far-infrared and submm view of galaxy evolution is presented, to bring attention to the results of the recent FIR missions and to prepare the science that will become feasible with the millimeter arrays planned for the next decade. One chapter is devoted to diagnostics obtained from quasar absorption lines, covering the Lyman alpha forest, metal absorbers and damped Lyman alpha systems. On the theoretical side, the techniques of population synthesis which allow the comparison of galaxy formation models to observational data are extensively discussed. Finally, three chapters cover the galaxy properties in the Local Universe and the derived clues on galaxy formation: elliptical galaxies, spirals, and dark matter.

Each chapter comprises extensive graphical data, including pre-publication materials, and concludes with a long reference list.

Ismael Pérez-Fournon, Marc Balcells,
Fernando Moreno-Insertis & Francisco Sánchez
Instituto de Astrofísica de Canarias
September 2001

Acknowledgements

The organizers express their gratitude to the lecturers for their dedication to the School and for the preparation of the proceedings, as well as all the participants for making the School a success.

We acknowledge the financial assistance from the European Commission through the "Improving Human Research Potential and the Socio-economic Knowledge Base" Programme, from the Spanish Ministerio de Educación y Cultura and from the *Cabildo* of the Island of Tenerife.

The dedication of many persons in the IAC staff was essential to the School. The School Secretariat, Lourdes González and Nieves Villoslada, together with Mónica Murphy, ensured that everything ran smoothly. Begoña López and Carmen del Puerto of the press group put together the interviews to the professors that were published in the special issue of the *IAC Noticias* Newsletter. Jesús Burgos provided invaluable help with all the issues concerning the EC funding. Gotzon Cañada drafted the school poster, and Ramón Castro played an important role improving the quality of some figures for the book. Gabriel Gómez contributed his LATEX skills to formatting the chapters. Their dedication and enthusiasm is an essential ingredient to the Canary Islands Winter School program, and is kindly acknowledged.

Galaxy Formation and Evolution: Recent Progress

By RICHARD ELLIS

California Institute of Technology MS 105-24, Pasadena, CA 91125 USA

In this series of lectures, I review recent observational progress in constraining models of galaxy formation and evolution highlighting the importance advances in addressing questions of the assembly history and origin of the Hubble sequence in the context of modern pictures of structure formation.

1. Introduction

These are exciting times to be working on any aspect of studies of galaxies at high redshift whether observational or theoretical. Most would agree that the current period represents something of a *golden era* in the subject. Figure 1 shows the increasing extent to which articles concerned with galaxy evolution dominate the published literature over the past 25 years (gauged xenophobically I'm afraid by keyword statistics only in two North American journals).

To try and understand the cause for this prominence in the subject, the dates associated with the commissioning of some major observational facilities have been marked. The progress appears to have been driven largely by new kinds of optical and near-infrared data: faint counts and searches for primaeval galaxies in the late 1970's and early 1980's (Peterson et al. 1979, Tyson & Jarvis 1979, Kron 1980, Koo 1985), faint galaxy redshift surveys made possible by multi-object spectrographs in the late 1980's and early 1990's (Lilly et al. 1995, Ellis et al. 1996, Cowie et al. 1996, Cohen et al. 2000), the launch of Hubble Space Telescope (HST) and its revelation of resolved galaxy images to significant redshifts (Griffiths et al. 1994, Glazebrook et al. 1995, Brinchmann et al. 1998), the remarkable Hubble Deep Field image (Williams et al. 1996) and the plethora of papers that followed (Livio, Fall & Madau 1998) and the arrival of the Keck telescopes bringing a new wave of faint Lyman-break galaxy spectroscopy at unprecedented redshifts (Steidel et al. 1996, Steidel et al. 1999)†.

One often hears claims that a subject undergoing spectacular progress is one that is nearing completion (c.f. Horgan 1997). After all, the rise in Figure 1 clearly cannot continue indefinitely and fairly soon, it could be argued, we will then have solved all of the essential problems in the subject. As if anticipating this, a theoretical colleague gave a recent colloquium at my institute entitled *Galaxy Formation: End of the Road*!

Consider the evidence. Observationally we may soon, via photometric redshifts, have determined the redshift distribution, luminosity evolution and spatial clustering of sources to unprecedented limits. If one accepts photometric redshifts are reliable, the rate of progress in the traditional pursuit of $N(m, color, z)$ is limited solely by the field of view of the telescope and the exposure times adopted. Panchromatic data matching that obtained with optical and near-infrared telescopes from SIRTF, FIRST, and ALMA will

---

† A correlation was also made with three key international conferences (Larson & Tinsley 1978, Frenk et al. 1989, Livio, Fall & Madau 1998) but I was horrified to see that these appeared to have had a *negative* effect on the community's output! I assume this arose from a much-needed period of post-conference reflection!

also enable us unravel the cosmic star formation history $\rho_{SFR}(z)$ to unprecedented precision (Madau et al. 1996, Blain et al. 1999). It has already been claimed that the above data, e.g. $N(m, color, z)$ and $\rho_{SFR}(z)$, can be understood in terms of hierarchical models of structure formation where galaxies assemble through the cooling of baryonic gas into merging cold dark matter halos (CDM, Kauffmann et al. 1994, Baugh et al. 1998, Cole et al. 2000a).

The word 'concordance' was recently coined astrophysically in an article reconciling different estimates of the cosmological parameters (Ostriker & Steinhardt 1996). Such concordance in our understanding of galaxy evolution is a natural consequence of semi-analytical theories whose sole purpose is to explain the 'big picture' as realised with the extant galaxy data. In this series of lectures I want to show that we have our work cut out for some considerable time! Exciting progress is definitely being made, but observers must rise to the challenge of testing the fundamentals of contemporary theories such as CDM and theorists must get ready to interpret qualitatively new kinds of data that we can expect in the next decade.

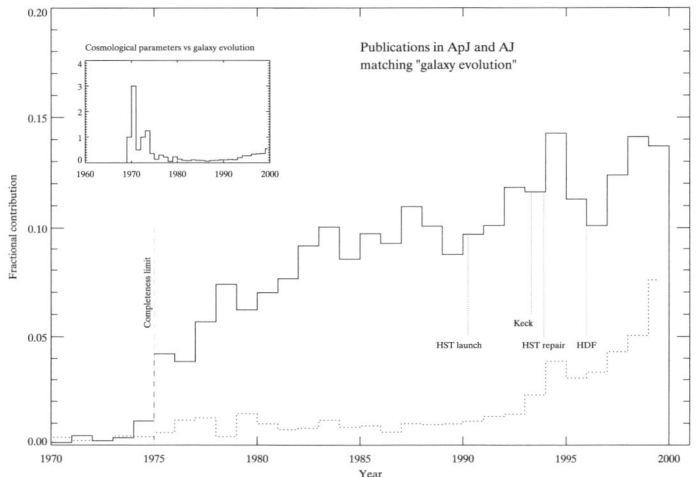

Figure 1: *The remarkably rapid growth in galaxy evolution studies: the fraction of the ApJ and AJ literature containing the key word 'galaxy evolution' over the past 25 years. The inset shows the marked decline in the use of galaxies as probes of the cosmological parameters during 1970-1980 (after Brinchmann, Ph.D. thesis 1999).*

These lectures are intended for interested graduate students or postdocs entering the field. There is an obvious observational flavor although I have tried to keep in perspective an ultimate goal of comparing results with recent CDM predictions. The bias is largely to optical and near-infrared applications; there is insufficient space to do justice to the rapidly-developing contributions being made at sub-millimetre, radio and X-ray wavelengths which other contributors at this winter school will cover in detail.

2. Galaxy Formation and Cosmology

Traditionally faint galaxies were studied in order to constrain the cosmological world model (Sandage 1961); their evolution was considered just one more tedious correction (the so-called *evolutionary correction*) in the path to the Holy Grail of the deceleration parameter q_0 ($\equiv \Omega_M/2$ in Λ=0 Friedmann models). The most useful galaxies in this respect were giant ellipticals in rich clusters. Tinsley (1976) demonstrated how sensitive the derived q_0 was to the assumed main sequence brightening with look-back time in these populations.

The traditional view for the formation history of an elliptical followed Eggen, Lynden-Bell & Sandage (1962). Monolithic collapse and rapid star formation leads to a subsequent track known as 'passive evolution' (i.e. without further star formation). Tinsley showed that main sequence brightening in such a stellar population is largely governed by the rate at which stars evolve off the main sequence, i.e. the slope $x(\simeq 1)$ of the initial mass function at the typical turnoff mass 0.4–$1 M_\odot$. Whence:

$$E(z,t) = dM_v/d \ln t \sim 1.3 - 0.3\,x \qquad (2.1)$$

and, in terms of its bias on q_0:

$$\Delta\,q_0 = 1.4 (H_0\,t_o)^{-1}\,d M_v/d \ln t = 1.8 - 0.42\,x \qquad (2.2)$$

Tinsley argued that one would have to know the evolutionary correction to remarkable precision get a secure value of q_0. In fact, noting that the difference in apparent magnitude for a standard candle at $z=1$ between an empty and Einstein-de Sitter Universe is only $\simeq 0.5$ mag, the relative importance of cosmology and evolution can be readily gauged.

Despite the above, it is always a mystery to me why several of our most eminent astronomers (Kristian et al. 1978, Gunn & Oke 1975) continued to pursue the Hubble diagram as a cosmological probe using first-ranked cluster galaxies, in some cases for several years after the challenge of resolving the evolutionary correction became known. Tammann (1985) estimated about 400 nights nights of Palomar 200-inch time was consumed by the two competing groups whose resulting values of q_0 fundamentally disagreed. Recently Aragón-Salamanca (1998) showed, in a elegant summary of the situation, how the modern K-band Hubble diagram is most likely complicated further by the fact that first-ranked cluster galaxies are still assembling their stars over the redshift interval $0 < z < 1$, offsetting the main sequence brightening (Figure 2).

In the late 1970's therefore, the motivation for studying faint galaxies became one of understanding their history rather than using them as tracers of the cosmic expansion (see inset panel in Figure 1). This is not to say that uncertainties in the cosmological model do not affect the conclusions drawn. The connection between cosmology and source evolutions remains strong in three respects:

(*a*) We use our knowledge of stellar evolution to predict the past appearance of stellar populations in galaxies observed at high redshift. However, stellar evolution is baselined in *physical time* (the conventional unit is the Gyr: 10^9yr), whereas we observe distant sources in *redshift* units. The mapping of time and redshift depends on the world model. Broadly speaking there is less time for the necessary changes to occur in a high Ω_M universe and consequently evolutionary trends are much stronger in such models.

(*b*) Many evolutionary tests depend on the *numbers* of sources, the most familiar being the number-magnitude count which is remarkably sensitive to small changes in source luminosity. However, the relativistic volume element $dV(z)$ depends sensitively on curvature being much larger in open and accelerating Universes than in the Einstein-de Sitter case.

(*c*) Predictions for the mass assembly history of a galaxy in hierarchical models depend also on the cosmological model in a fairly complex manner since these models jointly satisfy constraints concerned with the normalisation of the mass power spectrum via the present abundance of clusters (e.g. Baugh et al. 1998). Figure 3 illustrates one aspect of this dependence (Kauffmann & Charlot 1998a); structure grows more rapidly in a dense Universe so the decline with redshift in the abundance of massive spheroidal galaxies, which are thought in this picture to forms via mergers of smaller systems, is much more marked in high density models than in open or accelerating Universes.

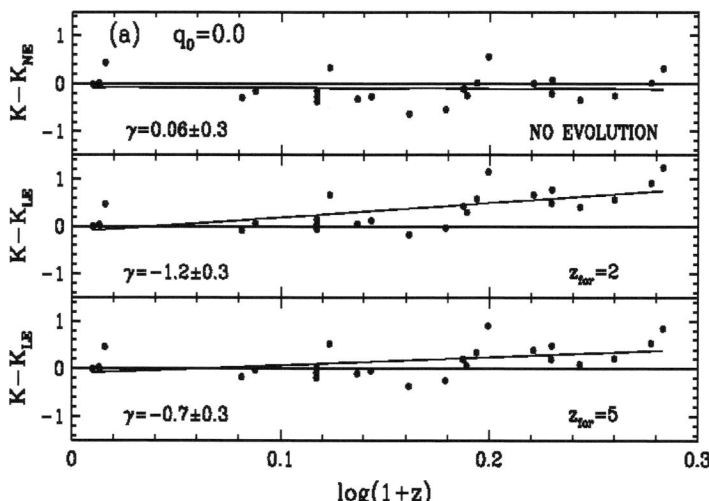

Figure 2: *A recent appraisal of the prospects of securing cosmological constraints from the Hubble diagram of brightest cluster galaxies (Aragón-Salamanca et al. 1998). Luminosity evolution is parameterised as $L = L(0)(1+z)^\gamma$. For $q_0=0$, the top panel shows residuals and best fit trend applying k-correction and luminosity distance effects only; no luminosity evolution is seen. The middle and bottom panels show the residuals when evolution is modeled for single burst populations formed at $z_F=2$ and 5, respectively. High z galaxies are* less *luminous than expected, presumably because they are still accreting material. Quantitatively, the effect amounts to a factor of 2-4 less stellar mass depending on the assumed q_0 (c.f. van Dokkum et al. 1999).*

Fortunately, we are making excellent progress in constraining the cosmological parameters from independent methods, the most prominent of which include the angular fluctuation spectrum in the microwave background (de Bernardis et al. 2000, Balbi et al. 2000), the Hubble diagram of distant Type Ia supernovae (Garnavich et al. 1998, Perlmutter et al. 1999), the abundance of rich clusters at various epochs (Bahcall & Fan 1998) and the redshift-space distortion in large redshift surveys such as 2dF (Peacock et al. 2001).

Given it matters, how then should we respond to the widely-accepted *concordance* in the determination of H_0, Ω_M, Λ from various probes (Ostriker & Steinhardt 1996, Bahcall et al. 1999)? The claimed convergence on the value of Hubble's constant (Mould et al. 2000) is not so important for the discussion below since most evolutionary tests are primarily concerned with *relative* comparisons at various look-back times where H_0 cancels. The most bewildering aspect of the concordance picture is the resurrection of a non-zero Λ, the evidence for which comes primarily from the Hubble diagram for Type Ia supernovae.

As a member of the Supernova Cosmology Project (Perlmutter et al. 1999) I obviously take the supernova results seriously! However, this does not prevent me from being surprised as to the implications of a non-zero Λ. The most astonishing fact is how readily the community has apparently accepted the resurrection of Λ - a term for which there is no satisfactory physical explanation (c.f. Wang et al. 2000). To one poorly-understood component of the cosmic energy density (non-baryonic dark matter), we seem to have added another (vacuum energy). It seems a remarkable coincidence that all three significant constituents $(\Omega_B, \Omega_{DM}, \Omega_\Lambda)$ are comparable in magnitude to within a factor of 10, and hardly a step forward that only one is physically understood!

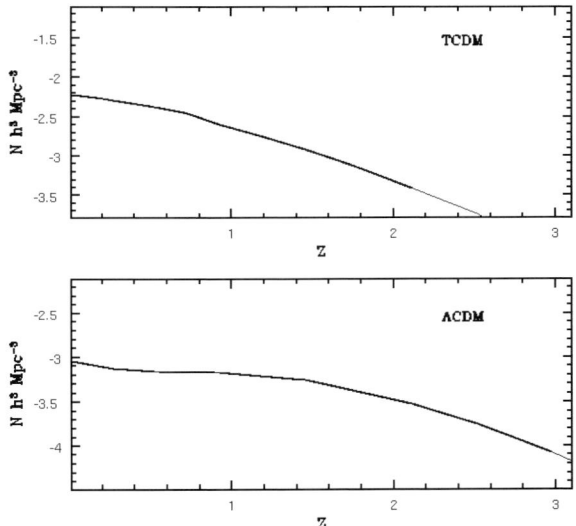

Figure 3: *The abundance of massive ($>10^{11} M_\odot$) systems as a function of redshift in two hierarchical models (Kauffmann & Charlot 1998a) showing the strong decline in a high density (τCDM) model c.f. that in a low density accelerating model (ΛCDM).*

The lesson I think we should draw from the *cosmic concordance* is similar to the comment I made in §1 when we discussed some theorists' triumphant reconciliation of their theories with faint galaxy data (a point we will debate in detail in §3). In both cases, the hypothesis certainly reproduces a wide range of observations but note it takes, as input, parameters for which there is not yet a clear physical model. One should not, therefore, regard a concordant picture as anything other than one of many possible working hypotheses. In the case of the cosmological models, we need to invest effort into understanding the physical nature of dark matter and vacuum energy. In the case of galaxy evolution our goal should be to test the basic ingredients of hierarchical galaxy formation.

3. Star Formation Histories

One of the most active areas of relevance to understanding the rate at which galaxies assemble is concerned with determining the cosmic star formation history. The idea is simple enough. A systematic survey is conducted according to some property that is sensitive to the on-going rate of star formation. The volume-average luminosity density is converted into its equivalent star formation rate averaged per unit co-moving volume and the procedure repeated as a function of redshift to give the cosmic star formation history $\rho_*(z)$. In this section we will explore the uncertainties and also the significance of this considerable area of current activity in terms of the constraints they provide on theories of galaxy formation.

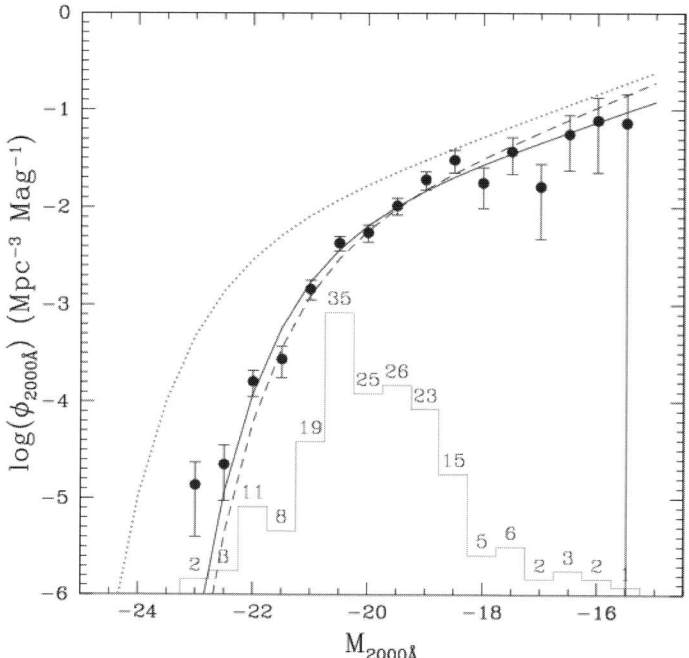

Figure 4: *The luminosity function for galaxies selected at 2000 Å from the recent survey of Sullivan et al (2000). The histogram and associated numbers indicate the absolute magnitude distribution observed which is corrected by volume and k-correction effects to give the data points. The dotted curve illustrates the considerable effect of extinction as gauged by Balmer decrements determined individually for those galaxies with emission lines. Such uncertainties translate in factors of two uncertainty in the local UV luminosity density.*

The joint distribution of luminosity L and redshift z, $N(L, z)$, for a flux-limited sample permits the construction of the luminosity function $\Phi(L)$ according to procedures which are reviewed by Efstathiou, Ellis & Peterson (1988) and compared by Ellis (1997). The luminosity function is often characterised according to the form defined by Schechter (1976), viz:

$$\Phi(L) \, dL/L^* = \Phi^* \, (L/L^*)^{-\alpha} \, \exp(-L/L^*) \, dL/L^* \qquad (3.3)$$

in which case the integrated number of galaxies per unit volume N and the luminosity density ρ_L then become:

$$N = \int \Phi(L) \, dL = \Phi^* \, \Gamma(\alpha + 1) \qquad (3.4)$$

and

$$\rho_L = \int \Phi(L) \, L \, dL = \Phi^* \, L^* \, \Gamma(\alpha + 2) \qquad (3.5)$$

and the source counts in the non-relativistic case, applicable to local catalogs, is:

$$N(<m) \propto d^{*\,3}(m) \int dL\, \Phi(L)\, (L/L^*)^{\frac{3}{2}} \propto \Phi^* L^{*\,\frac{3}{2}} \Gamma(\alpha + \frac{5}{2}) \qquad (3.6)$$

Frequently-used measures of star formation in galaxies over a range of redshift include rest-frame ultraviolet and blue broad-band luminosities (Lilly et al. 1995, Steidel et al. 1996, Sullivan et al. 2000), nebular emission lines such as Hα (Gallego et al. 1995, Tresse & Maddox 1998, Glazebrook et al. 1999), thermal far-infrared emission from dust clouds (Rowan-Robinson et al. 1997, Blain et al. 1999) and, most recently, radio continuum emission (Mobasher et al. 1999).

Since only a limited range of the luminosity function centered on L^* is reliably probed in flux-limited samples, a key issue is how well the integrated luminosity density ρ_L can be determined from such surveys. In the Schechter formalism, equations [3.4] and [3.5] show that whilst N would diverge for $\alpha < -1$, the luminosity density is convergent unless $\alpha < -2$.

Figure 4 shows the local rest-frame ultraviolet (2000 Å) luminosity function from Sullivan et al. (2000) whose faint end slope $\alpha = -1.6$ is markedly steeper than that found for samples selected in the near-infrared (Mobasher et al. 1993, Gardner et al. 1997, Cole et al. 2001) (where $\alpha \simeq -1$). This contrast in the luminosity distribution of young and old stellar populations is an important result which emphasizes the relatively weak connection between stellar mass and light and implies there may be significant uncertainties in the estimation of integrated luminosity densities for star-forming populations.

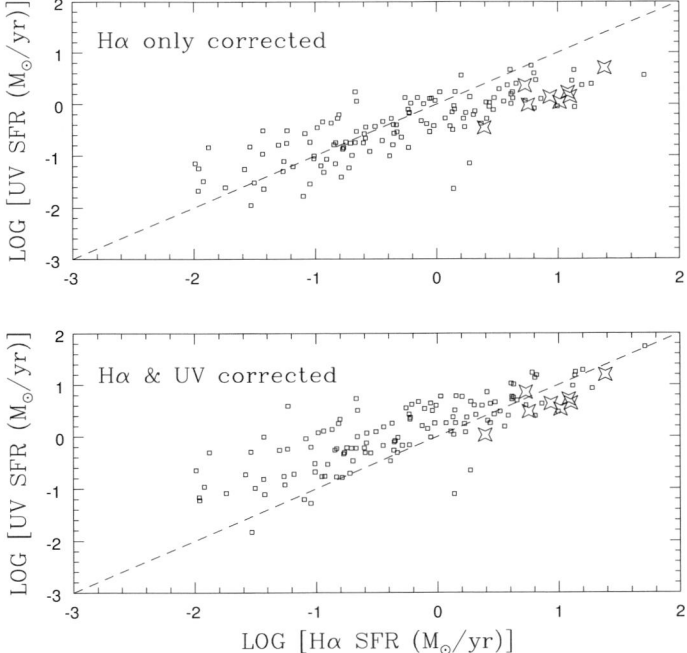

Figure 5: *Star formation rates derived from UV (2000 Å) continua versus those derived from Hα fluxes from the local survey of Sullivan et al. (2000, open squares) and the $z \simeq 1$ samples of Glazebrook et al. (1999, large stars). For the Sullivan et al. sample, extinction corrections were derived from individual Balmer decrements assuming Case B recombination and applied to the Hα fluxes in the upper panel and both estimates in the lower panel.*

Kennicutt (1998) carefully reviewed the relationships between the various observational diagnostics listed above and the star formation rate. Clearly a major uncertainty in any transformation based on the ultraviolet/optical continuum or nebular emission line measures is the likely presence of absorbing dust (Figure 4). Other uncertainties include the form of the initial stellar mass function and the nature of the star formation history itself.

Some of these uncertainties are quite imponderable and the only way to estimate their effect in typical populations is to undertake a comparison of the various diagnostics for the same sample. Sullivan et al. (2000) compared UV and Hα-based estimators for their local balloon-based UV-selected sample and Glazebrook et al. (1999) undertook a similar comparison for a restricted incomplete sample of high redshift galaxies (drawn from a I-selected sample). Bell & Kennicutt (2001) independently examined some of Sullivan et al's conclusions based on a smaller local sample with satellite UV fluxes. The comparison analysed by Sullivan et al. is shown in Figure 5. Although an overall linear relation is observed the scatter is quite considerable, greater than accountable from observational errors. The uncertainties would appear to be alarming in view of the fairly modest trends claimed in $\rho_{SFR}(z)$ (see below).

In addition to the scatter arising from extinction (accounted for via individual Balmer emission line decrements), Sullivan et al suggest that some fraction of their UV-selected population must be suffering star formation which is erratic in its time history. In such a situation, different diagnostics will be sensitive to bursts of activity for different periods, corresponding to the time over which the contributing stars remain on the main sequence. $H\alpha$ flux arises from recombination photons linked to those emitted below the Lyman limit from main sequence stars with lifetimes $\simeq 10^6$ years. The UV and blue continua persist for much longer periods ($\simeq 10^8 - 10^9$ years).

Depending upon how widespread star formation histories of this kind may be, two forms of error may arise in estimating cosmic star formation histories. Firstly, the star formation rate derived for an individual galaxy will be a past time average, smoothing over any erratic behavior, rather than a true instantaneous value. More importantly however, particularly at high redshift, galaxies may be preferentially selected only if their star formation history is erratic, for example in $H\alpha$ surveys where some threshold of detectability may seriously restrict the samples.

Figure 6 shows a recent estimate of the cosmic star formation history drawn from various surveys (Blain 2001). There appears to be a marked increase in activity over $0 < z < 1$ with a possible decline beyond $z > 2$. Although, inevitably perhaps, attention has focused on the case for the high redshift decline, even the strong rise to $z \simeq 1$ remains controversial. Originally proposed independently by Lilly et al. (1995) and Fall et al. (1995), revised estimates for the local luminosity density (Sullivan et al. 2000) and independent surveys (Cowie et al. 1999) have challenged the rapidity of this rise. Part of the problem is that no single survey permits a self-consistent measurement of ρ_{SFR} over more than a very limited range in z. Most likely, therefore, much of the scatter in Figure 6 is simply a manefestation of the kinds of uncertainties discussed above in the context of Sullivan et al.'s survey.

Beyond $z \simeq 2$, the available star formation rates have been derived almost exclusively from UV continua in Lyman break galaxies selected by their 'dropout' signatures in various photometric bands (Madau et al. 1996, Steidel et al. 1996, Steidel et al. 1999) and from currently scant datasets of sub-mm sources interpreted assuming thermal emission from dust heated by young stars (Blain et al. 1999, Barger et al. 1999b). There has been much discussion on the possible disparity between the estimates derived from these two diagnostics (which other lecturers will address). Two points can be made: firstly, the measured UV luminosity densities will clearly underestimate the true values given likely extinctions. Secondly, the sample of sub-mm sources with reliable redshifts remains quite inadequate for luminosity density estimates in the sense described above. Most of the constraints arise from modelling their likely properties in a manner consistent with their source counts and the integrated far-infrared background.

Have we become over-obsessed with determining the cosmic star formation history? Observers are eager to place their survey points alongside others on the overall curve and different groups defend their methods against those whose data points disagree. We should consider carefully what role this cosmic star formation history plays in understanding how galaxies form.

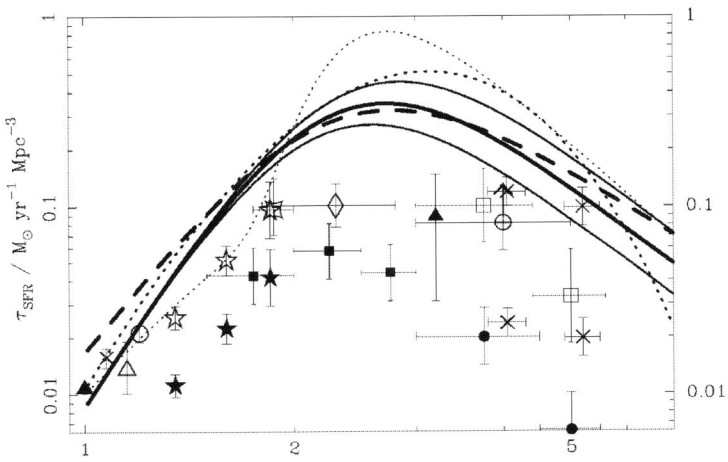

Figure 6: *The history of recent star formation from the recent compilation of Blain (2001). Data points are taken from a variety of sources referenced in that article. Thick solid and dashed lines represent trends expected from simple luminosity evolution and hierarchical models, respectively. It is clear there is considerable observational scatter at all redshifts, not just beyond $z \simeq 1$ as often assumed.*

Clearly, the prime conclusion we can draw from Figure 6 is that the stars which make the galaxies we see today formed continuously over a very wide redshift range. This may seem such an obvious deduction that it hardly merits stating but it is important to stress the absence of any obvious detectable 'epoch of star formation' as was once imagined (Eggen, Lynden-Bell & Sandage 1962, Frenk et al. 1989). Hierarchical modelers were quick to point out (e.g. Baugh et al. 1998) that they predicted extended star formation histories as early as 1990 (White & Frenk 1991). It is certainly true that a continuous assembly of galaxies is a major feature of these models and thus one supported by the data.

However, what about the *quantitative* form of Figure 6 which remains so difficult to pin down: does the shape of the curve really matter? Firstly, we should recognise that the luminosity density integrates over much detailed astrophysics that may be important. A particular ρ_{SFR} at a given redshift could be consistent *either* with a population of established massive sources undergoing modest continous star formation *or* a steep luminosity function where most of the activity is in newly-formed dwarf galaxies. In terms of structure formation theories, these are very different physical situations yet that distinction is lost in Figure 6.

Secondly, theoretically, the cosmic star formation history is not particularly closely related to how galaxies assemble. It is more sensitive to the rate at which gas cools into the assembling dark matter halos, a process of considerable interest but which involves a myriad of uncertain astrophysical processes (Figure 7) which are fairly detached from the underlying physical basis of say the hierarchical picture. In support of this, we should note that Baugh et al. (1999) were able, within the same Λ-dominated CDM model, to 'refine' their earlier prediction to match new high redshift datapoints revealing a much less marked decline beyond $z \simeq 2$.

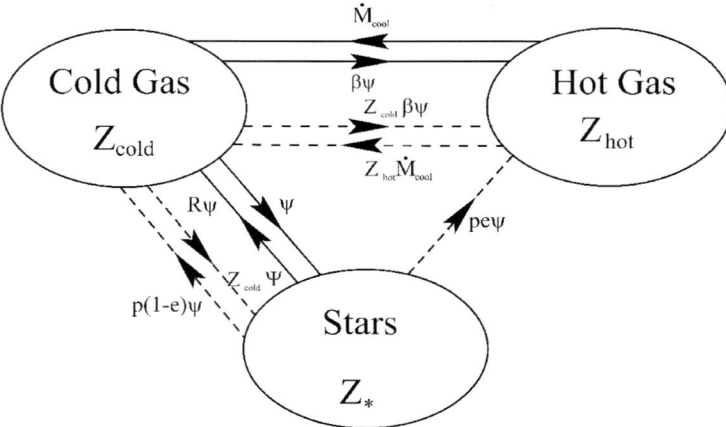

Figure 7: *An illustration of the complex physical processes governing the star formation rate of a young galaxy (courtesy of Carlos Frenk). Star formation is governed by the rate at which baryonic gas cools and falls into dark matter halos and this is inhibited by heating, e.g. from supernovae. The precise form of the cosmic star formation history gives us more insight into the interplay between these processes, integrated over all star-forming galaxies, than in distinguishing between various forms of structure formation (e.g. hierarchical vs. monolithic).*

4. Morphological Data from HST

As we discussed in §1, one of the most exciting new datasets that arrived in the mid-1990's was the first set of resolved images of galaxies at significant look-back times from HST. Much of the early work was conducted in rich clusters (Couch et al. 1994, Dressler et al. 1994, Couch et al. 1998, Dressler et al. 1998) where the well-known 'Butcher-Oemler' effect (Butcher & Oemler 1978) - a surprisingly recent increase in the fraction of blue cluster members - was found to be due to a dramatic shift in the morphology-density relation (Figure 8). As recently as $z \simeq 0.3 - 0.4$ ($3 - 4$ Gyr ago), cluster S0s were noticeably fewer in proportion, their place apparently taken by spirals, many of which showed signs of recent disturbances, such a distorted arms and tidal tails.

The physical origin of this transformation from spirals to S0s remains unclear and is currently being explored by detailed spectroscopy of representative cluster members (Barger et al. 1996, Abraham et al. 1996, Poggianti et al. 1999). A key diagnostic here is the interplay between the changing morphologies, the presence of nebular emission lines (such as [O II] 3727 Å – Hα is generally redshifted out of the accessible range) and

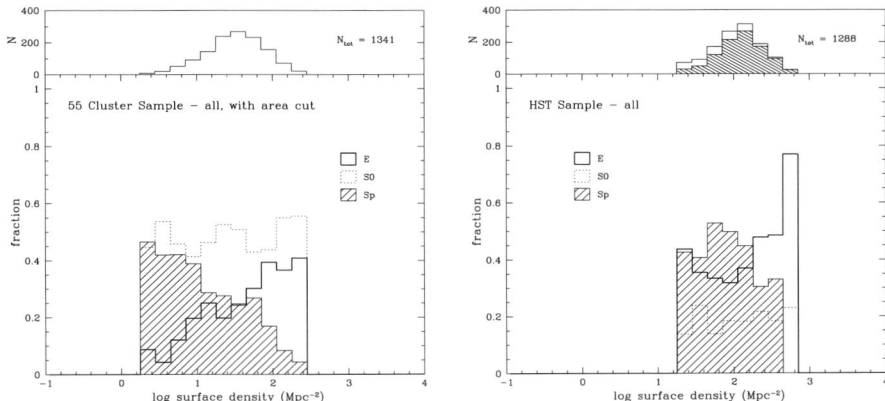

Figure 8: *Evolution in the morphology density relation from the 'Morphs' survey of Dressler et al. (1998). (Left) The fraction of E/S0/Spirals as a function of projected galaxian surface density for Dressler's 55 local cluster sample. (Right) As left, for all distant HST clusters with $0.3 < z < 0.55$. The comparison refers to the same cluster core radius ($< 0.6 Mpc$, $h = 0.5$) and includes galaxies to the same rest-frame V luminosity ($M_V = -20.0$). Note the dramatic decline in the S0 population and the marked increase in the spiral fraction for environments with high projected density.*

Balmer absorption lines (such as Hδ 4101 Å). The latter lines are prominent in main sequence A stars which linger for $\simeq$1 Gyr after any enhanced starburst activity.

Barger et al. (1996) proposed a simple cycle where an unsuspecting galaxy undergoes some perturbation, perhaps due to a merger or its first encounter with the intracluster gas, subsequently becomes morphologically-distorted and spectrally-active before subsiding to a regular spheroidal with a decaying Balmer absorption line. Whereas such a cycle can explain the *proportion* of unusual objects, it has difficulty matching their *luminosities*. A galaxy should be rendered more luminous during a burst and thus blue examples cannot easily be the precursors of the equally-luminous red *post-burst* cases. A controversy has since arisen over the fractions of objects seen in the various spectrally-active classes (Balogh et al. 1999) suggesting much work is needed in this area, both in quantifying cluster-cluster variations and also radial variations in the responsible processes.

Although the cluster work discussed above represents something of a digression in our overall theme, the realisation that galaxies can so easily be transformed morphologically has profound implications for our understanding of galaxy formation. Much of the early work explaining the Hubble sequence (Tinsley 1977) assumed galaxies evolve as isolated systems, however the abundance of morphologically-peculiar and interacting galaxies in early HST images (Griffiths et al. 1994) has been used to emphasize the important role that galaxy mergers must play in shaping the present Hubble sequence (Toomre & Toomre 1972, Barnes & Hernquist 1992). Merger-induced transformations of this kind are a natural consequence of hierarchical models (Baugh, Cole & Frenk 1996). Early disk systems are prone to merge during epochs when the cosmic density is high and the peculiar velocity field is cold, forming bulge-dominated and spheroidal systems which may then later accrete disks.

The possibility that galaxies transform from one class to another is a hard hypothesis to verify observationally since, as we have seen, traditionally observers have searched for redshift-dependent trends with subsets of the population chosen via an observed property (color, morphology, spectral characteristics) which could be transient. Moreover, experience ought to teach us that the outcome of tests of galaxy formation rarely come down simply to either Theory A *or* Theory B; usually it is some complicated mixture or the question was naive in the first place! Fortunately, the late formation of massive regular galaxies in the hierarchical picture (Figure 9) seems a particularly robust prediction and one in stark contrast to the classical 'monolithic collapse' picture (Tinsley 1977, Sandage 1986). The distinction is greatest for ellipticals presumed to form at high redshift with minimum dissipation (their central density reflecting that of the epoch of formation). Studying the evolutionary history of massive ellipticals is thus an obvious place to start.

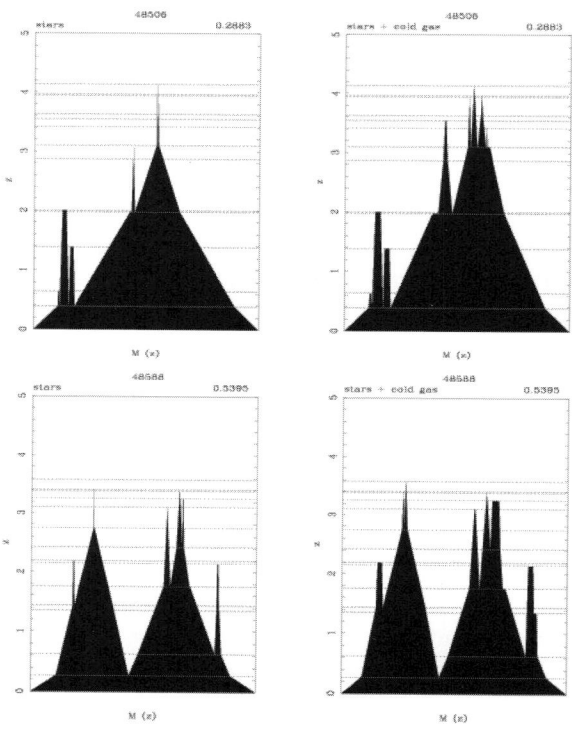

Figure 9: *The important role of late merging in a typical CDM semi-analytical model (Baugh et al. 1996). The panels show the redshift-dependent growth for the stellar mass (left) and that of all baryonic material (right), as indicated by the thickness of the black area at a given epoch, for two present-day massive galaxies. The top system grows gradually and is thought to represent a present-day spiral. The bottom system suffers a late equal-mass merger thought to produce a present-day elliptical. Note the remarkably late assembly; most of stellar mass in both cases assembles in the interval $0 < z < 1$.*

An oft-quoted result in support of old ellipticals is the remarkable homogeneity of their optical colors (Sandage & Visvanathan 1978, Bower et al. 1992). The idea is simple: the intrinsic population scatter in a color sensitive to recent star formation, such as $U - B$, places a constraint either on how synchronous the previous star formation history must have been across the population or, if galaxies form independently, the mean age of their stellar populations. By combining cluster data at low redshift (Bower et al. 1992) with

HST-selected samples at intermediate redshift (Ellis et al. 1998), the bulk of the cluster elliptical population was deduced to have formed its stars before $z \simeq 2$, in apparent conflict with hierarchical models. Similar conclusions have been drawn from evolution of the mass/light ratio deduced from the fundamental plane (Ziegler & Bender 1997, van Dokkum et al. 1998).

Unfortunately, one cannot generalize from the results found in distant clusters. In hierarchical models, clusters represent early peaks in the density fluctuations and thus evolution is likely accelerated in these environments (Kauffmann 1995) plus, of course, there may be processes peculiar to these environments involving the intracluster gas. It is also important to distinguish between the history of *mass* assembly and that of the *stars*. Recent evidence for widespread merging of ellipticals in clusters (van Dokkum et al. 1999) lends support to the idea that the stars in dense regions were formed at high redshift, in lower mass systems which later merged.

For these reasons, attention has recently switched to tracking the evolution of *field* ellipticals. The term *field elliptical* is something of a misnomer here since a high fraction of ellipticals actually reside in clusters. What is really meant in this case is that we prefer to select ellipticals systematically in flux-limited samples rather than concentrate on those found in the cores of dense clusters†.

The study of evolution in field ellipticals is currently very active and I cannot possibly do justice, in the space available, to the many complex issues being discussed. Instead let me summarise what I think are the most interesting results.

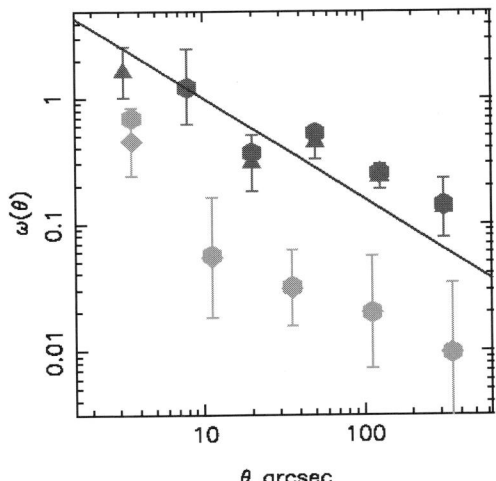

Figure 10: *Evidence for a clustered population of red objects in a 1000 arcmin² area of the ongoing Las Campanas Infrared Survey (McCarthy et al. 2000). The angular correlation function for objects with $I - H > 3.5$ (top set of data points) is substantially above that for all H-selected galaxies (bottom set) and consistent with a high fraction of the red objects being clustered ellipticals with $\overline{z} \simeq 1.0 - 1.5$.*

- Searches for a population of faint intrinsically red objects, representing the expected $z > 1$ precursors of passively-evolving ellipticals which formed their stars at high redshift have been conducted both with and without HST morphological data (Zepf 1997, Barger et al. 1999a, Menanteau et al. 1999, Daddi et al. 2000, McCarthy et al. 2000).

† A major concern in all the work relating to the evolution of galaxies in clusters is precisely how the clusters were located.

However, only recently have substantial areas of sky been mapped. This is because such searches are most sensitive to high z sources when conducted using optical-near infrared colors and access to large infrared arrays is a recent technical development. Both Daddi et al (2000) and McCarthy et al. (2000) (Figure 10) claim strong angular clustering in their faint red populations and there is limited evidence that the abundance is consistent with a constant comoving number density, in contrast to the hierarchical predictions. However, without confirmatory spectroscopy neither the redshift range nor the nature of these red sources is yet clear. Even if it later emerges, as was claimed on far less convincing data (Zepf 1997, Barger et al. 1999a, Menanteau et al. 1999), that there is a shortage of intrinsically red objects beyond $z \simeq 1$, only a modest amount of residual star formation is needed to substantially bluen a well-established old galaxy (Jimenez et al. 1999). Even when redshift data is secured, color alone may be an unreliable way to track a specific population.

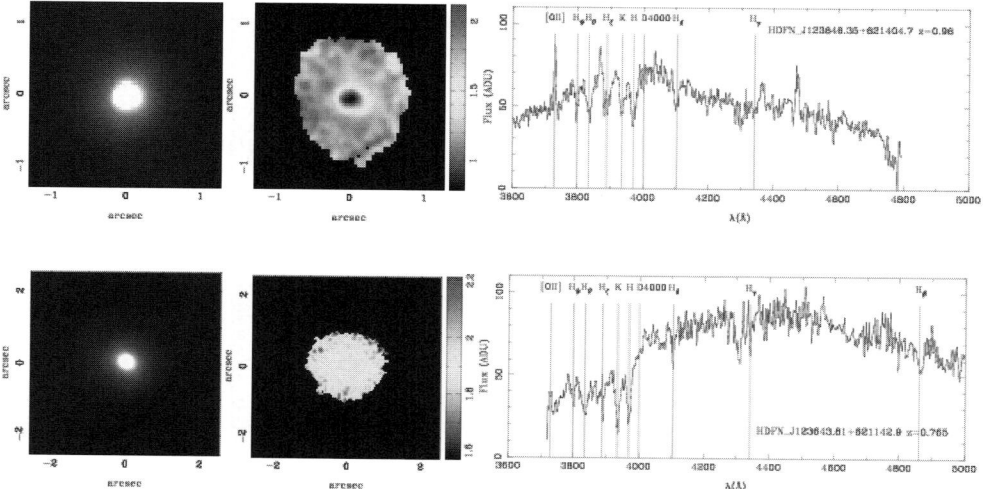

Figure 11: *Color inhomogeneities in HDF field ellipticals suggest continued star formation is occurring, possibly as a result of hierarchical assembly. Each row displays the I HST image, a V − I color image and the Keck LRIS spectrum. The top set refers to a z=0.92 elliptical with a blue core; its spectrum shows features indicative of active star formation ([O II] emission and deep Balmer absorption lines). The bottom set refers to a quiescent example at z=0.966 whose spectrum is consistent with an old stellar population. The amount of blue light can be combined with the depth of the spectral features to statistically estimate the amount and timescale of recent star formation.*

- At brighter magnitudes, systematic redshift surveys with associated HST data give constraints on the luminosity function and colors of morphologically-selected ellipticals (Brinchmann et al. 1998, Schade et al. 1999, Menanteau et al. 1999, Im et al. 2001). Unfortunately, because of the disparity in field of view betweeen WFPC-2 and the ground-based multislit spectrographs, the samples remain small and hence the conclusions are subject to significant field-to-field clustering uncertainties. However, no substantial decline in the volume density of ellipticals is yet observed to $z \simeq 1$, although there is some dispute as to the fraction which may deviate in color from the passive track (Im et al. 2001 c.f. Menanteau et al. 1999). Current spectroscopic surveys may not be quite deep enough to critically test the expected evolution in the hierarchical models, particularly if $\Lambda \neq 0$.

- A completely independent method of determining whether field ellipticals form continuously as expected in hierarchical models is possible in the Hubble Deep Fields (Menanteau et al. 2001). Here, the imaging signal/noise is sufficient to permit an examination the *internal* colors of ellipticals with $I < 24$, a subset of which have redshifts. Menanteau et al. (2001) find about 25% of the HDF ellipticals show blue cores and other color inhomogeneities suggestive of recent star formation, perhaps as a result of the merger with a gas-rich low mass galaxy. Keck spectroscopy (Ellis et al. 2001b) supports this suggestion: galaxies with blue cores generally show emission and absorption line features indicative of star formation (Figure 11). The amount of blue light seen in the affected ellipticals can be used to quantify the *amount* of recent star formation and the associated spectrum can be used to estimate the *timescale* of activity through diagnostic features of main sequence stars. Only modest accretion rates of $\simeq 10\%$ by mass over $\simeq 1$ Gyr are implied, albeit for a significant fraction of the population. This continued growth, whilst modest c.f. expectations of hierarchical models, is noticeably *less* prominent in rich clusters (Menanteau et al. 2001).

What evolution is found in the properties of other kinds of galaxies? Brinchmann et al. (1998) secured HST images for a sizeable and statistically-complete subset of CFRS and LDSS redshift survey galaxies and found the abundance of spirals to $I = 22$ – a flux limit which samples $0.3 < z < 0.8$ – is comparable to that expected on the basis of their local abundance if their disks were somewhat brighter and bluer in the past as evidenced from surface photometry (Lilly et al. 1999). In practice, however, the detectability of spiral disks is affected by a number of possible selection effects (Simard et al. 1999, Bouwens & Silk 2000) and it may be some time before a self-consistent picture emerges.

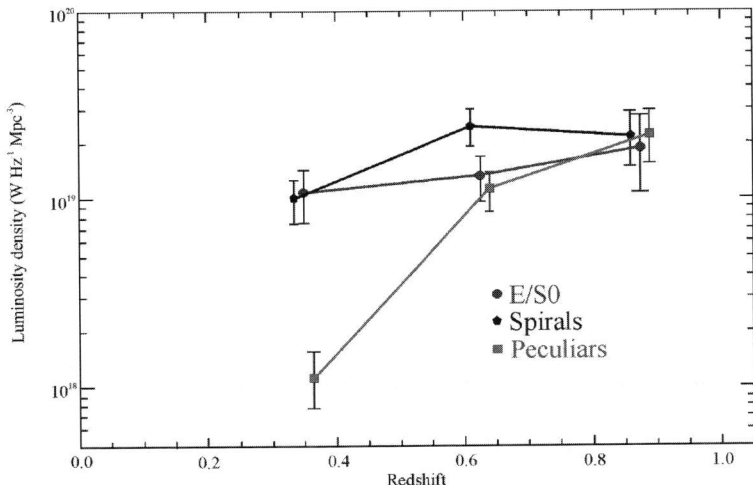

Figure 12: *The morphological dependence of the blue luminosity density from that subset of the CFRS/LDSS redshift survey imaged with HST (Brinchmann et al. 1998). The marked decline in the luminosity density of galaxies with peculiar morphology over $0 < z < 1$ is the primary cause for steep slope in the blue faint galaxy counts.*

A less controversial result from Brinchmann et al. (1998) claimed in earlier analyses

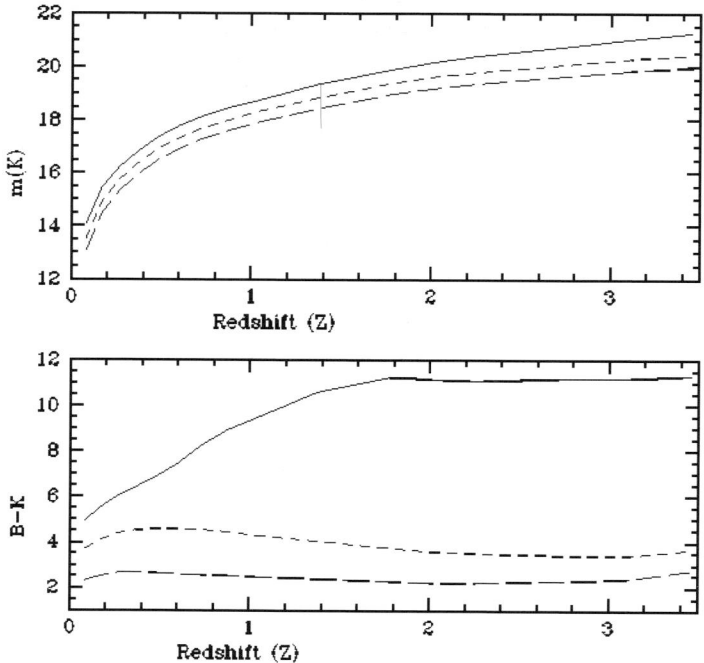

Figure 13: *The K-band luminosity is a good measure of the underlying stellar mass irrespective the past star formation history (Kauffmann & Charlot 1998b). The curves show the observed K magnitude as a function of the redshift at which such an object is selected for a system containing $10^{11} M_\odot$ produced according to a variety of star formation histories. Even across extreme cases (single burst at $z = \infty$, solid line, to a constant star formation rate to the epoch of observation, short-dash), the K-band output remains the same to within a factor of $\simeq 2$. The lower panel shows, how different the observe optical-infrared color would be in these cases.*

without redshift data (Glazebrook et al. 1995, Driver et al. 1995) is the remarkably high abundance of morphologically-peculiar galaxies in faint HST data. Brinchmann et al. quantified this in terms of the luminosity density arguing that a substantial fraction of the claimed decline in the blue luminosity density since $z \simeq 1$ (c.f. Figure 6) arises from the demise of this population (Figure 12).

Given our earlier concerns with over-interpreting the cosmic star formation history, should we be cautious in drawing conclusions from Figure 12? Although Brinchmann et al's redshift sample is small, the basic result is consistent with the HST morphological number counts where much larger samples are involved. Whereas early skeptics argued that morphologically-peculiar galaxies represent regular systems viewed at unfamiliar ultraviolet wavelengths, recent NICMOS imaging (see Dickinson's lectures) suggests such 'morphological bandshifting' is only of minor consequence. In quantitative detail, as before, uncertain corrections must be made for the effects of the flux limited sample and of course extinction is a major uncertainty. However, it seems inescapable that the bulk of the decline in blue light (the so-called *faint blue galaxy problem*, Ellis 1997) arises from the demise of a population of late-type and morphologically-peculiar systems. A key question therefore is what happened to this population? We will address this problem in the next section.

5. Constraining the Masses of Distant Galaxies

A recurring issue arises from the discussions in the earlier sections. Whilst observers are, with some restrictions, able to measure distant galaxy properties such as rest-frame colors, luminosities and star formation rates, these may be poor indications of the underlying stellar and total masses predicted most straightforwardly by contemporary models of structure formation. Either we put our faith in the forward modelling of the readily-available observables (i.e. we invest a lot of effort in understanding the complexities of feedback, Figure 7), or we consider how to measure galactic masses.

Ideally we seek methods for determining the *total* mass (baryonic plus the dark matter halo) but this seems out of reach for the moment except for local systems with tracers of the larger halo in which galaxies are thought to reside. Useful tracers here include the dynamical properties of attendant dwarf galaxies (Zaritsky et al. 1998) and globular clusters. A promising route in the future might be galaxy-galaxy gravitational lensing (Blandford & Narayan 1992). Here a foreground population is restricted in its selection, perhaps according to morphology or redshift, and the statistical image distortions in a background population analysed. Early results were based on HST data, for cluster spheroidals (Natarajan et al. 1999) and various field populations (Griffiths et al. 1996), however with extensive panoramic data from the Sloan Digital Sky Survey, convincing signals can be seen with ground-based photometry (Fischer et al. 2000). Again, photometric redshifts will be helpful in refining the sample selection and in determining the precise redshift distribution essential for accurate measures on an absolute scale.

Unfortunately, promising though the technique appears, the restrictions of galaxy-galaxy lensing are numerous. It only gives mass estimates for statistical samples: the signal is too weak to be detected in individual cases, unless a strong lensing feature is seen (Hogg et al. 1996). Moreover, the redshift range and physical scale on which the mass is determined is defined entirely by geometrical factors and, ultimately, one may never be able to apply the method to galaxies beyond $z \simeq 1$.

Extensive dynamical data is becoming available for restricted classes of high redshift galaxy, via linewidth measures (Koo et al. 1995), resolved rotation curves (Vogt et al. 1997) for sources with detectable [O II] 3727 Å emission, and via internal stellar velocity dispersions for absorption line galaxies such as spheroidals. Under certain assumptions, these give mass estimates and have enabled the construction of the fundamental plane for distant spheroidals (Treu et al. 2000) and the Tully-Fisher relation for high redshift disk galaxies (Vogt et al. 1997). The greatest progress in the former has been in constructing the fundamental plane in rich clusters (van Dokkum et al. 1998) where slow evolution in the inferred mass/light ratio for cluster ellipticals is consistent with a high redshift of formation (see §5). For the emission line studies it is not straightforward to convert data obtained over a limited spatial extent into reliable masses even for regular well-ordered systems. For compact and irregular sources, the required emission lines may come from unrepresentative components yielding poor mass estimates (Lehnert & Heckman 1996).

The prospects improve significantly if we drop the requirement to measure the *total* mass and are willing to consider only the *stellar mass*. In this case, the near-infrared luminosity is of particular importance. Broadhurst et al. (1992) and Kauffmann & Charlot (1998b) have demonstrated that the K (2.2μm) luminosity is a good measure of

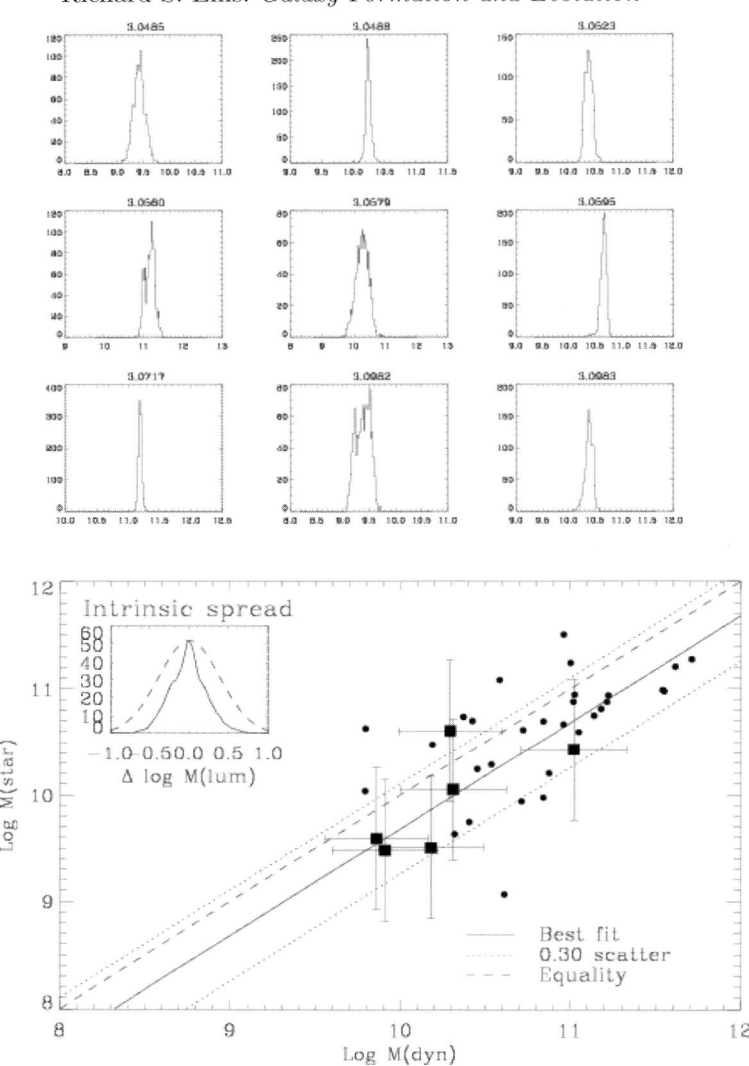

Figure 14: *The infrared method for determining the stellar mass of a distant galaxy (Ellis et al. 2001a). The technique fits the observed SED for a galaxy of known redshift in the context of evolutionary synthesis models where the stellar mass is the fitted variable. (Top) Likelihood functions for the derived logarithmic stellar mass for sample galaxies in the CFHT/LDSS redshift survey (Brinchmann, Ph.D. thesis 1999); a typical uncertainty of 30-50% is secured at I≃22. (Bottom) Correlations of stellar and dynamical mass for both low z (circles) and high z (squares with error bars) galaxies from the analysis of Brinchmann & Ellis (2000).*

its underlying stellar mass *regardless of how that mass assembled itself* (Figure 13). This remarkable fact arises because K-band light in all stellar populations (whether induced in bursts or continuous periods of activity) arises from long-lived giants whose collective output mirrors the *amount* of past activity, smoothing over its production timetable.

A deep K-band redshift survey thus probes the very existence of massive systems at early times. A slightly incomplete survey to $K=20$ (Cowie et al. 1996) and a complete photometric survey to $K=21$ (Fontana et al. 1999) indicates an apparently shortfall of luminous K objects beyond $z \simeq 1 - 1.5$ c.f. pure luminosity evolution models. Unfortunately, small sample sizes, field-to-field clustering, spectroscopic incompleteness and untested photometric redshift techniques beyond $z\simeq 1$ each weaken this potentially important conclusion. An important goal in the immediate future must be to reconcile these claims with the apparently abundant (and hence conflicting) population of optical-infrared red objects to $K \simeq 19 - 20$ (Daddi et al. 2000, McCarthy et al. 2000).

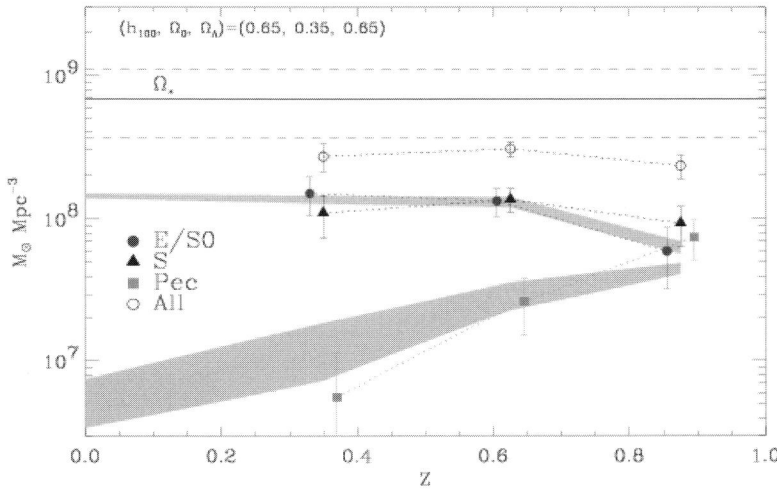

Figure 15: *Evolution of the stellar mass density $\rho_{stars}(z,T)$ from the analysis of Brinchmann & Ellis (2000). A remarkable decline with time in stellar mass density is seen for the morphologically-peculiar class which argues against a truncation of their star formation activity as the primary cause for their demise. Brinchmann & Ellis argue that this population must be transforming, possibly via mergers, into the regular classes. A simple model which implements a likely redshift-dependent merger rate (LeFevre et al. 2000) with elliptical products can broadly reproduce the trends observed (shaded area of the plot).*

The precision of the technique introduced by Kauffmann & Charlot (1998b) can be improved if the optical-infrared color is available as an extra parameter (Ellis et al. 2001a). In this way a first-order correction can be made for the past star formation history and hence the effect of the spread in the lower panel of Figure 13 can be used to improve the mass estimate. Importantly, such a technique for determine accurate stellar masses can then be applied to *all* galaxies, regular or peculiar, irrespective of their dynamical state and over a range in redshift (providing the data is sufficiently precise). The technique can be considered as a modification of that frequently utilised in estimating photometric redshifts. The observed optical-infrared SED for an object of known redshift is used to optimally fit the *stellar mass* rather than the redshift in the framework of an evolutionary synthesis code. Stellar masses can be derived to within a random uncertainty of 30-50% by this technique although at present there is no reliable way to verify the results except by comparison with independent dynamical measures (Figure 14).

6. Origin of the Hubble Sequence

The availability of stellar masses for *all* types enables the construction of a powerful evolutionary plot, analogous to Figure 6, involving the *stellar mass density*, $\rho_{stars}(z,T)$, as a function of morphology T. Whilst the stellar mass density can *grow* by continued star formation, unlike the *UV luminosity density*, ρ_{UV}, it is difficult to imagine how it can *decline*. As we saw earlier ρ_{UV} can decline significantly in only 1-2 Gyr because of an abrupt truncation of activity. However, such a change would have very little effect on the infrared output as illustrated in Figure 13.

Brinchmann & Ellis (2000) secured K luminosities and optical-IR SEDS for over 300 galaxies in the CFRS/LDSS and Hawaii survey fields and derive $\rho_{stars}(z,T)$ (Figure 15). Estimating the integrated stellar mass density is prone to all of the difficulties reviewed earlier for the luminosity density and there is the added complication that the redshift surveys in question are *optically-selected* and thus must miss some (red) fraction of a true K-limited sample. Accordingly, the mass densities derived are lower limits to the true values.

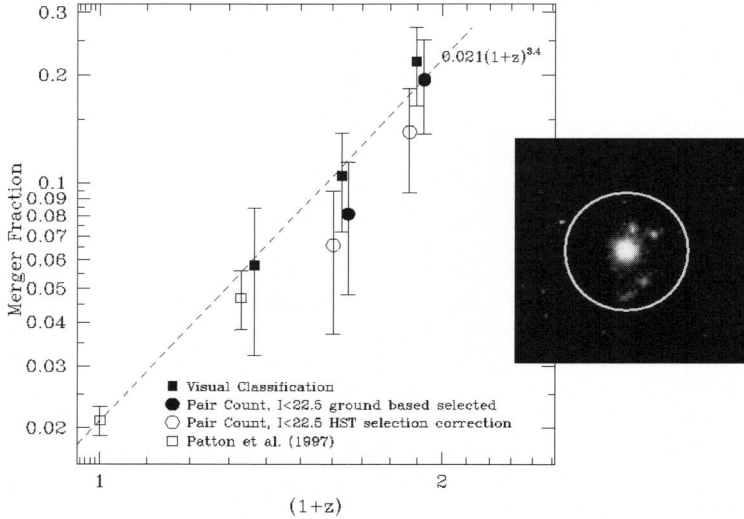

Figure 16: *An increase in the merger fraction as a function of redshift from the HST analysis of LeFevre et al. (2000). Galaxies of known redshift were examined for satellites brighter than a fixed rest-frame luminosity within a projected radius of $20h^{-1}$ kpc and corrections made for unrelated line-of-sight contamination. This redshift-dependent merger rate was adopted by Brinchmann & Ellis (2000) in Figure 15.*

Remarkably, $\rho_*(z,T)$ is a declining function for the intriguing population of morphologically-peculiar galaxies. Whereas the declining UV luminosity density could imply a fading population, such an explanation cannot be consistent with Figure 15 which argues, instead, that the objects are genuinely disappearing into other systems. The most logical explanation for their declining contribution to the stellar mass density is that morphologically-peculiar objects are being transformed, e.g. by mergers, into regular objects.

Merging has been an attractive process governing the evolution of galaxies for many years (Toomre & Toomre 1972, Rocca-Volmerange & Guiderdoni 1990, Broadhurst et al. 1992) and of course is fundamental to the hierarchical formation picture. However it has been extremely difficult to determine the observed rate at intermediate redshift. The fundamental problem is that we observe galaxies at various look-back times via discrete 'snapshots' without ever being able to *prove* two associated systems are destined to merge on a particular timescale. Using the CFRS/LDSS HST dataset referred to earlier, LeFevre et al. (2000) undertook a quantitative survey of the *fraction* of luminous galaxies with satellites brighter than a fixed absolute magnitude within a $20h^{-1}$ kpc metric radius and, after allowance for projection effects, determined the merger *fraction* increases with redshift as $\propto (1+z)^{3.4\pm0.5}$ – a result consistent with earlier ground-based efforts. Sadly, it is not straightforward to convert the proportion of galaxies with associated sources into a physical merger rate or, as ideally required, a mass assembly rate without some indication of the dynamical timescale for each merger and the mass of each satellite. Moreover, there are several annoying biases that affect even the derived merger fraction.

Brinchmann & Ellis (2000) attempted to reconcile the decline of the morphologically-peculiar population, the redshift dependence of the LeFevre et al. merger fraction and associated evidence for continued formation of ellipticals (Menanteau et al. 2001) into a simple self-consistent picture. They transferred the dominant population of morphologically-irregular galaxies, via the z-dependent merger rate, into a growth in the regular galaxies (shaded area of Figure 15). This is clearly a simplistic view, but nonetheless, gives a crude empirical rate at which regular galaxies are assembling. If correct, how does this agree with mass assembly histories predicted, say in ΛCDM?

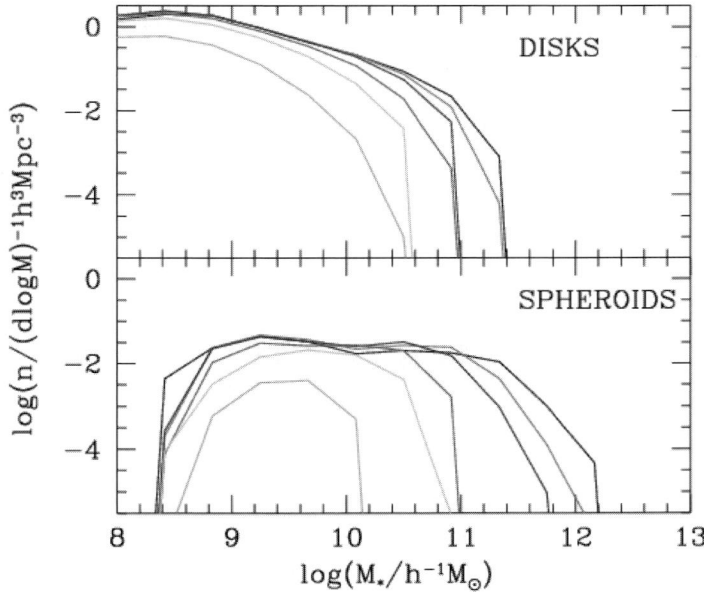

Figure 17: *Predicted evolution in stellar mass functions for disk and spheroidal populations in a ΛCDM hierarchical model (Frenk, priv. comm). The curves define mass functions as a function of redshift ($z=0,0.5,1,2$, from right to left). Modest growth over $0 < z < 2$ is expected for disk galaxies but significant growth is predicted for massive spheroidals.*

Figure 17 shows a recent prediction of the assembly history of spheroids and disks (Frenk, private communication). Although there are some discrepancies between this and its equivalent prediction from Kauffmann & Charlot (1998a, Figure 3), the trends

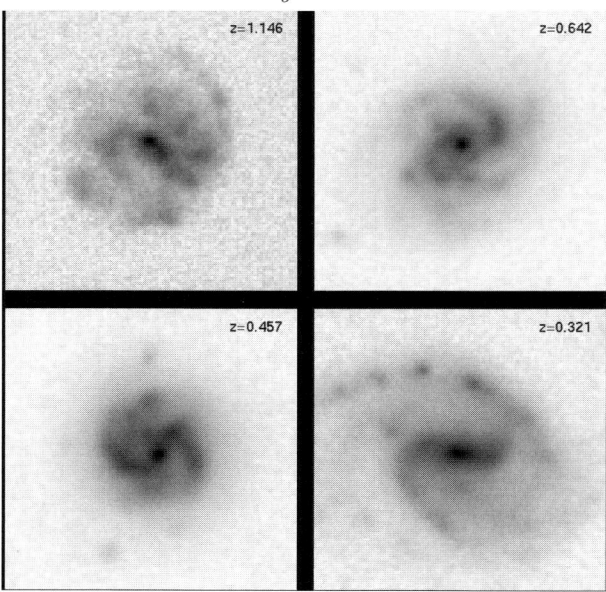

Figure 18: *Face-on barred spirals of known redshift in the Hubble Deep Field. Abraham et al. (1999) claim that all such systems should easily be recognised to $z\simeq 1$ in HDF-quality data whereas, beyond $z\simeq 0.6$, there appears to be a marginal paucity of such systems compared to their non-barred counterparts. If supported by further data, this could indicate an epoch corresponding to the 'dynamical maturing' of stellar disks.*

are clear. The strongest evolutionary signal is expected in terms of a recent assembly of massive spheroids; the equivalent growth rate in stellar disks is more modest. To the extent it is currently possible to test this picture, the qualitative trend is supported by the data. Field ellipticals are certainly still assembling (Menanteau et al. 2001) but perhaps more slowly than expected according to Figure 17; unfortunately deeper samples with redshifts are needed for a precise statement. Brinchmann (in prep.) has examined the stellar mass growth rate in disks using the infrared-based method over $0 < z < 1$ and finds only modest changes. This is very much a developing area and one that would benefit from significantly enlarged HST datasets chosen to overlap the growing faint redshift survey databases.

The HST data, particularly that in the Hubble Deep Fields (HDF), is an astonishingly rich resource which is still not completely exploited. As an indication of what might be possible with future instrumentation, I will close with some remarks on the important role that bulges and bars may play in the history of the Hubble sequence.

50% of local spirals have bars which are thought to originate through dynamical instabilities in well-established differentially-rotating stellar disks. If we could determine the epoch at which bars begin appearing, conceivably this would shed some light on how recently mature spirals came to be. Via careful simulations based on local examples, Abraham et al. (1999) showed that face-on barred galaxies should be recognisable to $z \simeq 1$ in the HDF exposures. In fact, many are seen (Figure 18) but tantalisingly the barred fraction of face-on spirals appears to drop beyond a redshift $z\simeq 0.6$. The effect is marginal but illustrative of a powerful future use of morphological imaging with the Advanced Camera for Surveys.

The story with bulges is also unclear, although potentially equally exciting. Tradi-

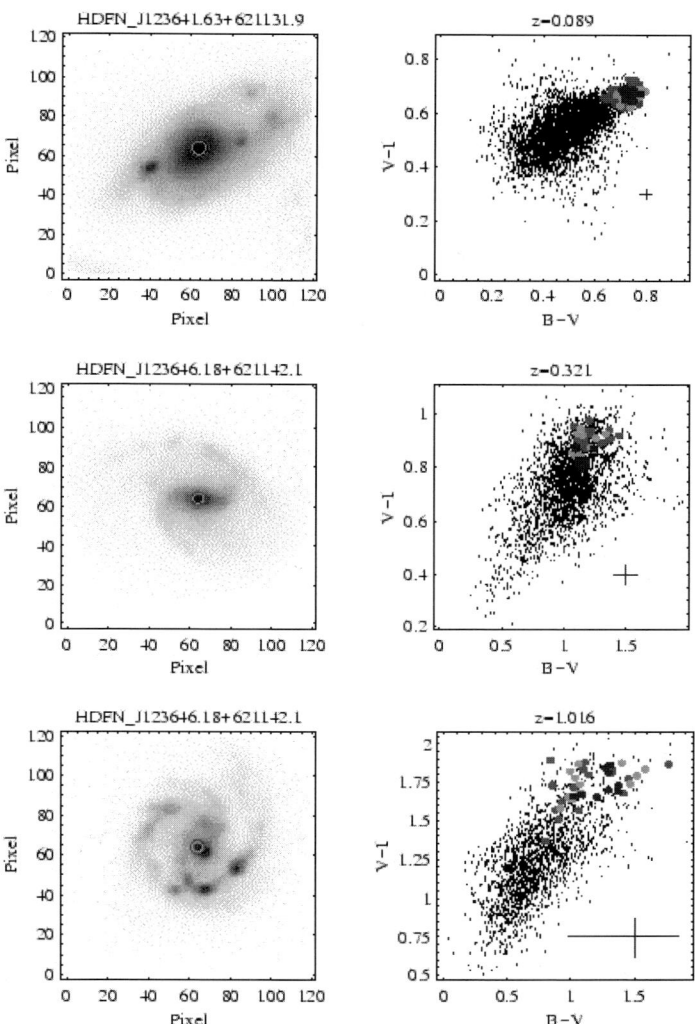

Figure 19a: *The remarkable diversity of intermediate redshift spiral bulges in the Hubble Deep Fields as revealed in the analysis of Ellis et al. (2001a). Selected face-on spirals in the HDF with pixel-by-pixel BVI color distributions. The marked points represent various aperture selections which serve to define the mean bulge color; in each case the bulge remains the reddest part of the spiral galaxies.*

tionally, bulges were thought to represent miniature ellipticals which formed passively at high redshift (Eggen, Lynden-Bell & Sandage 1962). Detailed studies of local examples, including the Galactic bulge, have shown a considerable diversity in properties, both in integrated color and even in their photometric structure (Wyse 1999). There is some evidence of a bimodality in the population; prominent bulges in early type spirals share surface brightness characteristics of ellipticals, whereas those in late-type spirals are closer to exponential disks. This might indicate two formation mechanisms, one pri-

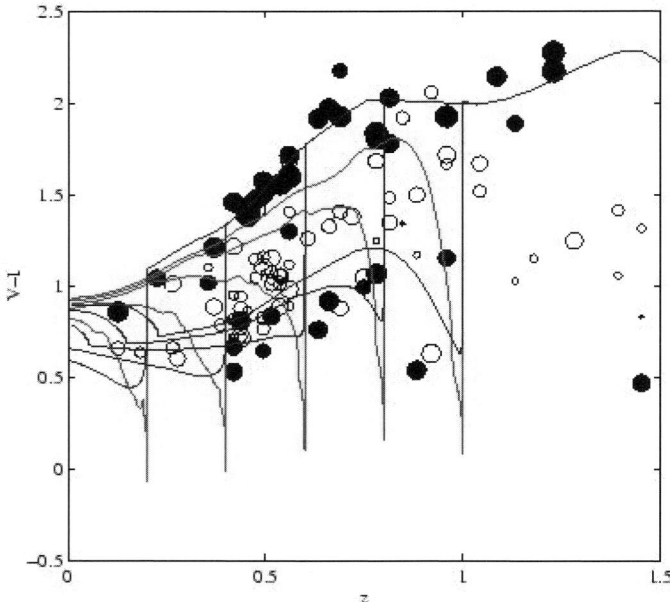

Figure 19b: *The remarkable diversity of intermediate redshift spiral bulges in the Hubble Deep Fields as revealed in the analysis of Ellis et al. (2001a). $V - I$ aperture color for bulges (open circles) and integrated color for ellipticals (filled circles) versus redshift. Bulges are generally more diverse with a mean color bluer than their elliptical counterparts. Curves illustrate that a continued infall of 5% by mass over 1-2 Gyr timescales could explain the observed trends.*

mordial (as in the traditional picture), the other related perhaps to the merging assembly history or via disk instabilities through what is termed 'secular' evolution.

Taking advantage of the HDF images, including those from NICMOS, Ellis et al. (2001a) have examined the color distribution for a large sample of spirals bulges of known redshift and compared these colors with their integrated equivalent for the HDF ellipticals. If bulges are miniature ellipticals formed at high redshift, one would expect similar trends. Interestingly, in the hierarchical picture, one expects bulges to be *older* and presumably redder than ellipticals (since the latter predominantly form from merged disk systems which most likely contain bulges as early merger remnants). Ellis et al. (2001a) find intermediate redshift bulges are the reddest part of a typical spiral but, surprisingly, they are often bluer than their elliptical counterparts and far less homogeneous as a population. Contamination from disk light is an obvious concern though simulations suggest only modest bias arises to redshifts where these trends become prominent. What could be responsible for this puzzling behavior? Evolutionary synthesis modelling suggest only a modest amount of star formation corresponding to continued infall of $\simeq 5\%$ by mass would be needed to explain the bluing.

7. Conclusions

In summary, despite the frantic increase in publication rate in this field, there is an enormous amount of work still to be done, both observationally, in exploiting the connection between resolved images from HST and ground-based spectroscopy, and theoretically, in predicting more accurately the expected evolutionary histories of resolved components. In my opinion the subject suffers too much from a satisfaction with simply replicating, according to a particular theory, a range of observations. This is particularly dangerous when the observables are luminosities, colors and star formation rates since the parameters involved are numerous. The challenge will be to overcome the obvious limitations we presently face in determining galactic masses for complete samples of galaxies viewed at various look-back times, as well as integrating the growing body of data being obtained in the far infrared and sub-mm spectral regions.

I thank my students, past and present, and collaborators at Cambridge, Caltech and elsewhere for allowing me to present the results of unpublished work undertaken with them. I also thank Marc Balcells, Ismael Pérez-Fournón and Francisco Sánchez for inviting me to Tenerife to give these lectures and for their remarkable patience in waiting for this written version.

REFERENCES

ABRAHAM, R.G. MERRIFIELD, M.R., ELLIS, R.S. et al. 1999, MNRAS, 308, 569.
ABRAHAM, R.G., SMECKER-HANE, T.A., HUTCHINGS, J.B. et al. 1996, ApJ, 471, 694.
ARAGÓN-SALAMANCA, A., BAUGH, C.M. & KAUFFMANN, G. 1998, MNRAS, 297. 427.
BAHCALL, N. & FAN, X. 1998, ApJ, 504, 1.
BAHCALL, N., OSTRIKER, J.P., PERLMUTTER, S. & STEINHARDT, P.J. 1999, Science, 284, 1481.
BALBI, A., ADE, P.A.R., BOCK, J.J. et al. 2000 ApJ, 545, L1.
BALOGH, M.L., MORRIS, S.L., YEE, H.K.C., CARLBERG, R.G. & ELLINGSON, E. 1999, ApJ, 527, 54.
BARGER, A., ARAGÓN-SALAMANCA, A., ELLIS, R.S. et al. 1996, MNRAS, 279, 1.
BARGER, A., COWIE, L.L., SMAIL, I. et al. 1999b, AJ, 117, 2656.
BARGER, A., COWIE, L.L., TRENTHAM, N. et al. 1999a, AJ, 117, 102.
BARNES, J. & HERNQUIST, L. 1992, ARA&A, 30, 705.
BAUGH, C.M., BENSON, A., COLE, S.M., FRENK, C.S, & LACEY, C.G. 1999, in *Photometric Redshifts*, in press (astro-ph/99007054).
BAUGH, C.M., COLE, S.M. & FRENK, C.S. 1996, MNRAS, 283, 1361.
BAUGH, C.M., COLE, S., FRENK, C.S. & LACEY, C.G. 1998, ApJ, 498, 504.
BELL, E.F. & KENNICUTT, R. 2001, ApJ, 548, 681.
BLAIN, A. 2001, in *Starburst Galaxies: Near and Far*, eds. L. Tacconi & D. Lutz, (Berlin:-Springer), 303 (astro-ph/0011387).
BLAIN, A., SMAIL,I., IVISON, R.J. & KNEIB, J-P. 1999, MNRAS, 302, 632.
BLANDFORD, R. & NARAYAN, R. 1992, ARA&A, 30, 311.
BOUWENS, R. & SILK, J. 2000, ApJ, accepted (astro-ph/0002133).
BOWER, R.G., LUCEY, J.R. & ELLIS, R.S. 1992, MNRAS, 254, 589.
BROADHURST, T.J., ELLIS, R.S. & GLAZEBROOK, K. 1992, Nature, 355, 55.
BRINCHMANN, J. 1999, Ph.D. Thesis, University of Cambridge.
BRINCHMANN, J., ABRAHAM, R.G., SCHADE, D. et al. 1998, ApJ, 499, 112.
BRINCHMANN, J. & ELLIS, R.S. 2000, ApJ, 536, L77.
BUTCHER, H. & OEMLER, A. 1978, ApJ, 219, 18.
COHEN, J.G., HOGG, D.W. & BLANDFORD, R. et al. 2000, ApJ, 538, 29.
COLE, S.M., LACEY, C.G., BAUGH, C.M. & FRENK, C.S. 2000, MNRAS, 319, 168.

COLE, S.M., NORBERG, P., BAUGH, C.M. et al. 2001, MNRAS, 327, 1297.
COUCH, W.J., BARGER, A., SMAIL, I. et al. 1998, ApJ, 497, 189.
COUCH, W.J., ELLIS, R.S., SHARPLES, R.M. & SMAIL, I. 1994, ApJ, 430, 121.
COWIE, L.L., SONGAILA, A. & BARGER, A.J. 1999, AJ, 118, 603.
COWIE, L.L., SONGAILA, A., HU, E.M. & COHEN, J.G. 1996, AJ, 112, 839.
DADDI, E., CIMATTI, A., POZZETTI, L. et al. 2000, A&A, 361, 535.
DE BERNARDIS, P., ADE, P.A.R., BOCK, J.J. et al. 2000, Nature, 404, 955.
DRESSLER, A., OEMLER, A., BUTCHER, H. & GUNN, J.E. 1994, ApJ, 430, 107.
DRESSLER, A., SMAIL, I., POGGIANTI, B. et al. 1998, ApJS, 122, 51.
DRIVER, S.P., WINDHORST, R.A. & GRIFFITHS, R.E. 1995, ApJ, 453, 48.
EFSTATHIOU, G.P., ELLIS, R.S. & PETERSON, B.A. 1988, MNRAS, 232, 431.
EGGEN, O., LYNDEN-BELL, D. & SANDAGE, A.R. 1962, ApJ, 136, 748.
ELLIS, R.S. 1997, ARA&A, 35, 389.
ELLIS, R.S., ABRAHAM, R.G. & DICKINSON, M.E. 2001a, ApJ, 551, 111
ELLIS, R.S., COLLESS, M., BROADHURST, T.J. et al. 1996, MNRAS, 280, 235.
ELLIS, R.S., SMAIL, I., DRESSLER, A. et al. 1998, ApJ, 483, 582.
ELLIS, R.S., VAN DOKKUM, P., ABRAHAM, R. & MENANTEAU, F. 2001b, in preparation.
FALL, S.M., CHARLOT,S. & PEI, Y.C. 1995, ApJ, 464, 43.
FISCHER, P. et al. 2000, AJ, 120, 1198.
FONTANA,A.,D'ODORICO, S., POLI, F. et al. 1999, AJ, 120,2206.
FRENK, C.S. et al. 1989, *The Epoch of Galaxy Formation*, (London:Kluwer).
GALLEGO, J., ZAMORANO, J., ARAGÓN-SALAMANCA, A. & REGO, M. 1995, ApJ, 455, L1.
GARDNER, J.P., SHARPLES, R.M., FRENK, C.S. & CARRASCO, B.E. 1997, ApJ, 480, L99.
GARNAVICH, P., KIRSHNER, R.P., CHALLIS, P. et al. 1998, ApJ, 493, L53.
GLAZEBROOK, K., BLAKE, C., ECONOMOU, F. et al. 1999, MNRAS, 306, 843.
GLAZEBROOK, K., ELLIS, R.S., COLLESS, M. et al. 1995, MNRAS, 275, L19.
GRIFFITHS, R., CASERTANO, S., IM, M. & RATNATUNGA, K.U. 1996, MNRAS, 282, 1159.
GRIFFITHS, R., CASERTANO, S., RATNATUNGA, K. et al. 1994, ApJ, 435, L19.
GUNN, J.E. & OKE, J.B. 1975, ApJ, 195, 255.
HOGG, D.W., BLANDFORD, R., KUNDIC, T. et al. 1996, ApJ, 467, 73.
HORGAN, J. 1997, *The End of Science*, Abacus.
IM, M., SIMARD, L., FABER, S.M. et al. 2001, ApJ, accepted (astro-ph/0011092).
JIMENEZ, R., FRIACA, A.C.S., DUNLOP, J.S. et al. 1999, MNRAS, 305, L16.
KAUFFMANN, G. 1995, MNRAS, 274, 153.
KAUFFMANN, G. & CHARLOT, S. 1998a, in *The Birth of Galaxies*, Xth Blois Conference, in press (astro-ph/9810031).
KAUFFMANN, G. & CHARLOT, S. 1998b, MNRAS, 297, 981.
KAUFFMANN, G., GUIDERDONI, B. & WHITE, S.D.M. 1994, MNRAS, 267, 981.
KENNICUTT, R. 1998, ARA&A, 36, 189.
KOO, D.C. 1985, AJ, 90, 418.
KOO, D.C., GUZMAN, R., FABER, S.M. et al. 1995, ApJ, 440, L49.
KRISTIAN, J., SANDAGE, A.R. & WESTPHAL, J.A. 1978, ApJ, 221, 383.
KRON, R.G. 1980, ApJS, 43, 305.
LARSON, R.B. & TINSLEY, B.M. 1978, *Evolution of Stellar Populations*, Yale University Press.
LE FÈVRE, O., ABRAHAM, R.G., LILLY, S.J. et al. 2000, MNRAS, 311, 565.
LEHNERT, M.D. & HECKMAN, T. 1996, ApJ, 472, 546.

Lilly, S.J., Schade, D.J., Ellis, R.S. et al. 1999, ApJ, 500, 75.
Lilly, S.J., Tresse, L., Hammer, F. et al. 1995, ApJ, 455, 108.
Livio, M., Fall, S.M. & Madau, P. 1998, *The Hubble Deep Field*, STScI Symposium Series (New York:Cambridge University Press), 11.
Madau, P., Ferguson, H., Dickinson, M.E. 1996, MNRAS, 283, 1388.
McCarthy, P., Carlberg, R., Marzke, R. et al. 2000, in *Deep Fields*, ESO Publications in press (astro-ph/0011499).
Menanteau, F., Abraham, R.G. & Ellis, R.S. 2001, MNRAS, 322, 1.
Menanteau, F., Ellis, R.S. & Abraham, R.G. 1999, MNRAS, 309, 208.
Mo, H.J., Mao, S. & White, S.D.M. 1998, MNRAS, 295, 319.
Mobasher, B., Cram, L., Georgakakis, A. & Hopkins, A. 1999, MNRAS, 308, 45.
Mobasher, B., Sharples, R.M. & Ellis, R.S. 1993, MNRAS, 263, 560.
Mould, J.R., Huchra, J.P., Freedman, W. et al. 2000, ApJ, 529, 786.
Natarajan, P., Kneib, J-P, Smail, I. & Ellis, R.S. 1999, ApJ, 499, 603.
Ostriker, J.P. & Steinhardt, P.J. 1996, Nature, 377, 600.
Peacock, J.A. et al. 2001, Nature, 410, 169.
Perlmutter, S., Aldering, G., Goldhaber, G. et al. 1999, ApJ, 517, 565.
Peterson, B.A., Ellis, R.S., Kibblewhite, E.J. et al. 1979, ApJ, 233, L109.
Poggianti, B., Smail, I., Dressler, A. et al. 1999, ApJ, 518, 576.
Rocca-Volmerange, B. & Guiderdoni, B. 1990, MNRAS, 247, 166.
Rowan-Robinson, M., Mann, R.G., Oliver, S.J. et al. 1997, MNRAS, 289, 490.
Sandage, A.R. 1961, ApJ, 134, 916.
Sandage, A.R. 1986, A&A, 161, 89.
Sandage, A.R. & Visvanathan, N. 1978, ApJ, 225, 742.
Schade, D., Lilly, S.J., Crampton, D. et al. 1999, ApJ, 525, 31.
Schechter, P.L. 1976, ApJ, 203, 297.
Simard, L., Koo, D.C., Faber, S.M. et al. 1999, ApJ, 519, 563.
Steidel, C.C., Adelberger, K.L., Giavalisco, M. et al. 1999, ApJ, 519, 1.
Steidel, C.C., Giavalisco, M., Pettini, M. et al. 1996, ApJ, 462, L17.
Sullivan, M., Treyer, M., Ellis, R.S. et al. 2000, MNRAS, 312, 442.
Tammann, G. 1985, in Trieste review
Tinsley, B.M. 1976, ApJ, 203, 63.
Tinsley, B.M. 1977, ApJ, 211, 621.
Toomre, A. & Toomre, J. 1972, ApJ, 178, 623.
Tresse, L. & Maddox, S.J. 1998, ApJ, 495, 691.
Treu, T., Stiavelli, M., Casertano, S. et al. 2000, MNRAS, 308, 1037.
Tyson, A.J. & Jarvis, J.F. 1979, ApJ, 230, L153.
van Dokkum, P.G., Franx, M., Fabricant, D. et al. 1999, ApJ, 530, L95.
van Dokkum, P.G., Franx, M., Kelson, D.D. & Illingworth, G. 1998, ApJ, 504, L17.
Vogt, N., Phillipps, A.C., Faber, S.M. et al. 1997, ApJ, 479, L121.
Wang, L., Caldwell, R.R., Ostriker, J.P. & Steinhardt, P.J. 2000, ApJ, 530, 17.
White, S.D.M. & Frenk, C.S. 1991, ApJ, 379, 52.
Williams, R., Blacker, B., Dickinson, M.E. et al. 1996, AJ, 112, 1335.
Wyse, R. 1999, in *The Formation of Galactic Bulges*, eds. Carollo, C.M., Ferguson, H. C., & Wyse, R. F. G., (Cambridge:Cambridge University Press), 195.
Zaritsky, D., Smith, R., Frenk, C.S. & White, S.D.M. 1998, ApJ, 478, L53.

ZEPF, S.E. 1997, Nature, 390, 377
ZIEGLER, B.L. & BENDER, R. 1997, MNRAS, 291, 527.

Galaxies at High Redshift

By MARK DICKINSON

Space Telescope Science Institute, 3700 San Martin Dr., Baltimore MD 21218, USA

We study galaxies at very high redshift because (1) we can, (2) it's fun (which keeps us motivated!), and (3) we can actually *watch* cosmic history unfolding, seeing galaxies form and evolve. We are now able to find and study galaxies out to $z \approx 6$, throughout most of cosmic time. In these lectures I review some of the observational techniques and important surveys that are used to find and study high redshift galaxies, and discuss some aspects of what we have learned by doing so. These include the evolution of galaxy morphologies, spectrophotometric properties, the nature of star–forming Lyman break galaxies at $z \approx 3$, galaxy clustering at very large redshift, and our (as yet) limited knowledge of the population of galaxies at $z > 5$.

1. Introduction

In recent years, our observational knowledge of galaxies in the early universe has made advances which are so dramatic that they resemble a sort of phase transition. Ten years ago, the first large–scale faint galaxy redshift surveys were just beginning to explore galaxies at redshifts of a few tenths, spanning the last few billion years of cosmic history, and occasionally venturing out toward $z = 1$. Virtually our only direct observations of the universe at $z > 1$ came from studying rare and highly unusual objects such as quasars and powerful radio galaxies, with indirect (but highly valuable) evidence from QSO absorption line systems and the Lyman α forest. Today, we have spectroscopically confirmed redshifts for more than a thousand relatively ordinary, typical galaxies at $1 < z < 4$. We can reliably identify and study thousands of other galaxies at similar redshifts from their photometric properties, and have imaged large samples with the exquisite angular resolution of the Hubble Space Telescope (*HST*). With such data, we have begun to carry out extensive statistical surveys of galaxy properties which were previously possible only for the local universe. Galaxy luminosity functions, color distributions, star formation rates, chemical abundances, kinematics, morphologies and clustering can all be measured at high redshift. New facilities are probing the distant universe at radio, mid– and far–infrared, optical and X–ray wavelengths. A handful of galaxies and quasars are now known at $z > 5$, and the most distant confirmed objects are at $z > 6$, with unconfirmed candidates at still higher redshift. The large majority of the cosmic timeline is now open to direct scrutiny, and we are beginning to piece together the story of galaxy formation and evolution from observations, radically transforming a field that was previously the subject of theoretical speculation.

In these lectures, I will review some of the key observations that have lead to this dramatic progress, and describe some of the things we have learned about galaxy evolution from this work. To a large extent, I will restrict the scope of discussion here to galaxies at redshifts $z > 1$. This is arbitrary but practical. First, $z = 1$ still represents an approximate boundary between the redshift range that has been well–probed by traditional, magnitude-limited redshift surveys ($0 < z < 1$) and redshifts where most progress has come from specialized survey methods and selection techniques (e.g., color–selection methods such as the Lyman break technique, "extremely red objects" (EROs), radio sources and other types of AGN, etc. Second, the review by Richard Ellis in this volume deals extensively with the properties of ordinary field galaxies, drawing upon results from large redshift surveys and complementary follow–up observations, and thus paints a fairly comprehensive picture of galaxy evolution at $z < 1$. The $z = 1$ division will not be strict, and I will at times discuss galaxy properties (e.g., using results from the Hubble Deep Fields) at all redshifts, but I will concentrate on the subject of finding and studying the most distant galaxies. Also, keeping in mind the other reviews in this volume, I will generally steer away from the important and highly relevant subjects of obscured star formation and far–infrared sources (see the lectures by Alberto Franceschini) the intergalactic medium and QSO absorption lines (covered by Jill Bechtold).

In this review, I will discuss how to find high redshift galaxies, and what to do with them once you've found them. In § 2, I will review some of the ways in which we identify galaxies at high redshift. In § 3, I will briefly describe one of the most important and observational surveys from *HST*, the Hubble Deep Fields (HDFs). § 4 discusses some aspects of galaxy properties across a broad range of redshifts in the HDF and elsewhere. § 5 concentrates on the properties of Lyman break galaxies, the best–studied population of galaxies at $z > 2$. § 6 reviews galaxy clusters and clustering at $z > 2$, and § 7 briefly discusses our limited information about the universe of galaxies at $z > 5$. I finish with some notes about future observational prospects in § 8.

2. How to Find High Redshift Galaxies

How do we find galaxies at high redshift, and especially at $z \gg 1$? Astronomers have employed a wide variety of methods, from sheer brute force (e.g., magnitude limited redshift surveys of hundreds or thousands of objects) to specialized, targeted searches designed to identify the most distant objects based on special signatures which make them recognizable among the multitude of foreground galaxies. An excellent review by Stern & Spinrad (1999) discusses search techniques for high redshift galaxies in more detail than can be presented here. Here I will just summarize some of the most important and interesting methods, and in particular concentrate on certain approaches which will be relevant to other discussions later in this article.

2.1. *Look very hard*

A traditional approach to studying distant galaxies is to carry out a redshift survey of every object brighter than some magnitude limit. Major surveys such as the Canada–France Redshift Survey (CFRS, Lilly et al. 1995), the AUTOFIB survey (Ellis et al. 1996), the Hawaii redshift surveys (Songaila et al. 1994), or the Caltech Deep Redshift Survey (Cohen et al. 1999, 2000) have all contributed enormously to our understanding of galaxy evolution by measuring redshifts for large and reasonably complete galaxy samples down to faint optical and/or infrared limiting magnitudes. When searching for very high redshift galaxies, however, this method has the disadvantage that the large majority of galaxies with magnitudes that are readily accessible even to 8–10m class telescopes are at relatively low redshifts. Figure 1 shows the redshift–magnitude distribution for galaxies in the HDF–North using photometric redshifts where necessary. In order to find a significant number of galaxies at $z > 1$ requires spectroscopy to $I_{814} > 22.5$, and $z > 2$ requires working fainter than $I_{814} > 23.5$. In any case, the high redshift objects will be greatly outnumbered by foreground "interlopers." Thus while magnitude–limited samples may be valuable for many purposes, there are more efficient means of isolating just the most distant objects.

It may seem obvious now that the most distant galaxies are faint and outnumbered by more nearby objects, but this was not always considered a certainty. In particular, many early models for galaxy evolution predicted a very luminous phase of tremendously rapid star formation during the initial collapse of the galaxy which would result in quite bright

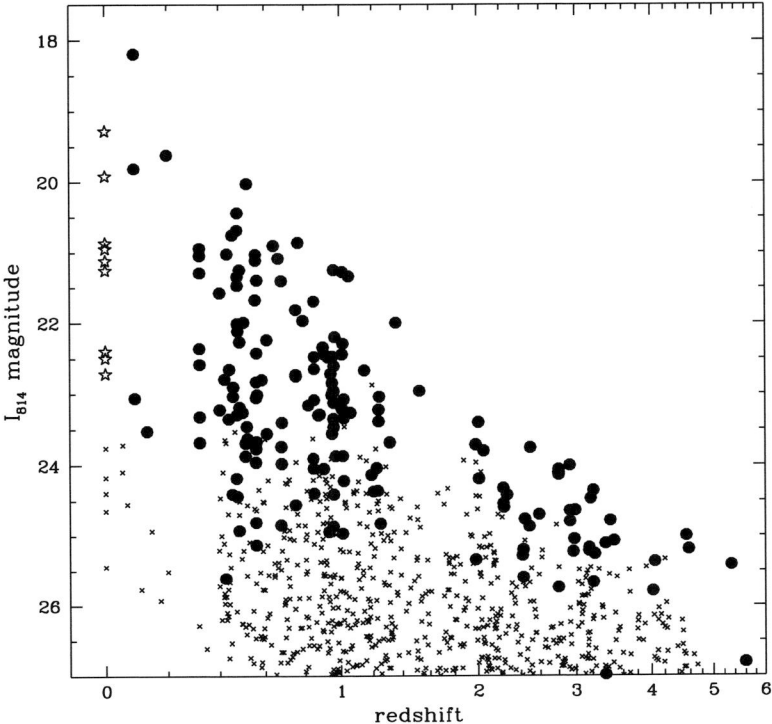

FIGURE 1. I_{814} magnitude vs. redshift for galaxies in the Hubble Deep Field North. Solid circles mark galaxies with spectroscopically confirmed redshifts and stars are confirmed galactic stars, while crosses indicate photometric redshifts derived from multi–band optical/infrared data from the *HST* WFPC2 and NICMOS instruments and ground–based K–band photometry.

magnitudes at high redshift. The large excess of "faint blue galaxies" seen even in deep photographic galaxy surveys was interpreted by some authors as evidence for this early population of bright, "primeval" galaxies. The fact that early redshift surveys did not find a substantial tail of objects extending to large redshifts motivated a revision of galaxy evolution models to suppress this early, bright phase, by invoking either gradual assembly from smaller subcomponents via hierarchical merging, or dust obscuration which might hide the UV/optical light from primeval galaxies. Today it seems that both processes may well be at work, with reasonable evidence (and a strong theoretical prejudice) for hierarchical assembly over a broad range of redshifts, and with the recent discovery of high redshift far–IR and submillimeter sources which may well represent the dust–obscured phases of early galaxy formation.

2.2. *Get some help (gravitational lensing)*

One way to ease the task of finding and spectroscopically confirming very distant galaxies is to look for objects whose light has been boosted by gravitational lensing. In this way, "ordinary" galaxies at high redshift, normally very faint, may be brought into range of even modest–aperture telescopes. A few particularly spectacular lensed galaxies at very large redshifts have been identified in this way. There are some particularly notable examples that have played an important role in the study of high redshift galaxies (here setting aside famous examples of lensed QSOs).

Lawrence et al. (1984) found that the radio source MG 2016+112 was a multiply-lensed active galaxy at $z = 3.273$, the first "non–QSO" found at $z > 2$. Recent Chandra observations (Chartas et al. 2001) have suggested that MG 2016+112 is a so–called "type 2 QSO," i.e., a quasar whose central engine is obscured from our sightline. Few such objects are known at high redshift (although arguably all high–z radio galaxies are actually type 2 quasars). F10214+4724 (Rowan–Robinson et al. 1991) is an ultraluminous infrared galaxy at $z = 2.29$ lensed by a foreground elliptical (Eisenhardt et al. 1996) and boosted above the IRAS detection limit. Although both lensing and AGN activity contribute to its remarkably bright far–IR flux, it was nevertheless the first known example of a powerful far–IR source at high redshift (of the sort now routinely studied by, e.g., SCUBA), and it has been the subject of extensive follow–up studies. cB-58 (Yee et al. 1996) is a galaxy at $z = 2.72$ amplified to $V = 20.5$ by the foreground cluster MS1512+36 (Seitz et al. 1998). This is an otherwise "ordinary" star forming galaxy similar to the Lyman break objects (see §§ 2.6.2 and 5), and gravitational lensing makes it bright enough to study in detail with high resolution spectroscopy (Pettini et al. 2000), infrared spectroscopy (Teplitz et al. 2000), and sub–mm/mm instruments (Baker et al. 2001; van der Werf et al. 2001). Warren et al. (1996, 1999) have found and studied an optical "Einstein ring," a galaxy at $z = 3.595$ lensed by a foreground elliptical galaxy. One last example, the $z = 4.92$ galaxy lensed by the cluster 1358+62 (Franx et al. 1997) was, for a brief time, the most distant object known.

These examples illustrate two of the virtues of gravitational lensing for finding and studying distant galaxies. (1) Sometimes the first examples of objects of a given type are insufficiently luminous to be detected by current technology without a boost from lensing. Thus F10214+4724 was the first object of its type known at high redshift, and no other examples would have been sufficiently luminous to be detected with IRAS without help from lensing. It may well be the case that the first galaxies at $z > 7$ (say) will be too faint to identify and confirm by ground–based infrared imaging alone, but that lensed examples could be found with a systematic infrared imaging program searching behind luminous galaxy clusters. (2) Even when examples can be found using more conventional techniques, gravitational lensing makes it possible to study some examples in greater detail or at higher S/N than would otherwise be possible. E.g., cB-58 may be a relatively ordinary Lyman break galaxy like thousands of others now known at similar redshift, but lensing makes it bright enough to be studied in ways which thousands of unlensed examples cannot.

More generally, cluster lensing has been used to boost sensitivity for statistical studies of faint, high redshift mid– and far–infrared sources from ISO (Altieri et al. 1999) and SCUBA (Smail, Ivison & Blain 1997). Lensing has proven to be particularly valuable when characterizing the number counts of infrared sources. The amplification makes it possible to count sources fainter than the normal detection limit of the instrument being used (although a good lensing model must be available to convert between lensed and intrinsic source fluxes). Moreover, lensing helps to overcome the effects of source confusion (common for mid– and far–IR observations with poor angular resolution) by "stretching" the background scene viewed behind the cluster.

2.3. *Get lucky*

Serendipity plays a surprisingly frequent role in the process of discovering very distant galaxies. Indeed, on several occasions, the most distant known objects have been found purely by chance. McCarthy et al. (1988) found a QSO at $z = 4.406$ that just happened to fall onto a long–slit observation of another, unrelated radio galaxy at lower redshift: this object nearly tied the redshift record set by another QSO at $z = 4.43$ found just a bit earlier in a large, systematic multicolor survey (Warren et al. 1987). Dey et al. (1998) broke the $z = 5$ barrier for the first time with the accidental discovery of "RD1," a galaxy at $z = 5.34$ that fell onto a slit being used to observe another galaxy with the "low" redshift of merely $z = 4.02$! Hyron Spinrad in particular has had a remarkably long and successful track record of serendipitous discoveries (as well as carefully planned ones!) of high redshift objects. The Spinrad group have turned this sort of good luck into a quantifiable process, searching for high–z, Lyman α emitters which fall into long–slit spectroscopic observations of other targets (cf. Stern & Spinrad 1999; Manning et al. 2000). Observers interested in using this method are well advised to collaborate with Hy.

2.4. Watch the fireworks

When we consider the 500 optically brightest objects in the sky, we find that they are almost all nearby: the sun, moon, planets, and relatively nearby stars. It is a remarkable fact that this is quite untrue at some other wavelengths. For example, the few hundred brightest radio sources in the sky, traced by the 3CR catalog, span the range from relatively nearby galactic sources (e.g., supernova remnants) out to $z = 2.474$, the radio galaxy 3C 257. This is largely due to the very strong cosmological evolution of the population of active galaxies, including radio sources: very powerful AGN were far more common (and perhaps more luminous as well) when the universe was young. Powerful radio sources once provided our only access to the universe of galaxies at $z > 1$, and have remained an important and interesting topic of study because they appear to occur in the rarest and, arguably, most massive galaxies in the universe, the giant ellipticals. The space density and physical properties of the most massive objects at any redshift are an important tracer of the process of galaxy evolution and of the power spectrum of density fluctuations. Thus, even today, when other methods (discussed below) have been used to find and study thousands of "ordinary," non–active galaxies at high redshifts, radio galaxies may yet offer important clues to early galaxy formation, and especially to the (perhaps ubiquitous) relation between massive black holes and the early history of galaxies. McCarthy (1993) provides a valuable review of the subject of high redshift radio galaxies, and the reader is pointed to Stern & Spinrad (1999) for a recent update.

Similarly, QSOs are tremendously bright "beacons" of UV and optical (and sometimes radio) emission, and it was a great surprise when spectra of some of the first identified quasars turned up objects at unprecedentedly large redshifts. Indeed, for many years QSOs held the redshift record, passing the $z = 1, 2, 3$ and 4 boundaries long before galaxies could be found at those redshifts. Only in the past five years have galaxies caught up to the quasars again: they greatly outnumber QSOs, making them easier to find if one can observe deeply enough, but this has required the capabilities of new 8-10m telescopes. Now galaxies and QSOs frequently trade places at the top of the redshift ladder, with QSOs having recently received a big boost from the Sloan Digital Sky Survey, which has found several objects at $z > 5$ and now even $z > 6$. It is ultimately important to understand both the populations of "ordinary" and "active" galaxies at high redshift, not only because the formation of giant black holes may be inextricably linked to the history of galaxy formation and evolution, but also because the energetic output from black holes at high redshift may be responsible for (or at least contribute strongly to) the reionization of the universe.

Finally, in recent years, gamma–ray bursts (GRBs) have joined the zoo of powerful, high–redshift exotica: they are now known to take place in galaxies at cosmological redshifts. Although the number of such objects with secure identifications and measured redshifts remains small, many are known or believed to be at $z > 1$, with examples known out to $z = 3.4$ (Kulkarni et al. 1998) and possibly even $z > 5$ (Fruchter 1999). Because we do not yet understand the physical nature of GRBs, their relevance to other aspects of galaxy formation or evolution is as yet unclear, but there are indications that they may be associated with massive star formation, perhaps explaining their prevalence at early cosmic epochs.

2.5. *Look next to something else*

Radio galaxies, quasars, and other powerful beacons at high redshift are not "typical" galaxies, and thus by themselves may offer a rather biased (if perhaps quite important) view of early galaxy evolution. However, because galaxies cluster, if one already knows where one high–z object is, then that is a good place to start looking for more. High redshift AGN are thus a natural place to begin searching for fainter cluster companions. In particular, various surveys (e.g., Yee & Green 1984, 1987; Yates et al. 1989; Hill & Lilly 1991; Ellingson, Yee & Green 1991; Allington-Smith et al. 1993; Wold et al. 2000; Blanton et al. 2000) have found that radio–loud AGN at $z \approx 0.5$ are frequently (but not always) situated in rich galaxy clusters. For this reason, many investigators have used distant quasars and radio sources as signposts in the search for rich clusters of more ordinary galaxies at high redshift.

This method of cluster hunting has several virtues. It restricts the search area to a manageable solid angle, whereas "blind" optical or X–ray surveys must cover rather large area to find rare clusters. It has a moderately high success rate at identifying clusters, and provides an *a priori* redshift (that of the radio galaxy or QSO) for any cluster candidate found. Its disadvantage is that it is difficult to use clusters found in this way for statistical studies, e.g., of the cluster luminosity or richness functions. It is primarily a good way to find individual examples of clusters whose properties (e.g., velocity dispersion, X–ray emission, etc.) or galaxy populations can then be studied in more detail.

At $0 < z < 1$, rich galaxy clusters are dominated by a population of early–type (E and S0) galaxies, whose red colors and spectral properties imply that they are dominated by an old population of stars, perhaps mostly formed at high redshift. Indeed, the study of the evolution of cluster ellipticals is a rich and active sub–field in itself, and may point the way toward the earliest epochs of galaxy formation. At $z > 1$, however, the red colors of elliptical galaxies introduce a strong bias against optical survey techniques, as k–correction effects dim the galaxies at optical wavelengths when the 4000Å break shifts through and beyond the reddest optical filters. For this reason, infrared imaging is extremely valuable and even essential for identifying clusters at $z > 1$. Many of the most distant galaxy clusters have been found using infrared imaging, even when other tracers such as X–ray emission first pointed the way to a particular spot on the sky (cf. clusters at $z \approx 1.25$ from Stanford et al. 1997 and Rosati et al. 1999). The same is true for targeted surveys around AGN. One of the first clusters at $z > 1$ identified in this way is associated with the radio galaxy 3C 324 ($z = 1.206$: Dickinson 1995a, b; Smail & Dickinson 1995). Many red, early–type galaxies are visible around the central radio galaxy in deep infrared images from the ground and from *HST*/NICMOS, and spectroscopy (see Dickinson 1997a, b, c) has shown that many share the redshift of the radio galaxy, although an additional foreground cluster or sheet of galaxies is also present at $z = 1.15$. Kajisawa et al. (2000a, b) have recently studied the 3C 324 cluster using deep near–infrared images from commissioning observations on the Subaru telescope.

By now, many other examples of likely clusters have been found associated with radio galaxies and QSOs at $0.8 < z < 1.4$. Some of the best candidates (albeit most still without spectroscopic confirmation) have been studied by Hutchings et al. (1993, 1995), Yamada et al. (1997), Tanaka et al. (2000), Chapman et al. (2000), Best (2000), and Haines et al. (2001). Few convincing (and no spectroscopically confirmed) examples of AGN–associated clusters have yet been found at $1.5 < z < 2$; the best candidates are probably two examples at $z \approx 1.5$ from Hall & Green (1998). Hall et al. (2001) have recently provided additional supporting evidence for some of these associations from narrow–band imaging of the redshifted Hα line and from photometric redshifts. In most of the cases cited above, the cluster candidates were found using infrared imaging, identified either as overdensities around the radio source or, especially, by the presence of a substantial excess of galaxies with atypically red optical–infrared colors, characteristic of early type galaxies at $z > 1$. Other than 3C 324, only a few other 3C radio galaxies at $z \approx 1$ have spectroscopic confirmation of associated clusters, such as those studied by Deltorn et al. (1997), Deltorn (1998). Liu et al. (2000) measured redshifts for four red galaxies at $z = 1.31$ associated with a MgII absorption system along the line of sight to a background QSO. It is possible, but not yet certain, that these are part of a virialized cluster.

Using narrow band imaging with a filter tuned to the wavelength of Lyman α in the field of the quasar PKS1614+051, Djorgovski et al. (1985) identified a companion object which was then spectroscopically confirmed to be at $z = 3.218$. This was the first unlensed galaxy identified at $z > 2$. Many other examples of companions to high redshift QSOs, radio galaxies, and damped Ly α systems have subsequently been identified in narrow–band imaging or spectroscopic programs. I will return to some of these, as well as a general discussion of galaxy clustering at $z > 2$, in § 6 below.

2.6. *Look smart*

Although all of the techniques described above have been used to find galaxies at $z \gg 1$, none are particularly efficient methods for a general search capable of turning up large numbers of objects. As noted in § 2.1, to a given magnitude limit, the highest redshift objects are always greatly outnumbered by a multitude of lower–redshift foreground galaxies. The most effective surveys for distant galaxies have employed special methods to "weed out" the foreground population and to isolate candidates for very high redshift objects. In recent years, these methods have opened the high redshift universe to active scrutiny, providing samples of thousands of galaxies at $z \gg 1$, and thus making it possible to carry out extensive statistical studies of their properties. Here I review a number of these methods, with particular emphasis on the Lyman break color selection technique.

2.6.1. *Emission line searches*

Young galaxies forming stars rapidly are expected to produce strong nebular line emission, and many investigators have sought to identify high redshift galaxies using observational techniques designed to detect objects by their emission line signatures. In particular, Lyman α can, in principle, be one of the strongest spectral features in young galaxies. In practice, Lyman α emission is quite easily destroyed: it is a resonant line, and therefore can undergo many scatterings through absorption and re–emission in the neutral Hydrogen ISM of a galaxy. This greatly increases the path–length it travels, and correspondingly the likelihood that a Ly α photon might encounter and be absorbed once and for all by a grain of dust. Despite this, there have been many Ly α searches for high redshift galaxies, and more recently also surveys using other emission lines which should be less subject to absorption.

Narrow band imaging surveys using interference filters or tunable filter systems are one of the most popular methods. Early searches (e.g., Pritchet 1994; Thompson & Djorgovski 1995) set only upper limits on the space density of such objects, although some individual examples were found in targeted surveys near QSOs, damped Lyman α systems, or other objects (e.g., Djorgovski et al. 1985; Lowenthal et al. 1991; Møller & Warren 1993; Giavalisco et al. 1994; etc.). Recent advances in telescope aperture and instrumentation have begun to yield more impressive results, however, pushing to fainter flux limits and surveying larger areas. Cowie & Hu (1998) and Hu, Cowie & McMahon (1998) used very deep exposures on the Keck telescope through a narrow filter tuned to Ly α at $z = 3.4$, detecting a significant number of line–emitting objects, some of which have been confirmed by spectroscopy. New, ongoing surveys on 4m telescopes (Rhoads et al. 2000; Stiavelli et al. 2001) are taking advantage of the wide fields of the new CCD mosaic cameras to survey larger total volumes, and are finding substantial numbers of candidates.

One disadvantage of narrow–band techniques compared to other methods such as broad–band color selection (e.g., the Lyman break method, see below) is that the narrow bandpass greatly restricts the co–moving volume that can be surveyed. The CCD mosaic surveys described above improve this somewhat, but one is still surveying only one thin slice of redshift space at a time. We now know that galaxies at high redshift are strongly clustered in large structures which can span transverse sizes of > 10 arcmin and many tens of co–moving Mpc (see § 6.2 below), making narrow band work potentially a hit–or–miss proposition. This can be turned to one's advantage, however, by surveying a region already known to be overdense. Some examples were given above in § 2.5. In one such program, Steidel et al. (2000) used very long exposures on the Palomar 200–inch with a filter tuned to Ly α at the redshift of a known overdensity of Lyman break galaxies (see below) at $z = 3.09$. They identified as many as 160 candidate members of this structure: some spectroscopically confirmed galaxies can even be recognized by Ly α *deficits* in the narrow–band data, due to the strong, damped Ly α absorption lines in their spectra.

Infrared narrow–band searches have begun to look for high redshift emission lines that are less subject to absorption than Ly α, such as Hα or the strong [OII] and [OIII] lines. Thus far, most successes have come from targeted surveys around QSOs, radio galaxies and damped Lyman α systems (e.g., Malkan, Teplitz & McLean 1995; Mannucci et al. 1998). The small fields of view of most infrared imagers have thus far limited the success rate of "blank field" infrared emission line searches.

Serendipitous or blind spectroscopic emission line searches have been carried out both with long–slit observations (e.g., Manning et al. 2000; Stern & Spinrad 1999; Hu et al. 1998) and with slitless observations, including infrared surveys with the *HST* NICMOS grisms (McCarthy et al. 1999; Yan et al. 1999; Hopkins, Connolly & Szalay 2000). Several groups are now pursuing creative techniques using either narrow band filters or "blind multislit" searches for emission lines that are redshifted into relatively dark "gaps" between the strong night–sky OH emission bands in the near–infrared, e.g., tuned to Ly α at $z = 6.5$ in the 9150Å airglow gap (e.g., Crampton & Lilly 1999; Stockton 2000).

2.6.2. Lyman break galaxies

In the absence of dust extinction, an actively star–forming galaxy should have a blue continuum at rest–frame ultraviolet wavelengths, nearly flat in f_ν units. Blueward of the 912Å Lyman limit, however, photoelectric absorption by intervening sources of neutral Hydrogen will sharply truncate the spectrum. This Hydrogen may be located in the photospheres of the UV–emitting stars themselves, in the interstellar medium of the distant galaxy, or along the intergalactic sightline between us and the object. When observed at some large redshift, the rest–frame Lyman limit of a galaxy shifts between

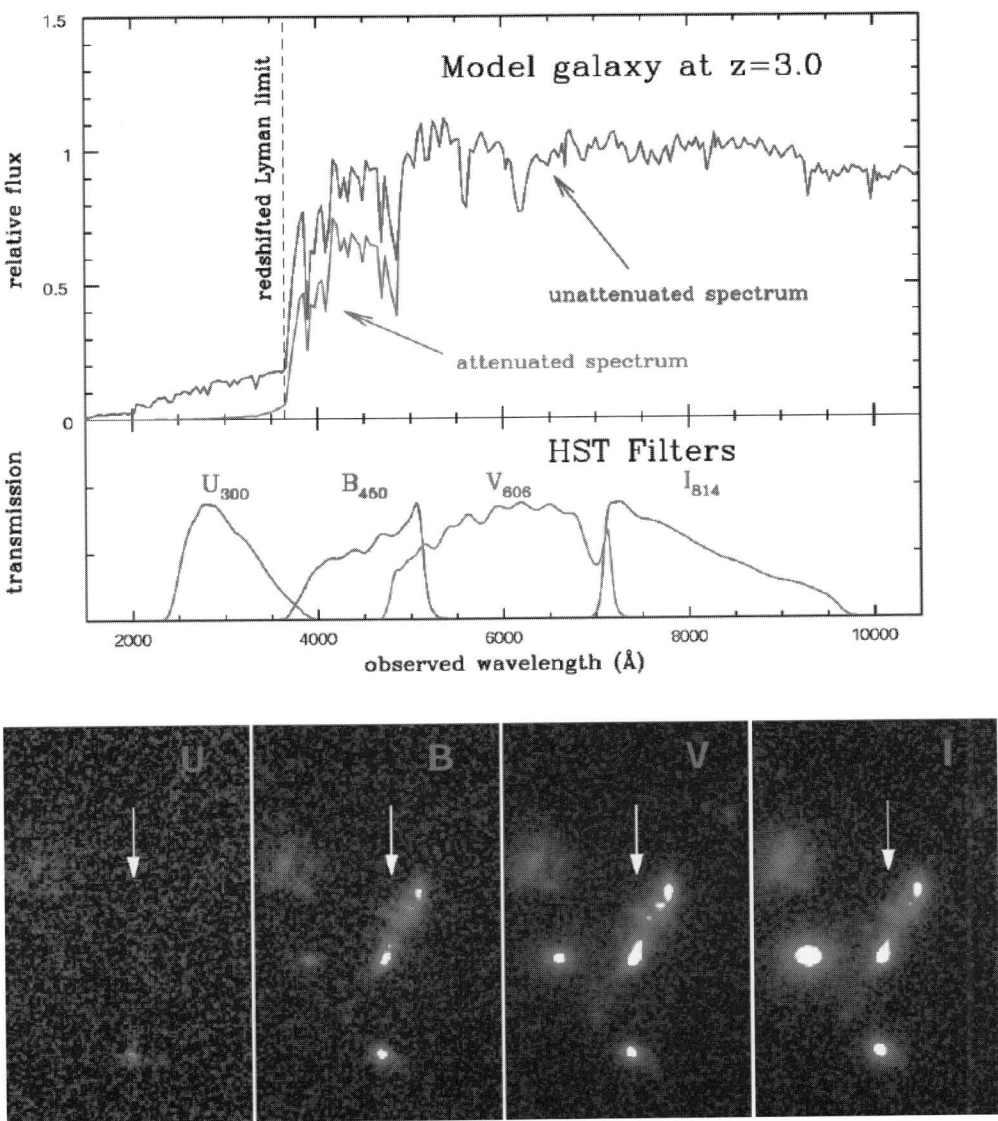

FIGURE 2. Illustration of the Lyman break technique as applied to the Hubble Deep Field. The upper panel shows a model spectrum of a star forming galaxy observed at $z = 3$. Its flat UV continuum is truncated by the 912Å Lyman limit, which is redshifted between the U_{300} and B_{450} filters (WFPC2 bandpasses shown below spectrum). In addition to photospheric absorption in the UV–emitting stars, the effects of intergalactic neutral Hydrogen further suppress the continuum in the U_{300} and B_{450} bands. At bottom, an HDF galaxy is shown in the four WFPC2 bandpasses. Clearly visible at I_{814}, V_{606} and B_{450}, it vanishes in the U_{300} image. This galaxy has been spectroscopically confirmed to have $z = 2.8$.

some pair of bandpasses (e.g. the WFPC2 U_{300} and B_{450} filters in Figure 2), and the galaxy "drops out" when viewed through the bluer filter because of the suppression of its flux. In addition to the Lyman continuum absorption, the cumulative effect of the Lyman α forest lines introduces an additional spectral break shortward of Lyman α at the emission redshift of the galaxy. This flux suppression is increasingly strong at higher redshifts as the forest thickens, and introduces its own color effects, particularly for galaxies at $z > 3$. Figure 2 demonstrates the principle of the Lyman break color selection technique using a galaxy from the HDF as an example. Figure 3 illustrates how the method is applied in practice, showing a two–color diagram for the HDF–N constructed from $U_{300}B_{450}V_{606}$ photometry. High redshift, star forming galaxies form a prominent "plume" of galaxies rising nearly vertically from the zero color point (i.e. flat spectrum galaxies) up toward very red $U_{300} - B_{450}$ colors as the Lyman break moves through the F300W bandpass, shifting them into a portion of color–color space which is unpopulated by low redshift objects.

Color selection based on the effects of the Lyman limit and Lyman α forest has been used for many years in surveys for distant QSOs (e.g. Warren et al. 1987). The method was applied to the study of distant galaxies by Guhathakurta et al. (1990) and Songaila, Cowie & Lilly (1990), who used it set limits on the number of star–forming galaxies at $z \approx 3$ in faint galaxy samples. Steidel & Hamilton (1992) and Steidel, Pettini & Hamilton (1995) reported the detection of significant numbers of high redshift galaxy candidates using this method. Spectroscopic confirmation of their redshifts was first presented by Steidel et al. (1996), and WFPC2 images of select examples were published by Giavalisco et al. (1996). To date, the majority of Lyman break selected galaxies come from the $U_n G \mathcal{R}$ survey of Steidel et al., who have spectroscopically confirmed more than 1000 galaxies at $z \approx 3$. The Lyman break technique, carefully applied, is remarkably efficient: it is not difficult to reach the faint magnitudes required with moderately long exposures on a 4m telescope, and the method probes a much larger redshift path–length and hence co–moving volume than do narrow–band methods searching for emission lines.

Madau et al. (1996) applied the Lyman break selection technique to search for both $U-$ and $B-$band "dropouts" in the HDF–North, objects at $z \approx 3$ and $z \approx 4$. That early work suggested a relative deficit of $z \approx 4$ objects compared to $z \approx 3$, and was interpreted to imply a decline in the co–moving volume density of UV luminosity and, by inference, global star formation at higher redshift. However, a much larger survey of $z \approx 4$ Lyman break galaxies by Steidel et al. (1999) (including spectroscopic redshifts measured for 48 galaxies) has found that the bright end of the LBG luminosity function is essentially unchanged over this redshift range, casting doubt on any strong evolution in their properties or the global star formation rate.

2.6.3. *Photometric redshifts*

Lyman break color selection requires three bandpasses (2 colors): one below the Lyman limit and/or Lyα forest breaks at the redshift of interest, and two above. Galaxies enter a color selection box at redshifts where one of these breaks reddens one color, and exit at some higher redshift where the breaks redden the other. In reality, galaxies do not all have the same intrinsic SEDs, nor is the opacity of the IGM (or the galaxies' ISM) uniform. This, along with photometric errors, introduces scatter so that objects move into and out of the box at somewhat different redshifts depending on their intrinsic SEDs (cf., Steidel et al. 1999). For these reasons, the method's selection efficiency is a function of redshift, intrinsic galaxy SED, and magnitude, i.e., $f(z, SED, m)$, and is not a uniform top–hat in z. The advantages of the 2–color selection method are simplicity of application and relative robustness. With a large enough sample of spectroscopic calibrators, as we now have from our ground–based survey, and with realistic Monte Carlo simulations to understand detection efficiencies and photometric errors, one can reliably model $f(z, SED, m)$, and derive the intrinsic distribution of galaxy colors and luminosities.

As an alternative, full–up photometric redshifts fit the SED of each galaxy using all

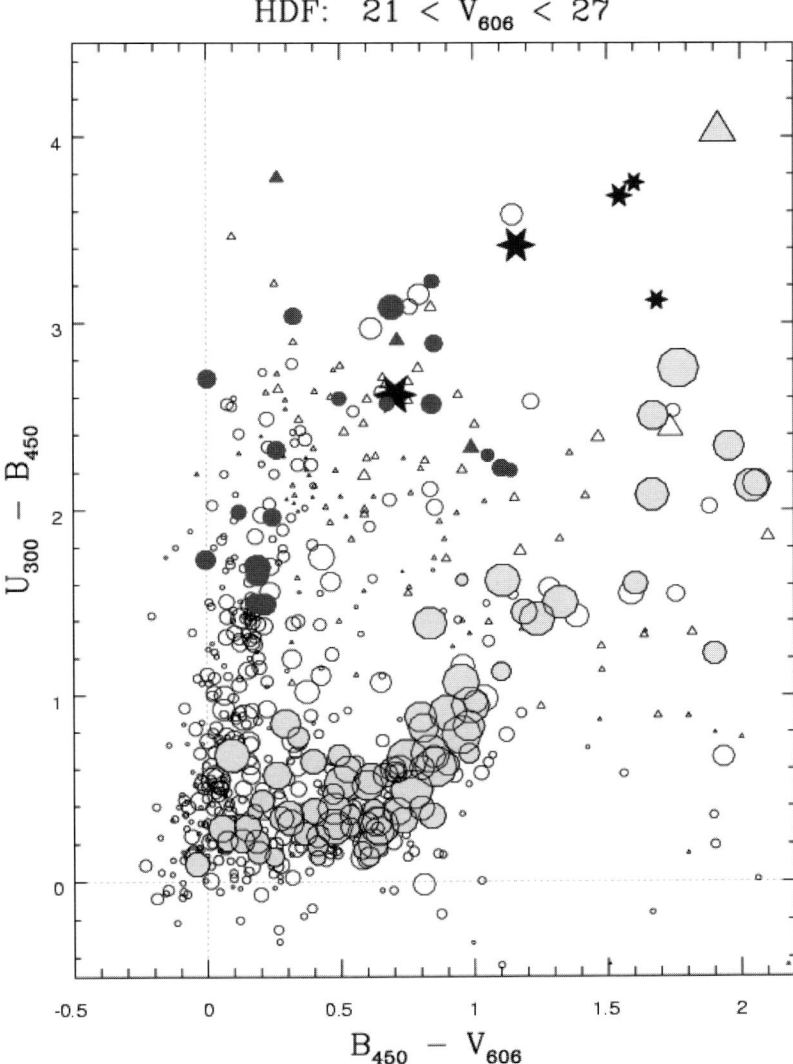

FIGURE 3. Color–color diagram of faint galaxies in the Hubble Deep Field, illustrating the "plume" of Lyman break objects rising from $U_{300} - B_{450} = B_{450} - V_{606} = 0$. These are nearly all galaxies at $z > 2$. Spectroscopically confirmed objects in this redshift range are shown as darker filled symbols; galaxies with measured redshifts $z < 2$ are shown as light filled circles, and stars are indicated by star–shaped points. Triangles mark lower limits (1σ to the $U_{300} - B_{450}$ color for objects undetected in U_{300}. Symbol size scales inversely with apparent V_{606} magnitude.

available bands. If they are well calibrated over the full range of galaxy types, phot–zs avoid the "hard edged" boundaries of the 2–color selection technique, i.e. there is no strict "redshift selection function." Phot–z fitting can also potentially identify objects whose intrinsic colors place them outside the conventional 2–color boundaries. Of course the photometric z's will have uncertainties, both random and occasionally systematic (i.e., due to poorly matched templates or multiple χ^2 minima in z_{phot}). Thus galaxies may scatter in with some error distribution $\delta z_{phot}(m, SED)$. Both color selection and photometric redshift fitting are occasionally subject to degeneracies, when objects at very different redshifts have similar observed colors, especially given real photometric errors. For example, galactic subdwarfs are the main contaminant for our ground–based $U_n G \mathcal{R}$ color–selected survey, while M dwarfs have have optical–to–near-IR colors which resemble those of $z \sim 4.5$ galaxies (cf. experience from QSO searches). Early type galaxies at $z \sim 0.8$ contaminate $G \mathcal{R} i$ selection for $z \sim 4$ galaxies (Steidel et al. 1999) and must be weeded out spectroscopically. In general, however, the successes of both methods have been impressive.

Photometric redshift estimation is now a booming industry, with several different methods (and many small variants) being used by a large number of groups. It is impractical to review all of these efforts here, or to discuss the different methods. The reader is directed to the proceedings of the OCIW workshop on photometric redshifts (Weymann et al. 1999) for many articles by the leading practitioners.

2.6.4. *Other forms of color selection and Extremely Red Objects (EROs)*

In addition to the Lyman break technique and generalized photometric redshifts, several other more "tuned" methods of color selection have been employed to identify interesting classes of high redshift objects.

Adelberger (2000) has developed a method of "Balmer break" selection designed to identify actively star–forming galaxies at $z \approx 1$ by the photometric signature of the 3646Å Balmer break in the same multicolor broad band data used for the Lyman break survey. This has been used to provide a sample of lower–redshift galaxies for comparison to the LBG population.

In any multicolor space, objects with the most extreme colors are often particularly interesting. The earliest imaging surveys using infrared arrays (e.g., Elston, Rieke & Rieke 1988) identified a population of unusually red galaxies with (e.g.) $R - K > 6$ (although this definition is not unique and many others have been used in the literature). This is much redder than colors of typical, faint field galaxies. Subsequent research has shown that these "extremely red objects" (EROs) fall into two rough categories. Some are early–type galaxies (E/S0s or early–type spirals) with old, red stellar populations whose k–corrected colors become very red at $z \gtrsim 1$. *HST* images of EROs clustered around powerful radio galaxies at $z \sim 1$ showed that they are mostly elliptical galaxies of this type (e.g., Dickinson 1995a,b). Other EROs, however, such as the famous example HR10 (Hu & Ridgway 1994; Graham & Dey 1996; Dey et al. 1999) at $z = 1.44$, appear to be dusty starburst galaxies, an interpretation borne out by sub–mm and mid–IR detections for some of them. An *HST* morphological study of EROs by Moriondo et al. (2000) finds that $\sim 15\%$ of the objects have irregular or interacting morphologies, 50–80% appear to be early–type galaxies, and the remainder are symmetric and compact but with disk–like morphology, possibly being early–type disk galaxies. The ERO population is also strongly clustered (Daddi et al. 2000; McCarthy et al. 2001), as would be expected for early–type galaxies. This adds complications for studying them with infrared imaging surveys, since it is difficult to survey a large area to sufficiently faint limits with present–day instruments, but smaller surveys suffer from dramatic variance in the surface density of EROs because of their clustering.

3. The Hubble Deep Fields

The Hubble Deep Field (HDF) observations (North and South) have been uniquely important both scientifically and in terms of scientific sociology. They offered the deepest, sharpest views of the distant universe, using very long observations from *HST* to image small patches of the sky that were carefully selected as clear sightlines out of our galaxy, but which were otherwise "unremarkable" – i.e., which provided a relatively unbiased sample of distant field galaxies.†

Perhaps even more important is the fact that the HDF data sets were non–proprietary, and fully reduced data products from the observations were generated at STScI and distributed to the astronomical community on a timescale of weeks. This has spurred an enormous amount of research activity, both in terms of analyzing the *HST* data themselves and in followup observations using a wide variety of ground– and space–based facilities. The HDFs have acted as a catalyst for research on galaxy evolution, and as a focus for subsequent observations. Virtually every large telescope operating at every wavelength, from X–ray through radio, has carried out its deepest and best observations on the HDFs, and many of these supporting observations have themselves been released to the community. More than 700 spectroscopic redshifts have been measured in the HDF–North region alone (Cohen et al. 2000), ~ 170 of which are within the 5 arcmin^2 area of the central WFPC2 HDF–N proper. Multicolor imaging from the UV through near–infrared provides photometry for thousands of galaxies and enables photometric redshift estimates for the many fainter galaxies which lack spectroscopic redshifts. Some of the deepest radio, X–ray, mid–infrared and sub–mm observations have all been carried out on the HDFs. The result is a truly rich collection of very deep multiwavelength observations, all directed on the same portions of the sky, providing a uniquely comprehensive resource for studying the evolution of galaxies, AGN, and even faint galactic stars.

The original HDF–North observations were carried out with the optical *HST* camera WFPC2, and are described in Williams et al. 1996. The HDF–South campaign used three *HST* instruments: WFPC2, the infrared camera NICMOS, and the imaging spectrograph STIS, which was used both to obtain spectra of the HDF–S QSO and to collect an extremely deep, unfiltered optical image of the field around the QSO. The HDF–S project is discussed in Williams et al. (2000), with papers by Casertano et al. (2000) and Gardner et al. (2000) describing the WFPC2 and STIS imaging observations, and papers in preparation by Fruchter et al., Ferguson et al., and Lucas et al. describing the NICMOS, STIS spectroscopy, and "flanking field" data sets.

It is impossible to discuss here all of the research programs based up the HDF data sets, or to list all of the follow–up observations that have been carried out. The review article by Ferguson, Dickinson & Williams (2000) provides the most useful summary of HDF research and observations, although already it is falling out of date, as more recent projects like the 1 megasecond Chandra observation of the HDF–North (Hornschemeier et al. 2000, 2001; Brandt et al. 2001) have been carried out and results have started to appear in the literature. I will therefore highlight only one type of HDF follow–up observation here, namely deep infrared imaging, in which I have been extensively

---

† Two possible caveats to the unbiased nature of the HDF sightlines should be noted. First, the HDF–N made some effort to steer away from relatively bright ($V \lesssim 18$) foreground galaxies. This may have introduced some bias against objects at $z \lesssim 0.3$, due to the fact that galaxies cluster and thus avoiding bright, low redshift galaxies may also bias the fields against fainter objects at similar redshifts. The co–moving volume of a WPFC2 field is extremely small at such low redshifts, however, so the value of the HDFs for studying such nearby galaxies is questionable anyway. Second, the HDF–South was selected to have a $z = 2.24$ QSO nearby, for which STIS obtained deep near–UV spectroscopic observations in order to probe gas along the sightline using absorption lines. Although the WFPC2 and NICMOS HDF–S fields are separated from the QSO sightline by several arcmin, it is quite conceivable that there could be some excess of galaxies in these fields associated with the QSO itself. No such objects have yet been identified, but to date there are relatively few published redshift measurements in and around the HDF–S.

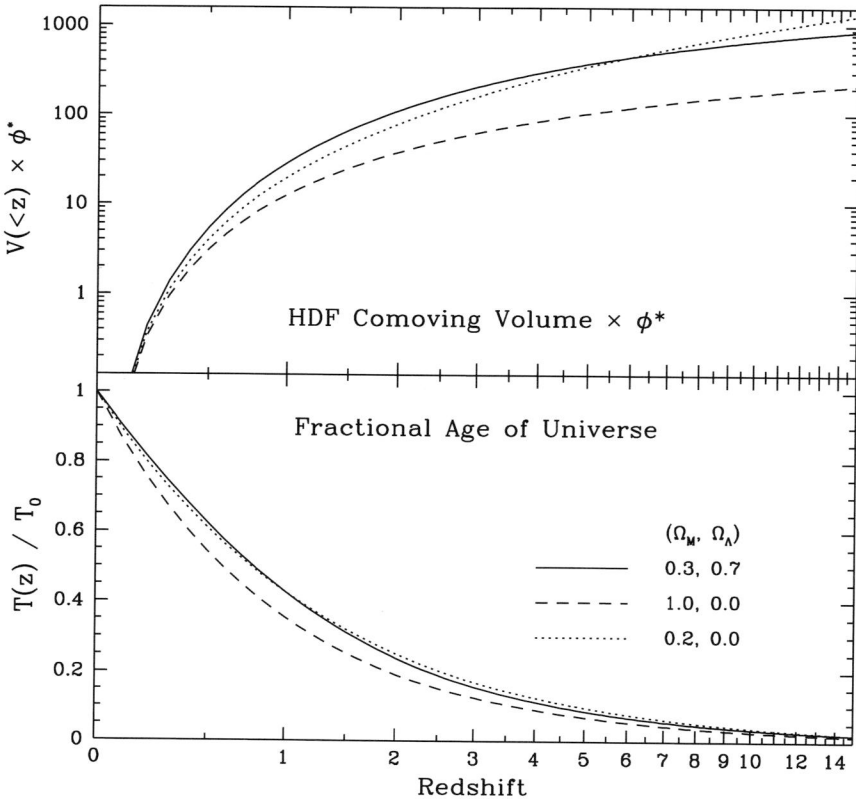

FIGURE 4. Volume and time in the HDF. The top panel shows the co-moving volume out to redshift z within one 5 arcmin^2 HDF field for several cosmologies. The volume is scaled by the present–day normalization of the galaxy luminosity function ϕ^* (here taken to be $0.0166h^3$ Mpc^{-3} from Gardner et al. 1997). This gives a rough measure of the number of "L^* volumes" out to that redshift. The bottom panel shows the fractional age of the universe versus redshift.

involved, and which will be important for some of the discussion which follows concerning the properties of high redshift galaxies.

It is always prudent to remember that although the HDF is extremely deep and thus detects galaxies over a long redshift path–length, its total *volume* is actually quite small, especially at low redshift. This is illustrated in Figure 4. Although most of cosmic time takes place at relatively low redshifts, $z < 1$, the total co-moving volume in the HDF at low redshifts is very small. At $z < 1$, we would only expect to find a handful of bright $(L > L^*)$ galaxies (or their progenitors) in the whole HDF, although this number increases to several hundred by $z = 5$. The vast majority of HDF galaxies are relatively faint, sub–L^* objects, even at fairly high redshift. It is worthwhile to keep this in mind when considering research projects with the HDF: small number statistics can limit many investigations, especially at $z < 2$. Moreover, we know that galaxies cluster very strongly even out to $z = 3$ (see § 6.2 below), which further complicates statistical studies. These were indeed strong motivations for having a second, southern HDF as a comparison to the first. However, even two HDFs cannot provide comprehensive and statistically robust answers to every question in high redshift galaxy evolution. Future *HST* instruments such as the Advanced Camera for Surveys (ACS) and Wide Field Camera 3 (WFC3, which will include an infrared channel) will greatly increase the efficiency of *HST* for deep imaging surveys over wider fields of view compared to WFPC2 and NICMOS, and will open new opportunities for a more thorough *HST* survey of the distant universe.

3.1. *Infrared observations of the HDFs*

Observing proposals and article introductions almost universally list a set of basic, interrelated themes that motivate deep near–infrared (NIR, here regarded as 1–3 μm) blank sky surveys:

- The integrated stellar spectra of normal galaxies peak in the NIR
- The stellar component of the extragalactic background peaks in the NIR
- NIR light measures familiar rest–frame optical wavelengths at high z
- Optical/NIR rest–frame light comes primarily from mid– to low–mass stars with long lifetimes relative to H_0^{-1}
- NIR light traces total stellar content/mass
- Evolutionary corrections are smaller and easier to model
- k–corrections are small or negative and similar for most galaxy types
- Effects of dust extinction are greatly reduced
- Access to $z > 6$, where galaxy light shifts beyond optical wavelengths.

The earliest, heroic efforts (Boughn, Saulson & Uson 1986; Collins & Joseph 1988) used single–element photometers to search for sky background fluctuations from primeval galaxies (PGs) whose light might be redshifted beyond the optical wavelength range. The field really came to life with the advent of array detectors, leading to the first faint near–infrared imaging surveys (Elston, Rieke & Rieke 1988; Cowie et al. 1988). It is interesting to read these and other early papers, where we find discoveries, concerns and hypotheses that have stayed with us ever since: PGs, extremely red objects ("EROs" as PG candidates, or high–z ellipticals, or dust–enshrouded galaxies), extremely blue objects (with rapid, cosmologically significant star formation), ERO clustering, UV–excess ellipticals, photometric redshifts, infrared number counts (to constrain space curvature and/or galaxy evolution), etc. While many of the issues remain the same today, the data quality has advanced tremendously, largely driven by progress in array technology, and most recently by the leap into space with *HST*/NICMOS (affording high angular resolution and far lower backgrounds). Survey sensitivities have improved by a factor of ~ 1000, source densities have increased ~ 200–fold, and we can now image the detailed morphologies of high redshift galaxies in their rest–frame optical light.

The *HST* WFPC2 and STIS observations of the Hubble Deep Fields (HDFs, North and South) are the deepest optical images of the sky, and correspondingly deep near–infrared observations of these areas are valuable for all the reasons outlined above. The HDF–N was observed from the ground in several different NIR programs (Hogg et al. 1997; Dickinson 1998; Barger et al. 1999; Hogg et al. 2000a,b), while the HDF–S has been imaged from the ESO NTT (da Costa et al. 1998) and more recently to with ISAAC on the VLT (Franx et al. 2000). The deepest observations at 1.1–1.6 μm have come from *HST* NICMOS imaging of the HDF–N (Thompson et al. 1999; Dickinson 1999) and for the HDF–S NICMOS field (Fruchter et al. 2001; this is distinct from the HDF–S WFPC2 field). The depth and angular resolution of our "wide–field" (only ~ 6 arcmin^2, smaller than the first NIR array surveys!) HDF–N/NICMOS program, combined with the great wealth of supporting imaging and spectroscopy at other wavelengths from the ground and from space, make this a premier resource for studying the near–infrared properties of galaxies at high redshift.

4. Properties of Distant Galaxies in the HDF and Elsewhere

4.1. *Morphologies*

One of the most striking results from early *HST* imaging was the abundance of irregular, peculiar–looking galaxies seen in deep images. The population of giant (as opposed to dwarf) galaxies in the nearby universe is dominated by ordinary, Hubble sequence spirals and ellipticals. Irregular galaxies are generally fairly low luminosity objects. Bright galaxies whose morphologies are distorted due to interactions or collisions are quite rare. Analyses of *HST* number counts divided by morphological type (e.g., Casertano et al. 1995; Driver et al. 1995; Glazebrook et al. 1995) have found that the abundances of faint spiral and elliptical galaxies are generally in line with what would be expected by extrapolation from the local galaxy population with only mild evolution out to $z \approx 1$. In the HDF, $\sim 40\%$ of galaxies at $I_{814} = 25$ fall into the irregular/peculiar/merging category (Abraham et al. 1996). However, the number of irregular galaxies greatly exceed such predictions, implying that such objects were more common and/or more luminous at high redshift.

From optical WFPC2 images alone it is difficult to assess the reasons for this over–abundance of faint irregular galaxies. Broadly speaking, there are two major reasons why galaxies might appear to be irregular. One is that they could be *structurally* disturbed, e.g., due to interactions, mergers, or simply due to youth, and have not had time to relax and settle into regular spiral or elliptical forms. Since hierarchical models predict a higher incidence of galaxy interactions and mergers in the past, this seems like a plausible explanation. However, it is not unique: "morphological k–correction" effects might also play a role. At high redshift, especially $z > 1$, optical instruments like WFPC2 are primarily viewing the rest–frame ultraviolet light from distant galaxies, light that is emitted by short–lived, massive stars, and modulated by dust extinction. Even in relatively ordinary nearby galaxies, the distribution of star formation can be asymmetric due to a certain degree of stochasticity in the process. Hence, the UV morphologies of star forming galaxies can appear to be irregular, even if the older stars that make up the bulk of the galaxy mass are in fact quite smooth and symmetric (e.g., O'Connell 1997; Kuchinski et al. 2001). At $R = 24$, the median redshift in the HDF (Cohen et al. 2000) is $\langle z \rangle = 1$, and at $I = 25$ (where Abraham et al. 1996 found that 40% of HDF galaxies are irregular), a substantial majority of objects should be at $z \gtrsim 1$, where WFPC2 samples rest frame ultraviolet light.

NICMOS on *HST* provided the opportunity to image faint, high redshift galaxies at rest–frame optical wavelengths with high angular resolution. The WFPC2+NICMOS HDF data set allows us to form images of galaxies at fixed rest frame wavelengths over a wide range of redshifts. Figure 5 illustrates several galaxies with peculiar WFPC2 morphologies, interpolating between bandpasses to form images at rest frame wavelengths 3000Å and 6500Å. The irregularities in these objects tend to be preserved across the UV–to–optical wavelength baseline: dramatic transformations, where peculiar objects are revealed to be comparatively ordinary galaxies at longer wavelengths, are comparatively rare. The structure of "chain galaxies" like 2-736.1 ($z = 1.355$, lower right in Figure 5) is almost entirely unchanged from 1300Å to 6800Å in the rest frame. Such structures are unlikely to be stable or to persist for long given the nominal dynamical time scales for these galaxies (e.g., Cowie, Hy & Songaila 1995). Broadly speaking, it appears that genuine *structural* disturbance, perhaps due to interactions and mergers, is primarily responsible for the irregular morphologies of high redshift galaxies, not band–shifting effects. In his review in this volume, Richard Ellis describes a mass census of galaxies by morphology out to $z \approx 1$ and its implications for the fate of these irregular galaxies.

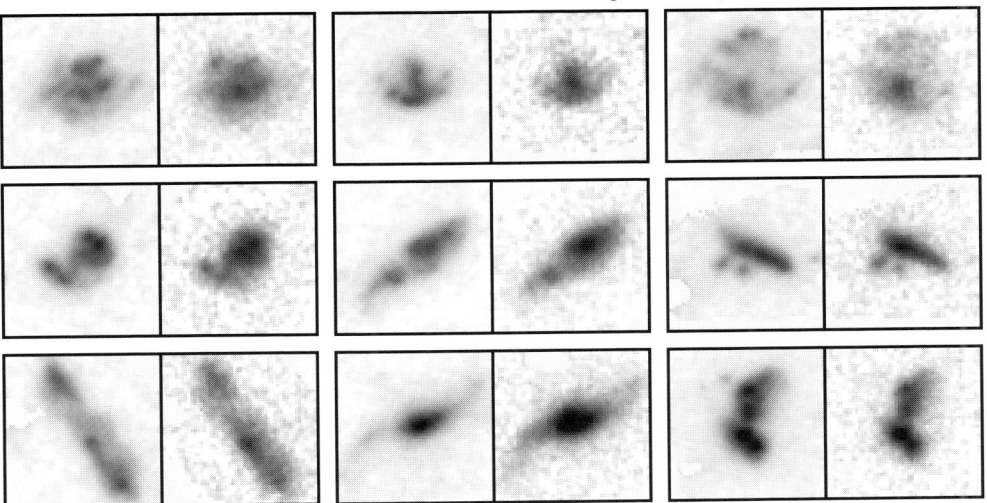

FIGURE 5. HDF–N objects with irregular morphologies at $0.75 < z < 1.36$, at rest frame 3000Å (left) and 6500Å (right). Box sizes are $\sim 32 h_{70}^{-1}$ kpc. These galaxies are large and quite luminous by comparison to typical irregular galaxies in the local universe. Although some of the galaxies are more centrally concentrated at longer wavelengths, in general the peculiar morphologies are preserved from over long wavelength baselines.

Giant spiral and elliptical galaxies, including ellipticals with very red, apparently old stellar populations, are present in the HDF and elsewhere in reasonably large numbers out to at least $z = 1.3$. The HDF-N/NICMOS data provides the only opportunity at present to assess the morphological mix at rest–frame optical wavelengths at higher redshifts. There are very few spectroscopic redshifts in the HDF in the "redshift desert" at $1.3 < z < 2$, where most strong spectral features shift out of the optical band (and where Lyα and the Lyman break have not yet shifted in from the UV). Therefore we must rely on photometric redshifts. Bearing that in mind, it does appear that the population of giant, easily recognizable spiral galaxies and red, "old" ellipticals does thin out considerably past $z_{phot} > 1.3$ (cf., Dickinson 2000a). There is one large and inherently quite luminous galaxy (HDF 4-948) with recognizable (if somewhat peculiar) spiral structure at $z > 1.3$, and which now has a tentative spectroscopic redshift of $z = 1.524$ reported by Dawson et al. 2001. Several other, fainter galaxies clearly have disk–like morphologies, albeit without clear spiral structure. There are at least two quite luminous, red, early–type galaxies with photometric redshift estimates $z \approx 1.7$. One of these (HDF 4-403.0) was the host of the supernova SN 1997ff (Gilliland, Nugent & Phillips 1999), now believed to be the most distant type Ia supernova known, and which has cited as demonstrating the transition between a decelerating and accelerating universe as predicted by world models with a cosmological constant or other forms of "dark energy" (Riess et al. 2001). The very existence of an apparently old, massive, red elliptical at $z \approx 1.7$ is itself cosmologically interesting, implying that some massive galaxies formed and assembled most of their stars at very high redshift. However, this galaxy (and the one other, nearby elliptical with a similar photometric redshift) seem to be the "final frontier" for red ellipticals in the HDF–N, at least: there are no good candidates for similar objects at higher redshifts. The HDF–S field also has several red ellipticals with photometric redshifts up to $z = 2$ (Stiavelli et al. 1999; Benítez et al. 1999), but few if any good candidates at higher redshift. Using earlier, pre–NICMOS data, several authors had suggested that the population of red, bright, early–type galaxies

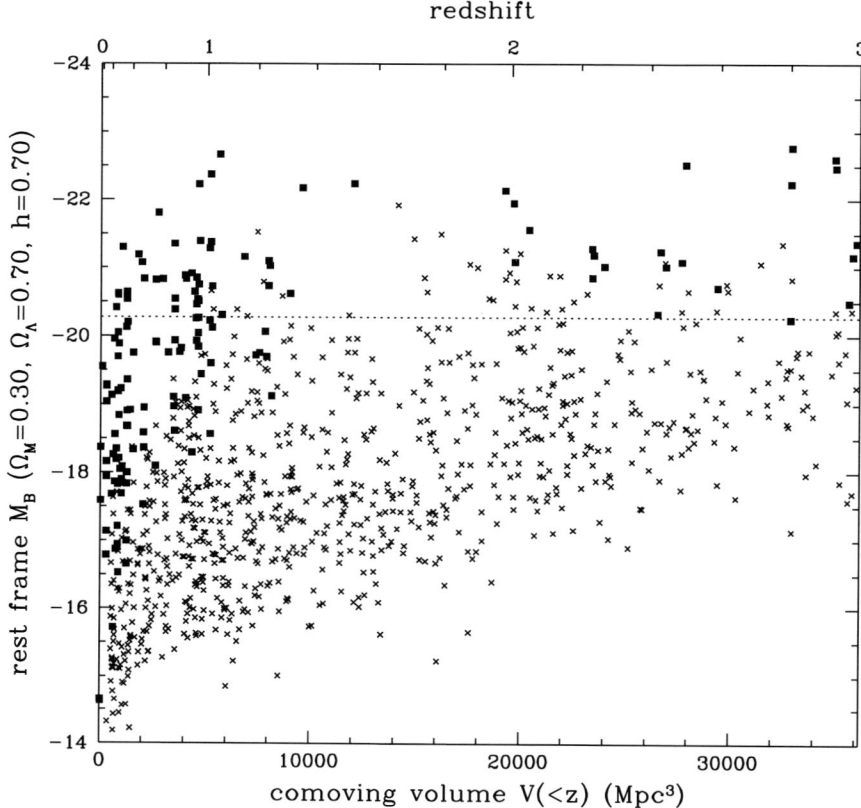

FIGURE 6. Rest–frame B–band absolute magnitudes for HDF–N galaxies vs. redshift (top axis). The horizontal scale is set to measure the co–moving volume out to redshift z (bottom axis). In this way, a constant density of points in the x–direction represents a constant space density of objects. A flat cosmology with $\Lambda = 0.7$ is assumed. The horizontal dotted line marks the B–band luminosity of a present–day L^* galaxy.

thins out at $z > 1$ (e.g., Zepf 1997; Franceschini et al. 1998; Barger et al. 1999), although this is not uncontroversial (e.g., Eisenhardt et al. 1998; Benítez et al. 1999; McCracken et al. 2000). From the limited fields surveyed by NICMOS, we can only add that there are few, if any, candidates for old, red ellipticals at $z > 2$.

4.2. Photometry

The availability of deep NICMOS photometry at 1.1 and 1.6μm enables us to observe the rest–frame B–band light from galaxies out to nearly $z = 3$. Given the redshift of an object (spectroscopic, or, when necessary, photometric), we can determine its photometric or morphological properties in the optical rest frame (see, e.g., Figure 5 above). Figure 6 shows the distribution of rest–frame B–band absolute magnitudes for HDF galaxies out to $z = 3$. The upper envelope of the luminosity distribution of galaxies remains fairly uniform out to $z = 3$. There are (coincidentally) a round 100 galaxies with $L_B \geq L_B^*(z = 0)$ in the HDF at $0 < z < 3$. This "100 brightest galaxies" sample is an interesting, volume–limited subset of objects to consider in some detail.

Figure 7 shows a montage of rest–frame B–band images for these galaxies ordered by

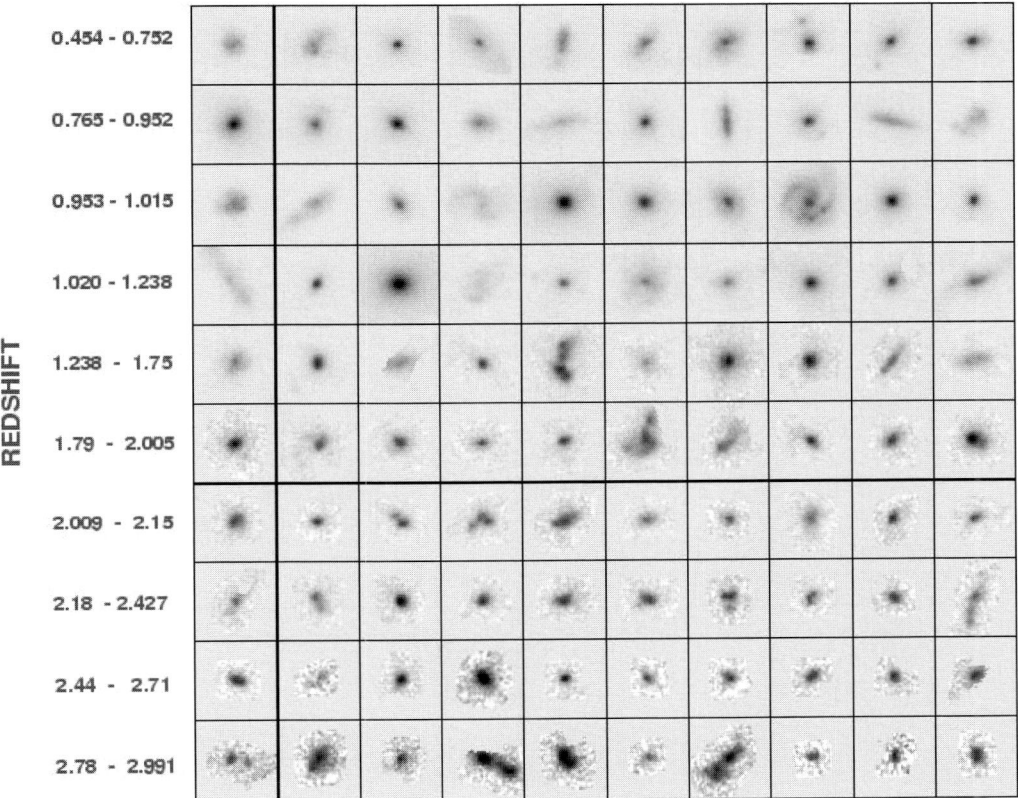

FIGURE 7. Rest–frame B–band images of the 100 brightest galaxies in the HDF (rest–frame B–band $L > L^*(z = 0)$) at $0 < z < 3$. The images were formed by pixel–wise wavelength interpolation between WFPC2 and NICMOS data, and have been scaled to have the same physical scale and rest–frame surface brightness.

redshift, all scaled to the same physical size and rest–frame surface brightness. Careful inspection shows the tendency for galaxies to become smaller and often more irregular at higher redshifts, and the disappearance of anything resembling face–on or thin, edge–on spiral disks beyond $z = 2$.

In Figure 8 we see the rest–frame $U - B$ colors of these galaxies versus redshift. There is a clear trend for the colors of these most luminous objects to get bluer with increasing redshift. In particular, nearly all of the galaxies at $2 < z < 3$ are bluer than almost *any* galaxy of comparable luminosity at $z < 1$. In this sample there is only one moderately red galaxy with a (photometric) redshift in this range, marked with an arrow and noted in the caption. The natural interpretation is that the very blue $U - B$ colors for the $z > 2$ galaxies are due to very active star formation in nearly all of these objects. We might have expected the Lyman break galaxies to have active star formation and blue colors, since they are *selected* based on their strong UV continuum and bluer UV colors (see § 2.6.2 above). What is new here is that almost all $z > 2$ galaxies in an *infrared–selected* sample, limited by rest–frame B–band luminosity rather than UV continuum, are as

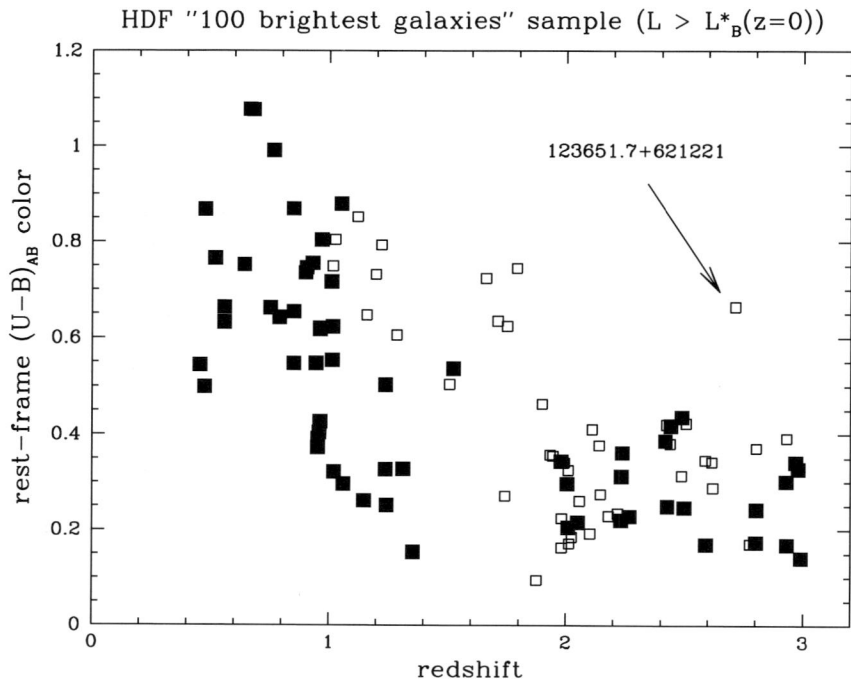

FIGURE 8. Rest–frame $U-B$ color vs. redshift for HDF–N galaxies with rest–frame $M_B < -20.3$ ($\Omega_M = 0.3$, $\Omega_\Lambda = 0.7$, $h = 0.7$). Filled and open symbols indicate galaxies with spectroscopic and photometric redshifts, respectively. There is a strong color trend with redshift; nearly all galaxies at $2 < z < 3$ are bluer than comparably luminous galaxies at $z < 1$. One redder object with $z_{\rm phot} = 2.7$ is marked; this object is also a radio (Richards et al. 1999) and hard X–ray source (Hornschemeier et al. 2000), and may be detected at 15 μm (Aussel et al. 1999) and 1.3 mm (Downes et al. 1999) as well.

blue as the Lyman break galaxies. Redder galaxies might be missed by the Lyman break selection process, but they should be picked up in an infrared catalog using photometric redshifts, and instead few are actually found. This also holds for HDF galaxies fainter than the limit considered here (see Dickinson 2000b and Papovich et al. 2001). As noted earlier, the HDF is a small volume and thus we cannot say with confidence that *no* red galaxies exist at $z > 2$, but it does seem that we may consider the properties of Lyman break–selected galaxies to be representative of those of the large majority of the bright galaxy population at these large redshifts.

5. Properties of Lyman Break Galaxies

5.1. *Photometric properties and star formation rates*

By selection, Lyman break galaxies are objects with strong ultraviolet continuum emission due to active star formation. Hot, massive OB stars produce UV continuum spectra

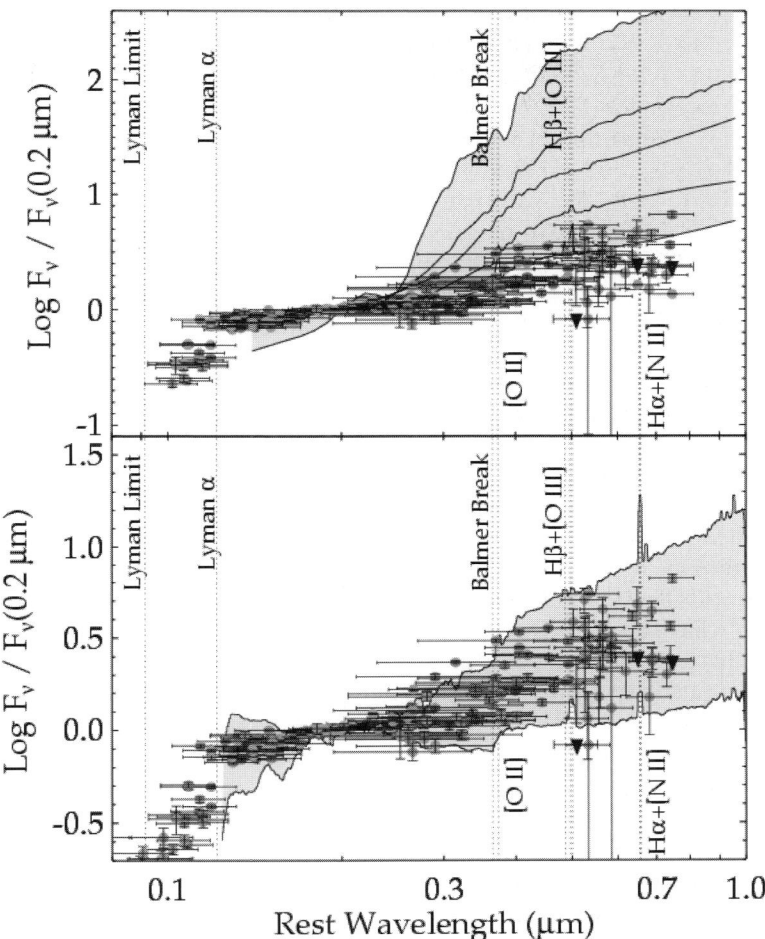

FIGURE 9. Rest–frame UV to optical spectral energy distributions for a collection of Lyman break galaxies in the HDF–North, normalized at 2000Å in the rest frame, from Papovich et al. 2001. In the top panel these are compared to template spectral energy distributions (SEDs) for ordinary Hubble sequence galaxies, ellipticals (reddest) through spirals and irregular galaxies (bluest), taken from Coleman, Wu & Weedman 1980. The LBGs are much bluer than present–day giant galaxies, but are well within the envelope of spectral shapes for local, UV–bright starburst galaxies (bottom) compiled by Kinney et al. 1996. The LBGs do show a rise in flux at optical wavelengths, redward of the Balmer and 4000Å region, indicating the presence of older starlight (A–type and later).

that are a roughly flat with wavelength (or somewhat bluer) in f_ν units. For reasonably continuous star formation, the UV continuum luminosity produced by OB stars is proportional to the star formation rate, and thus optical (rest–frame UV) photometry of LBGs has frequently been used to estimate their star formation rates, and even to infer the global star foramtion history of the universe (e.g., Madau et al. 1996; Madau, Pozzetti & Dickinson 1998). However, photometry and spectroscopy of actual Lyman break galaxies show that their observed UV continua are generally somewhat redder than flat spectrum, a fact which is generally interpreted as indicating the presence of some amount of dust extinction (cf. Meurer et al. 1997, 1999).

In Figure 8 above, we saw that LBGs are much bluer than most ordinary galaxies at lower redshift. We can see this again in Figure 9, which compares multiwavelength optical–to–infrared photometry (tracing the UV through optical rest–frame light from the galaxies) for a collection of HDF LBGs to template spectral energy distributions of nearby Hubble sequence spirals and ellipticals (top panel) and local, UV–bright starburst galaxies (bottom). The LBGs are much bluer than local spirals and ellipticals, demonstrating that the ratio of active to past star formation (i.e., newly formed stars to older stellar populations) is much larger than in galaxies today. Their SEDs are generally quite similar to those of local starbursts, however, with modest UV extinction.

Many authors have made estimates of extinction and star formation rates in Lyman break galaxies and other objects at high redshift, a topic which I will not review here in detail. There is very little direct and unambiguous data about extinction corrections or star formation rates for high–z LBGs and other objects. Such corrections are often made by assuming that local starburst dust attenuation laws (e.g., Calzetti et al. 2000) or empirical UV–to–far–IR correlations (Meurer et al. 1999) also hold for distant galaxies: a reasonable but largely untested hypothesis. The reader is directed to Adelberger & Steidel (2000) for a recent analysis incorporating a variety of multiwavelength constraints, and for discussion (not uncontroversial) of the relation between LBGs and high redshift sub–mm sources. Those authors find that the mean UV extinction for LBGs is a factor of 5 to 9, and that the objects with the highest star formation rates suffer the greatest extinction. This interpretation might naturally be extrapolated to the relatively rare but very luminous and highly obscured sub–mm sources, perhaps suggesting a continuity of properties rather than two distinct populations of objects. It might be hoped that infrared spectroscopy of nebular emission lines (especially Balmer lines) would provide an independent measure of star formation that is less subject to extinction than are rest–frame UV continuum measurements. However, Pettini et al. (2001) have obtained such data for ~ 20 LBGs, and found that star formation rates derived from Hβ are in general quite similar to those from the UV continuum. This is quite probably because ionized gas in HII regions suffers still greater extinction than does the light from less massive and longer–lived B–stars that dominate the UV continuum, thus partially or wholly cancelling the benefits of moving to longer wavelengths.

5.2. *UV and optical rest-frame spectra*

Rest–frame UV spectroscopy (i.e., from optical observations) of LBGs provides an additional basis of comparison to local starburst galaxies. Typical spectra of faint ($R \approx 25$) LBGs have low S/N, and only the strongest features can be seen clearly, even with 8–10m telescopes. These are generally interstellar metal absorption lines due to gas within the galaxy. Lyman α appears in *absorption* rather than emission for roughly half the objects. Although Ly α should be produced by star formation, it is easily absorbed by small amounts of dust due to resonant scattering within the galaxy ISM. A broad trough of damped Ly α is frequently seen. Lines such as CIV and NV frequently show P Cygni profiles, indicating the presence of strong stellar winds from OB and Wolf–Rayet stars. Actual stellar absorption features can only be seen for the brightest LBGs with the highest S/N spectroscopy, such as the lensed example, cB-58, discussed above in §2.2 and shown here in Figure 10.

The features seen in the UV spectra of LBGs are generally quite similar to those observed in space–based UV spectroscopy of nearby starburst galaxies. Pettini et al. (2000) (see also de Mello, Leitherer & Heckman 2000 and Leitherer et al. 2001) find that the

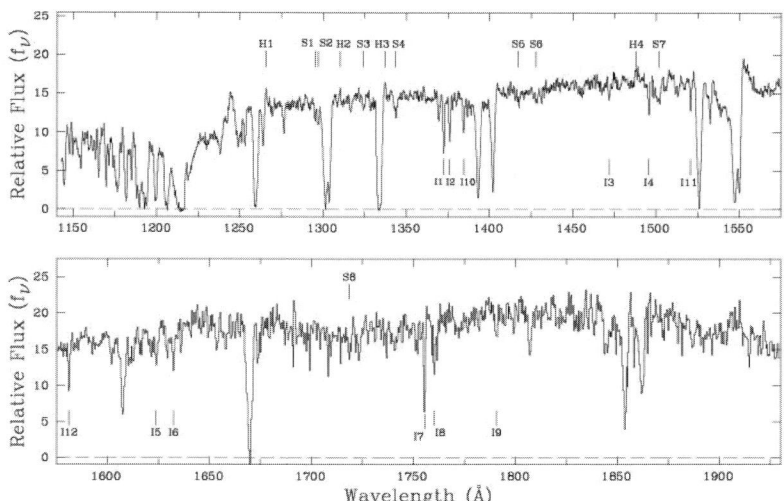

FIGURE 10. Rest–frame UV spectrum of the $z = 2.72$ galaxy cB-58, from Pettini et al. 2000. Gravitational lensing boosts the flux from this Lyman break galaxy, making it possible to observe it with high signal–to–noise ratios at moderately high dispersion. A large number of absorption features can be seen. Most of the strong absorption lines arise from the interstellar medium of the galaxy itself, and are due to species of S, Si, Al, O, C, N, Ni and Fe, as well as a broad, damped Ly α profile from neutral Hydrogen. Weaker stellar features (marked with "S" above the spectrum) can only be seen thanks to the high S/N enabled by the lensing. Features marked with "I" are metal absorption features from intervening material along the sightline to cB-58, like those seen in QSO spectra.

P Cygni profiles are well matched by continuous star formation from a stellar population whose IMF extends to $M > 50~M_\odot$ (less massive stars do not produce sufficiently strong winds). Analyses of both the P Cygni features and the interstellar absorption lines suggest metallicities ~ 0.2–$0.3 \times Z_\odot$. For cB–58, this estimate is supported by analysis of nebular emission lines measured with infrared spectroscopy (Teplitz et al. 2000).

Infrared nebular line spectroscopy of ordinary, unlensed LBGs is a daunting task even with 8–10m telescopes and new instruments such as Keck/NIRSPEC and VLT/ISAAC. In addition to providing another tracer of star formation (see above), measurements of rest–frame optical emission lines such as [OII], [OIII], Hβ, Hα and [NII] offer another means to measure chemical abundance in LBGs. Unfortunately, at $z \approx 3$, where most "U–dropout" LBGs are found, Hα and [NII] are redshifted beyond the limits of the K–band, and it is difficult to robustly measure abundances from the Oxygen and Hβ lines alone. The widely–used indicator

$$R_{23} \equiv (F_{5007} + F_{4959} + F_{3727})/F_{H\beta}$$

is sensitive to Oxygen abundance, but is generally double–valued at low metallicity. Pettini et al. (2001) have found that LBG abundances can generally be determined only to within one order of magnitude. Nevertheless, even their lower limits are interesting, showing that LBGs generally have abundances $> 0.1 Z_\odot$, i.e., greater than those of most damped Ly α (DLA) QSO absorption line systems at similar redshifts. Since DLAs are believed to contain most of the neutral gas in the universe at high redshift, this suggests that the active star formation in LBGs is spatially segregated from the bulk of the available gas reservoir at similar redshifts, which has yet to be consumed or heavily polluted by star formation.

An interesting pattern emerges when one compares LBG redshifts measured from nebular emission lines, interstellar UV absorption lines, and Ly α emission (Pettini et al. 2000, 2001). The ISM absorption lines are blueshifted by -200 to -400 km/s relative to the nebular emission lines, while the Ly α emission (when present) is redshifted by +200 to +1100 km/s. This is interpreted as evidence for galactic scale outflows or "superwinds," driven by star formation in the galaxies. The nebular emission from HII regions traces the systemic redshift, while the outflowing ISM is seen as blueshifted absorption. Ly α is wholly absorbed by the large column of foreground neutral gas, and can only be seen from the back of the expanding shell, behind the stars, where it is redshifted to velocities where foreground absorption does not affect it. These powerful outflows from LBGs may substantially contribute to the distribution of metals into the intergalactic medium, as observed in measurements of the Lyman α forest in QSO spectra.

5.3. Stellar populations

Papovich, Dickinson & Ferguson (2001) have carried out a detailed study of the stellar population properties of star–forming "Lyman break galaxies" (LBGs) from the HDF–N at $2 < z < 3.5$, using NICMOS data to extend previous work based on ground–based near–infrared photometry (Sawicki & Yee 1998). In addition to comparisons with empirical spectral templates of nearby galaxies (e.g., Figure 9 above), we have fit stellar population synthesis models (Bruzual & Charlot 1993, 2000; see also Gustavo Bruzual's contribution to this volume) to the galaxy photometry, varying their star formation histories and dust extinction. Even with high–quality *HST* optical/infrared photometry, we find only weak constraints on most parameters of the LBG stellar populations, with degeneracies between age, star formation time scale, metallicity and extinction. Perhaps the best constraints, however, are those on the total stellar mass $\mathcal{M}$ (see Figure 11). If the LBG star formation history $\Psi(t)$ is modeled by an exponential, $\Psi \propto e^{-t/\tau}$, with t and τ (and extinction) as free parameters, then with fixed assumptions about the IMF and metallicity, the typical 68% confidence interval on log $\mathcal{M}$ is approximately ± 0.25 dex. For LBGs with L^* UV luminosities (Steidel et al. 1999), the inferred stellar masses (assuming a Salpeter IMF, and varying the model metallicities) are 1 to $2 \times 10^{10} \mathcal{M}_\odot$ for $\Omega_M = 0.3$, $\Omega_\Lambda = 0.7$, $h = 0.7$. These are $\sim 1/10$th the stellar masses of L^* galaxies today (Cole et al. 2001). We may compare these masses to those derived from virial estimates using nebular line–widths and *HST*–measured sizes (Pettini et al. 2001), which are also $\sim 10^{10} \mathcal{M}_\odot$. This suggests that these kinematic measurements underestimate the full mass of the dark matter halo.

We may set an upper bound on the allowable stellar mass by considering how much light from a hypothetical, maximally old stellar population (formed at $z = \infty$) could be hidden beneath the glare of the young, star–forming population. On average, this upper bound is a factor of ~ 5 to $6\times$ larger than the mass derived for the "young" models. If this were generally the case, however, then virtually all galaxies with previous generations of star formation at $z \gg 3$ must *also* be forming stars rapidly at $2 < z < 3$. We see very few candidates for mature, red, non–star–forming galaxies in this redshift range, even with a NICMOS–selected sample where photometric redshifts should, in principle, readily identify such galaxies if they are present (see Figure 8).

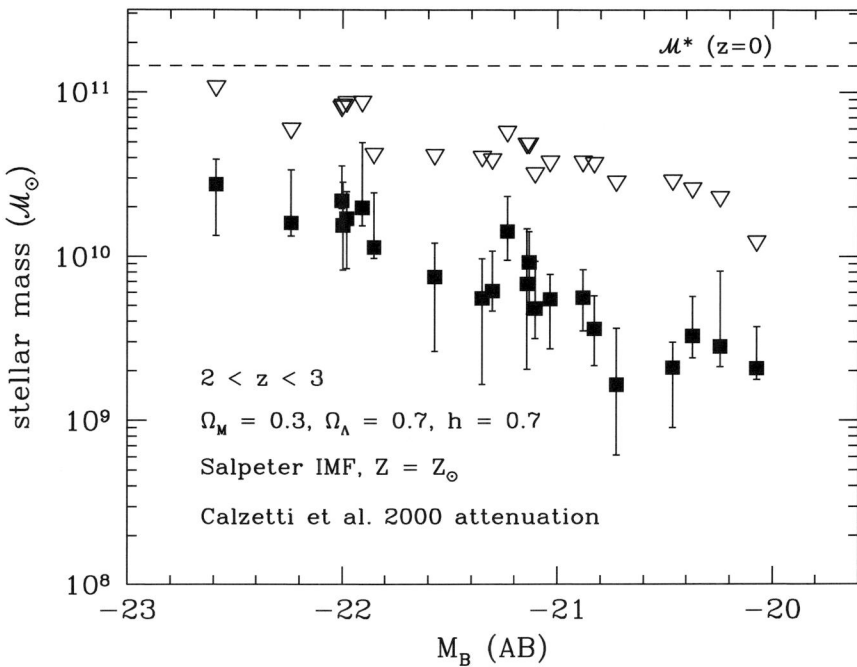

FIGURE 11. Stellar masses for HDF–N Lyman break galaxies derived from stellar population model fitting (Papovich et al. 2001). The filled points show best–fitting mass estimates using solar metallicity, Salpeter IMF models, and assuming exponential star formation histories, while the error bars show 68% confidence intervals. The downward triangles are upper mass limits allowing for the presence of an underlying, maximum $\mathcal{M}/\mathcal{L}$ stellar population formed at $z = \infty$. The dashed line shows the present–day characteristic $\mathcal{M}^*$ stellar mass (Cole et al. 2001) for the same IMF.

5.4. Stellar mass evolution

Using photometric and spectroscopic redshifts and NICMOS photometry for HDF–N galaxies, we may examine the rest–frame B–band luminosity distribution of galaxies at $2 < z < 3$ (Figure 12). Over the luminosity range shown here, this is not dissimilar to the local B–band luminosity function (LF) (Folkes et al. 1999). There appears to be an excess at bright magnitudes with respect to the local LF: this is confirmed by Shapley et al. (2001) using infrared photometry for a larger sample of brighter LBGs from the Steidel et al. ground–based survey. Galaxies with $2 < z_{\rm phot} < 3$ and $M_B < -18.75$ contribute a total blue luminosity density $\rho_B = 5.2 \times 10^{26}$ erg s^{-1} Mpc^{-3}. Without further correction for incompleteness or extrapolation to fainter magnitudes, this is $\sim 1.5\times$ that from the integrated 2dF LF, and nearly equal to that from the preliminary SDSS LF (Blanton 2001).

Although the optical luminosity densities are similar at $z = 0$ and $z = 3$, the implied stellar mass densities are quite different. Galaxies at $2 < z < 3$ are far bluer than local counterparts (Fig. 8), indicating much smaller mass–to–light ratios. From our modeling (Papovich et al. 2001, see Figure 11) for 22 HDF–N LBGs at $2 < z < 3$ we derive an average $\langle \mathcal{M}/\mathcal{L}_B \rangle = 0.10$ to 0.25 (in solar units), depending on assumptions about metallicity and IMF. These values refer to *emergent* luminosities, i.e., here L_B is *not* corrected for the effects of extinction – the *intrinsic* $\mathcal{M}/\mathcal{L}$ for the stellar populations are

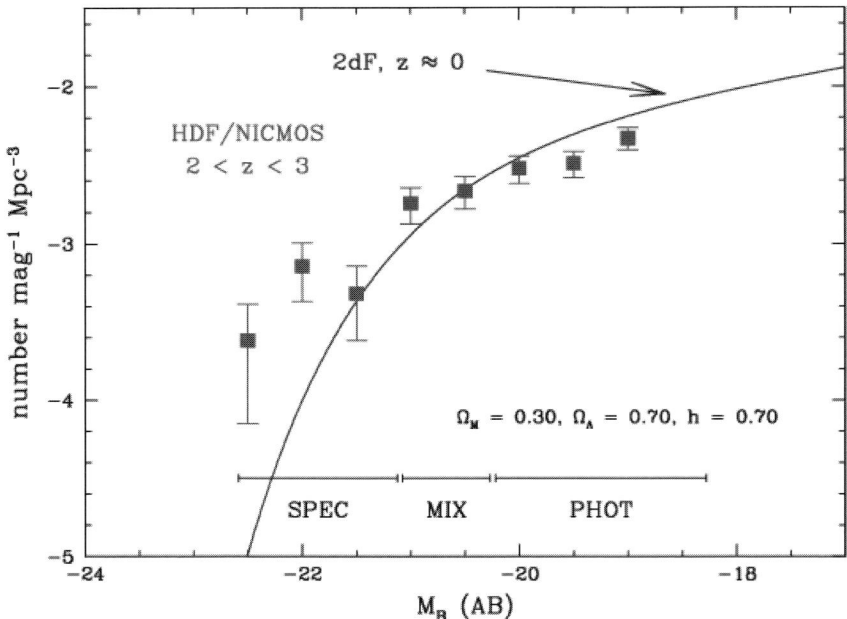

FIGURE 12. Rest–frame B–band luminosity distribution for galaxies at $2 < z < 3$ in the HDF–N. Horizontal bars indicate magnitude ranges where spectroscopic or photometric redshifts, or a mix of the two, have been used. The data have not been corrected for incompleteness, which may affect the fainter points. Error bars indicate Poisson uncertainties only. The local luminosity function from the 2dF survey (Folkes et al. 1999) is shown for comparison.

smaller. There is a trend of $\mathcal{M}/\mathcal{L}$ with rest–frame color, amounting to a factor of ~ 2 over the observed color range of the spectroscopic LBG sample, but we neglect this here and apply $\langle \mathcal{M}/\mathcal{L}_\mathcal{B} \rangle$ from the spectroscopic LBGs to the photometric redshift sample. The majority of galaxies with photometric redshifts $2 < z_{\rm phot} < 3$ have colors similar to or bluer than those in the spectroscopic sample (Fig. 8).

Restricting our attention to a Salpeter IMF, the starburst dust attenuation law (Calzetti et al. 2000), and model metallicities 0.2 to $1 \times Z_\odot$, we estimate a stellar mass density $\rho_* = 1.7$ to $2.9 \times 10^7 \mathcal{M}_\odot {\rm Mpc}^{-3}$ at these redshifts. Comparing this to the present–day stellar mass density computed from the 2dF+2MASS K–band luminosity function (Cole et al. 2001), using the same IMF assumptions, we find $\rho_*(z=2.5)/\rho_*(z=0) = 0.034$ to 0.058. It is also 8–14× smaller than the estimated mass density from bright galaxies at $z \approx 0.9$ (Brinchmann & Ellis 2000). This estimate of stellar mass density versus redshift, expressed as a cosmological density $\Omega_{\rm stars} h(z)$, is shown in Figure 13.

The estimate of the stellar mass density at $z = 2.5$ is probably a lower limit for several reasons. First, our assumed cosmology results in a nearly minimal luminosity density for currently acceptable values of the cosmological parameters. An Einstein–de Sitter model increases by the luminosity and mass densities by $\sim 80\%$. Second, we have made no corrections for incompleteness in the NICMOS–selected galaxy sample, nor any attempt to extrapolate to objects fainter than the HDF/NICMOS detection limit. Finally, as described in § 5.3, the galaxy masses may be larger if there were earlier generations of star formation. If we assign *every* LBG its *maximum* (at 68% confidence) stellar mass (Fig. 11) allowing for a older generation of stars formed at $z = \infty$, then for the adopted

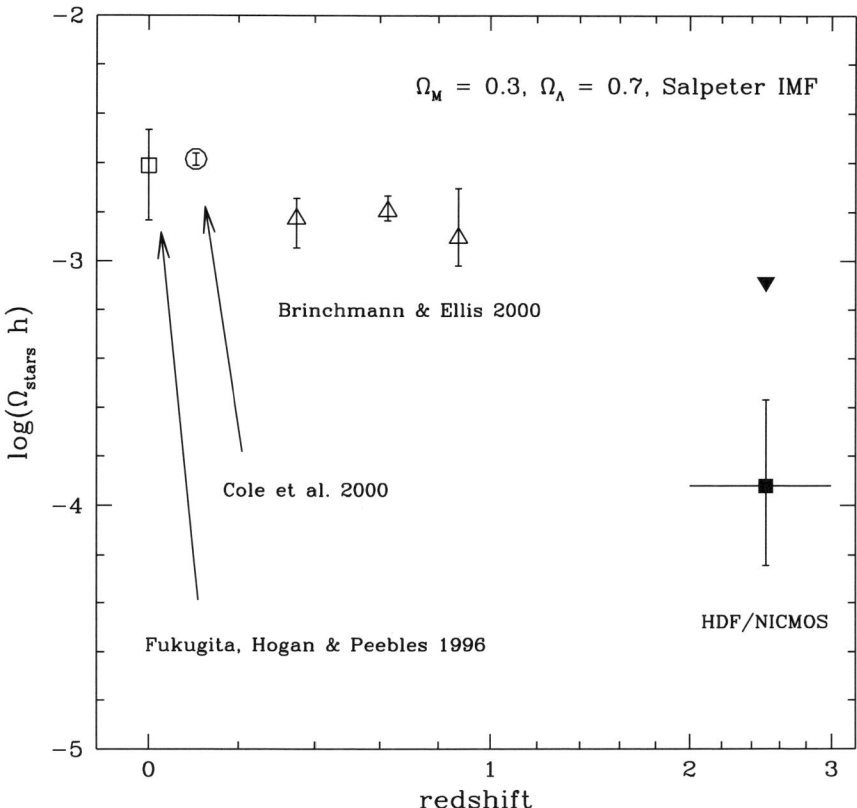

FIGURE 13. The co–moving stellar mass density, expressed as a fraction $\Omega_{\rm stars}h$ of the critical mass density, versus redshift. The points at low redshift come from to estimates of the local stellar mass density, e.g., from analysis of the K–band luminosity function. Points at $0.3 < z < 1$ are based on infrared measurements of distant field galaxies, and are lower limits because no attempt has been made to extrapolate over a luminosity function beyond the magnitude limit of the survey (see also Richard Ellis' contribution to this volume). The point at $z = 2.5$ comes from the analysis of Lyman break galaxies in the HDF-N/NICMOS data, described here. The large error bars reflect uncertainties due to degeneracies in the stellar population model fitting (although a constant Salpeter IMF has been assumed for all data points plotted here). The downward–pointing triangle represents an upper limit based on the assumption that every LBG has the *maximum* mass allowed by the photometry in an old stellar population which was formed at $z = \infty$.

cosmology we may set a conservative upper bound on the total stellar mass density contained within NICMOS–detected galaxies, $\rho_*(z = 2.5)/\rho_*(z = 0) < 0.30$.

This upper bound is barely consistent with the hypothesis that all stars in present–day galactic spheroids formed at $z > 3$ (Renzini 1999). In this scenario, all galaxies must have already formed most of their stars at $z \gg 3$, but must also be forming more stars at $z \approx 2.5$ (since there are are few candidates for evolved, non–star–forming HDF galaxies at that redshift). As described above, a more direct accounting for the mass present in NICMOS–detected galaxies at $z \approx 2.5$ implies a much smaller fraction, $\sim 5\%$, of the present–day stellar mass density.

6. Clustering at High Redshift

In § 2.5 we briefly discussed searches for high–redshift galaxies based on proximity to other, previously–known objects (e.g., radio galaxies or QSOs) at high redshift. Clustering is thus a tool for finding "ordinary" high redshift galaxies, but is also an interesting and important topic in itself, since the clustering of galaxies is intimately related to the power spectrum of mass fluctuations in the universe and the processes which drive galaxy formation and evolution. Here we briefly turn to observations of galaxy clustering at $z > 2$.

6.1. Cluster candidates at $z > 2$

At $z > 2$, there is an increasing number of examples of galaxy overdensities associated with radio galaxies and QSOs, but in few (if any) cases is there clear evidence for bound or virialized clusters *per se*. Perhaps the three best studied cases of spectroscopically confirmed associations are all found at $2.1 < z < 2.4$, a redshift range where Lyα imaging, infrared spectroscopy, and infrared multi–color selection can all be used effectively.

Francis et al. (1996, 1997, 1998, 2001) have made extensive observations of a system of galaxies and gas at $z = 2.38$ originally noted by the presence of several high column density Lyα absorption systems in the spectra of background QSOs. They have confirmed redshifts for several objects via Lyα emission and/or infrared spectroscopy, and argue that there is a very large reservoir of neutral gas out of which a rich cluster may form.

Pascarelle et al. (1996a,b, 1998) have studied a group of galaxies and AGN associated with the radio galaxy 53W002 at $z = 2.390$, using narrow– and intermediate–band Lyα imaging to identify a large number of candidates and spectroscopy to confirm redshifts for several. Pascarelle et al. (1998) emphasize the compact sizes of many of these galaxies in *HST* images, and suggest that they are sub–galactic "fragments" that will eventually merge to form more massive galaxies. Keel et al. (1999) surveyed a wider field around 53W002, identifying still more candidates, and studied their spatial distribution, suggesting that there was no clear evidence for a relaxed, virialized cluster, although it is clear that 53W002 is situated in a locally overdense region of space.

Finally, Pentericci et al. (1997, 2000) and Kurk et al. (2000) have been exploring the environment of the $z = 2.156$ radio galaxy PKS 1138-262, an object whose highly distorted radio morphology and extremely high radio rotation measure suggest the presence of a dense, gaseous environment. Using VLT narrow–band Lyα imaging, they have identified ~ 50 candidate emission line companion objects (as well as a giant Lyα emission halo around the radio galaxy itself), of which 15 have been spectroscopically confirmed. Although there is a substantial overdensity of objects at this redshift, their spatial distribution gives no strong indication that there is yet a truly relaxed, virialized cluster present.

There are many other suggestive examples in the literature of galaxy associations and cluster candidates at $z > 2$. Dickinson (1997b) presented a compendium of examples known up to 1997, and others have been found since that time. Clements (2000) has identified an excess of galaxies with red optical–infrared colors around a quasar at $z = 2.15$. Le Fèvre et al. (1996) spectroscopically confirmed several Lyα emitters around a radio galaxy at $z = 3.14$, and Le Fèvre et al. (2002) have used the Lyman break imaging technique to find a centrally concentrated group of galaxies around the same radio galaxy. Ivison et al. (2000) have noted an excess of sub–mm sources around the $z = 3.8$ radio galaxy 4C 41.17, suggesting that this may be a proto–cluster with massive starbursts in the process of forming cluster ellipticals. Djorgovski (1999) and collaborators have identified and confirmed galaxies associated with several QSOs at $z > 4$, and suggest that these are proto–cluster sites.

Clearly, AGN–marked cluster searches are an effective means of identifying galaxy associations at high redshift. At present, however, there are very few cases where we have enough supporting information to judge the evolutionary state of these proto–cluster candidates, especially at $z > 2$. It seems clear that even with the new generation of more powerful X–ray satellites, it will be long, hard work detecting, confirming, and deriving physically valuable information about the IGM for galaxy clusters at $z > 1$.

6.2. Galaxy clustering at $z > 2$

Although at $z > 2$ there is as yet little evidence for systems that most observers would agree are genuinely relaxed, virialized, massive clusters, the information on galaxy *clustering* at those redshifts is accumulating at an impressive rate. The first suggestions of clustering at such large redshifts came from studies of QSO absorption line systems, which found that the higher column density absorbers (e.g., CIV systems at $z \gtrsim 2$) were strongly clustered along the line of sight (cf. Sargent, Boksenberg & Steidel 1988; Quashnock, Vanden Berk & York 1996) with additional evidence for transverse clustering when multiple sightlines are available: e.g., the multiple absorption systems studied by Francis et al. (1996, 1997) as discussed above. QSOs themselves also exhibit clustering on large spatial scales (cf. Shaver 1984; Shanks et al. 1987; Shanks & Boyle 1994; Stephens et al. 1997; Roukema & Mamon 2000).

The most extensive evidence for galaxy clustering at very large redshift comes from the large survey of LBGs at $z \approx 3$ by Steidel et al. (1996, 1999). Giavalisco et al. (1998) detected angular correlation in the LBG population, and Steidel et al. (1998) identified a pronounced and highly significant "spike" at $z = 3.09$ in their spectroscopic survey of one field (SSA22). In fact, spikes and voids in the LBG redshift distribution are nearly ubiquitous in our survey, found along every sightline. Figure 14 shows one example, with an interesting twist related to the discussion of AGN clustering in § 2.5. In this field, we found 8 galaxies associated with a QSO at $z = 3.6$, despite the fact that our Lyman break color selection criteria should, in principle, be quite *ineffective* for identifying galaxies at $z > 3.5$. This suggests that this particular QSO may be in a remarkably rich galaxy environment.† These structures are large, with transverse dimensions that equal or exceed the scales of our imaging data ($\gtrsim 10$ Mpc).

Adelberger et al. (1998) have carried out a redshift–space counts–in–cells analysis of data from several fields to quantify the degree of clustering. Overall, the results demonstrate that bright LBGs cluster roughly as strongly as do present–day, massive ($\sim L^*$) galaxies, despite the fact that the overall clustering of *mass* in the universe must have been considerably smaller at these large redshifts. We interpret this as direct evidence for early galaxy biasing, in which young galaxies form preferentially in massive dark matter halos, which are expected to be more strongly clustered than the overall dark matter distribution (Bardeen et al. 1986).

In § 2.6.1, I discussed the deep, narrow–band Lyα images obtained by Steidel et al. (2000) of the $z = 3.09$ LBG overdensity in SSA22. Altogether, ~ 160 objects were identified that are likely members of the structure. Most remarkably, Steidel et al. found two enormous Lyα "blobs," each with physical dimensions $\gtrsim 100h^{-1}$ kpc and Lyα luminosities $\approx 10^{44}$ erg s^{-1}. These blobs are strikingly similar to the giant Lyα halos sometimes found around powerful radio galaxies (e.g., around the $z = 2.156$ galaxy PKS 1138+262, studied by Kurk et al. 2000 and discussed above). However, the SSA22

---

† It should never, however, be thought that *all* high redshift AGN are surrounded by a multitude of companion galaxies. We have also surveyed the field around the $z = 3.395$ radio galaxy 0902+34, but found only two objects (out of 42) with redshifts matching that of the radio galaxy.

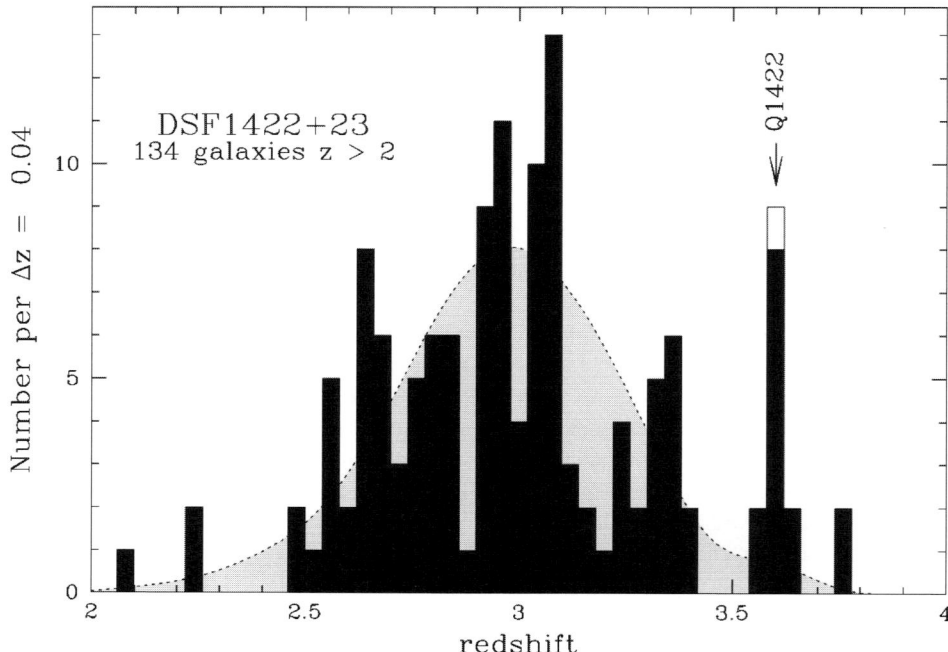

FIGURE 14. Redshift histogram for Lyman break galaxies in the Steidel et al. survey field DSF1422+23. The smooth shaded curve shows the redshift selection function of the Lyman break technique, derived from the average of many independent fields. The redshift distribution for individual sightlines, however, shows considerable structure, with strong over– and under–densities. This particular field has a "background" QSO at $z = 3.6$, and we have identified 8 other galaxies at similar redshift, despite the fact that the color selection is rather inefficient at that redshift.

blobs have no radio sources, and indeed are only loosely associated with UV–bright Lyman break galaxies found near their peripheries. One blob encompasses a red galaxy detected in deep K–band images. This blob is also situated within a larger–scale local overdensity in the distribution of objects within the spike. At this point, the nature of these blobs is unclear. We do not know what powers the Lyα emission, since there is no obvious source of UV radiation for photo–ionization. We may speculate that they trace gas condensing in a massive cooling flow from the surrounding overdense region, perhaps building galaxies in the central regions of a proto–cluster environment, although it is difficult to understand how the tremendous Lyα luminosity could be sustained through cooling alone without some mechanism to keep reionizing the recombining atoms. Most recently, Chapman et al. (2001) have detected sub–mm emission from both blobs, including a very powerful 20 mJy source associated with "blob 1" (the one with the central K–band object). This is one of the brightest high–z sub–mm sources known, and highlights the remarkable nature of these objects, although it is still unclear whether this source is an AGN that may be somehow photo–ionizing its environment (although that ionizing emission is invisible from our vantage point), or a site of tremendous star formation, perhaps induced by the cooling and inflow of the surrounding gaseous medium.

In general, the large transverse extent of the LBG spikes would suggest that they are more similar to the sheets and walls seen in local galaxy redshift surveys, rather than indicative of virialized clusters of galaxies. Given current estimates for the LBG bias relative to the mass distribution, the overdensity of these "spikes" relative to the background matter appears to be of order $\delta\rho/\rho \sim 2$ on ~ 10 Mpc scales, suggesting that they are close to turn–around. At low redshift we know that massive clusters tend to inhabit such sheets and walls, occurring particularly at their intersections. It is not unlikely that the structures we find at $z \approx 3$ are thus truly *proto–cluster* sites: the most overdense regions of the universe at that redshift, where galaxy formation is accelerated, and where large–scale motions and infall will eventually lead to the collapse and growth of Abell–type clusters. Simulations of the formation and evolution of galaxies and large scale structure also find that "mock LBGs" formed at $z \sim 3$ preferentially end up in overdense environments such as rich clusters at $z = 0$ (e.g., Governato et al. 1998). The Lyα blobs and the associated, powerful sub–mm emission may indicate the beginnings of the process of rich cluster formation: the early formation stages of massive, central cluster galaxies. Because many high redshift radio galaxies also have similar Lyα emission halos, and are also often detected at sub–mm wavelengths (Archibald et al. 2001), it is tempting to imagine that similar processes are at work in their vicinity, and that they also trace the sites of early cluster formation, much as radio–loud AGN frequently mark the locations of more mature galaxy clusters at $z \approx 1$ and below.

7. The Most Distant Galaxies

What lies beyond $z = 5$? To date, the most extensive and systematic results come from surveys for radio galaxies and QSOs. A moderate number of powerful radio sources are now known at $z > 4$ and even one or two at $z > 5$, but there is controversy about whether or not the space density declines with increasing redshift (see, e.g., Jarvis et al. 2001 and references therein). The Sloan Digital Sky Survey (SDSS) is now finding large numbers of color–selected QSOs at $4 < z < 6$, and early analysis (Fan et al. 2001) finds an apparently steep decline (by a factor of ~ 6) in the QSO space density from $z = 3.5$ to 5.

For "ordinary" galaxies, the situation is less clear. An increasingly large handful of individual galaxies at $z > 5$ has been found by a variety of means. The first object to break the $z = 5$ barrier, the galaxy RD1 at $z = 5.34$ (Dey et al. 1998) was found by pure chance when it fell on a spectrograph slit. Two galaxies in the HDF–N, initially selected as photometric redshift candidates (Fernández–Soto et al. 1999), have been spectroscopically confirmed at $z = 5.33$ and $z = 5.60$ (Spinrad et al. 1998; Weymann et al. 1998). Hu, McMahon & Cowie (1999) used narrow band imaging to identify a candidate Lyman α emitter which they confirmed spectroscopically to be at $z = 5.74$. Other candidates for objects at $z > 5$ and even $z > 6$ are floating about in the published literature and unpublished rumor mill.

Despite these individual successes, so far there have been few systematic surveys for galaxies at $z > 5$, and indeed there are few data sets capable of doing so. Lyman break or photometric redshift methods using optical data alone become ineffective at $z \gg 5$. At these redshifts, the dominant spectral feature leading to the "Lyman break" signature is no longer the Lyman limit itself, but instead the Lyman α forest break, which is redshifted into the I–band and beyond. Color selection methods optimally require at least two bandpasses redward of this break to measure a continuum color against which the break can be detected, and requiring near–infrared photometry at these large redshifts. The SDSS has detected QSOs up to $z = 6$ thanks to the availability of z–band data (effective wavelength ≈ 9000Å), but z–band CCD imaging is fairly inefficient from the ground because of the bright OH sky background and the declining quantum efficiency of CCD detectors. z–band surveys capable of detecting faint Lyman break galaxies have not yet been carried out. Space–based z–band surveys with the *HST* Advanced Camera may offer an important advantage in this area.

At $z > 6$, the optical light from galaxies is almost entirely suppressed, and true near–infrared photometry at $\lambda > 1\mu$m is required. Again, ground–based near–infrared surveys must fight against the bright sky background, and few infrared cameras have sufficiently wide fields of view to make very deep surveys over a sufficiently large field of view. *HST* NICMOS observations have the advantage of a very dark extra–atmospheric sky background, permitting excellent sensitivity, but the widest field of view available from NICMOS covered less than 1 arcmin^2, making it expensive to survey even one HDF–sized area. As described above in § 3.1, the HDF–N itself was completely covered with NICMOS to $AB \approx 26.5$ at 1.1 and 1.6μm, and still deeper images covering $\lesssim 1$ arcmin^2 were obtained on an HDF–N subregion and also the HDF–S NICMOS field. These are probably the only data sets suitable for making a systematic count of galaxy candidates at $z \gg 5$.

Using photometric redshifts fit to the deeper HDF–N GTO NICMOS field, Thompson, Weymann, & Storrie-Lombardi (2001) identified 8 galaxies with $5 < z_{phot} < 6.6$. Almost all are very faint, with seven having $26.5 < H_{AB} < 28.5$. No candidates at still higher redshifts were identified. From an analysis of the ultraviolet luminosity density, Thompson et al. concluded that the co–moving density of star formation is roughly constant from $3 < z < 5$, with a possible drop at $z > 6$, since their very deep NICMOS images should have been capable of detecting higher redshift galaxies if indeed they were present. Dickinson, Deltorn & Budavári (2001, see also Dickinson 2000b) have used the wide–field (but shallower) NICMOS observations covering the whole HDF for a similar investigation, identifying V– and I–band dropout objects and also considering template–based photometric redshift estimates. We used the measured UV luminosity function of LBGs at $z \approx 3$ and 4 from Steidel et al. 1999 and extensive Monte Carlo simulations to compare the number of NICMOS–selected candidates to the expectations for constant co–moving space density. Similar to the results from Thompson et al., we found that the number of V–band dropout candidates ($4.5 \lesssim z \lesssim 6.0$) is similar to predictions assuming an unevolving space density. However, only one or two reliable I–dropout candidates ($6.0 < z < 8.5$) were found, whereas ~ 8 would have been expected. Small number statistics limit the robustness of this result, but together with that of Thompson et al. it does suggest that galaxies at $z > 6$ are either fewer, fainter, or harder to see (e.g., due to intrinsically lower surface brightnesses) than their counterparts at $2 < z < 6$.

In a photometric redshift analysis of the HDF–S NICMOS field, Yahata et al. (2000) identified 21 objects with $z_{phot} > 5$. Of these, 14 have $z_{phot} > 6$, and 8 have $10 < z_{phot} < 16$. Several of the highest redshift candidates are remarkably bright, detected only in the relatively shallow NICMOS K–band (F222M) images with $23.6 < K_{AB} < 24.7$. Lanzetta, Yahil & Fernández–Soto (1998) had previously identified five similarly bright ($K_{AB} < 24$) candidates for galaxies at $z \gtrsim 10$ using ground–based near–infrared images of the HDF–N together with the WFPC2 data. From the HDF–N NICMOS images, Dickinson et al. (2000) detected one of these five objects at 1.6μm and confirmed the K–band detection with deeper Keck images. This object, HDFN-JD1, appears to be a "J–band dropout," invisible (or at best only marginally detected) at 1.1μm and shorter wavelengths. This remains a plausible candidate for an object (probably a galaxy, although it appears to be only barely resolved) at $z \approx 12.5$, although other explanations are certainly possible (e.g., a dust–obscured object at more modest redshift). The other four candidates from Lanzetta et al. (1998) were not detected with NICMOS.

If any of these unusually bright Lyman break or photometric redshift selected galaxy candidates at $z \gtrsim 10$ are real, it suggests a remarkable change in the galaxy population at very large redshifts. The deep NICMOS surveys find relatively few objects at $6 < z < 10$. However, these $z > 10$ candidates are considerably more luminous in their UV rest–frame than known Lyman break galaxies at $2 < z < 4.5$ selected from the wide–area ground–based survey of Steidel et al. It is possible that we are seeing an early epoch of early, rapid collapse and "primeval" star formation unobscured by dust. However, it is also possible that none of these candidates is actually at the very large redshifts that have been estimated for them. Unfortunately, spectroscopy to confirm the redshifts will be exceedingly difficult with any existing facility.

8. Future Directions

The observations described here have really just begun to address the most important questions about the evolution of galaxies, their mass assembly histories, and the formation of the present–day Hubble sequence. A new generation of instruments on the ground and in space will carry us further along down this road.

Spectrographs like VIRMOS and DEIMOS, soon to be commissioned on VLT and Keck, respectively, will vastly increase the multiplexing capabilities for redshift surveys on 8–10m telescopes. This will greatly enhance our ability to carry out statistical studies of high redshift galaxies (luminosity functions, clustering, etc.), but will also open the possibility for wholesale surveys of galaxy kinematics using moderately high dispersion spectroscopy and long exposures for thousands of objects.

Deep near–infrared surveys are still limited to very small solid angles and thus volumes at high redshift. New large format detectors will greatly improve this situation, although unfortunately very few cameras for 8–10m telescopes (whose aperture is really needed to study $z > 2$ galaxies) are being configured with wide fields of view. The first multi–object infrared spectrographs are just now coming on line; these will permit wholesale spectroscopy of distant galaxies at rest–frame optical wavelengths, offering a means of measuring kinematic masses and chemical abundances at high redshift.

On *HST*, the Advanced Camera for Surveys (ACS), to be installed in 2002, will offer a factor of nearly 10 "acceleration" for imaging surveys of high redshift galaxies compared to WFPC2, thanks to a larger field of view and greater throughput. Moreover, it should deliver better image quality thanks to finer pixel sampling. With ACS it becomes possible to image a significant fraction of a square degree with moderately deep, multicolor *HST* data, or to carry out a deep field going at least one magnitude fainter than the HDF. A few years later, the infrared channel of Wide Field Camera 3 will offer a big advance for infrared galaxy surveys, offering a field of view $5.5\times$ larger than that of NICMOS, with higher throughput and $1.7\times$ better pixel sampling. Deep 1.0–1.6 μm observations of HDF–sized regions will become routine, enabling wholesale studies of the rest–frame optical morphologies, luminosities and colors of galaxies at $z < 3$, and color–selected surveys for galaxies at $z > 6$.

Despite the promises usually made for near–infrared surveys (see § 3.1), our census of the stellar content of galaxies at $z > 2$ is still fundamentally limited by the wavelengths at which we can observe. The H and K–bands measure rest–frame B and V–band light at $z = 3$, leaving large uncertainties on estimates of stellar mass, ages, and other such parameters, while brave first attempts at 3–7 μm do not go deep enough to detect galaxies at $z > 2$ (Serjeant et al. 1997, Hogg et al. 2000b). The *SIRTF* Infrared Array Camera (IRAC), observing at 3.6–8.0 μm, can measure rest–frame K–band light from galaxies out to $z \approx 3$ and $\lambda_0 > 1$ μm emission out to $z = 7$. Extremely deep exposures will be required, however, to detect ordinary galaxies at such large redshifts. We will be carrying out a *SIRTF* Legacy Program, the Great Observatories Origins Deep Survey (GOODS), which will push observations at 3.6–24 μm to their limits in two fields (the HDF–N and Chandra Deep Field South) totaling approximately 330 arcmin^2. The survey goal is to provide multiwavelength data suitable for tracing the mass assembly history of galaxies and their energetic output from star formation and AGN activity out to the highest accessible redshifts. The *SIRTF* data, along with extensive supporting observations from ESO and NOAO facilities, will be distributed to the community, providing a rich archive for research and a pathfinder to future work with the Next Generation Space Telescope (*NGST*). *SIRTF* surveys from the guaranteed time programs will cover larger areas to shallower depths at 3.6–160μm, and another Legacy program, the *SIRTF* Wide–area InfraRed Extragalactic survey (SWIRE) will cover more than 50 deg^2.

NGST itself, currently planned to have a 6.5m mirror, will operate primarily at near–infrared wavelengths, from 0.6–5μm with a possible mid–infrared instrument extending the range to 30μm. One lofty goal for *NGST* is to see and study the first starlight in the universe: to capture the earliest moments of galaxy formation, when stars and quasars first began to shine. At the same time, multislit and/or integral field spectroscopy will make it possible to measure the rest–frame optical spectral properties of high–z galaxies with unprecedented sensitivity and angular resolution, opening the way for an extensive investigation of the chemical abundances, stellar populations, star formation rates and internal kinematics of galaxies in the young universe.

This is arguably the most interesting time to be working in this area of research. The "phase transition" mentioned in the introduction has taken place: most of the universe is now open to scrutiny. The basic cosmological parameters may already be fairly well known (even if the underlying physics of cosmological constants, quintessence, branes, or what-have-you is not!), and even the power spectrum of matter fluctuations will probably be pinned down with some accuracy in the near future from the combination of microwave background observations, large–scale structure surveys like 2dF and SDSS, and perhaps from weak lensing. In this context, the *background* for understanding galaxy formation is set. We must now go out and watch this evolution happen, observing galaxies over the broadest range of cosmic time, and interpreting what we see in terms of the physical processes at work. The tools (both observational and theoretical) are now available. The students who have attended this IAC winter school are here at just the right time to go forth and carry out this work, and I encourage you all to do so, and to have fun with it!

I would like to thank the organizers of this excellent winter school, Ismael Perez-Fournon and Marc Balcells, for inviting me, also to thank the IAC for their generous financial support. I must especially thank Marc and Ismael for their patience (which I severely tried) waiting for this manuscript. Support for this work was provided in part by NASA grant GO-07817.01-96A.

REFERENCES

ABRAHAM, R. G., TANVIR, N. R., SANTIAGO, B. X., ELLIS, R. S., GLAZEBROOK, K. & VAN DEN BERGH, S. 1996, MNRAS, 279, L47.

ADELBERGER, K. L. 2000, in *Clustering at High Redshift,* eds. A. Mazure, O. Le Fèvre, and V. Le Brun (San Francisco: ASP), 13.

ADELBERGER, K. L. & STEIDEL, C. C. 2000, ApJ, 544, 218.

ADELBERGER, K.L., STEIDEL, C. C., GIAVALISCO, M., DICKINSON, M., PETTINI, M. & KELLOGG, M. 1998, ApJ, 505, 18.

ALLINGTON-SMITH, J. R., ELLIS, R., ZIRBEL, E. L. & OEMLER, A. 1993, ApJ, 404, 521.

ALTIERI, B. *et al.* 1999, A&A, 343, L65.

ARCHIBALD, E. N., DUNLOP, J. S., HUGHES, D. M., RAWLINGS, S., EALES, S. A. & IVISON, R. J. 2001, MNRAS, 323, 417.

AUSSEL, H., CESARSKY, C. J., ELBAZ, D. & STARCK, J. L. 1999, A&A, 342, 313.

BAKER, A. J., LUTZ, D., GENZEL, R., TACCONI, L. J. & LEHNERT, M. D. 2001, A&A, 372, L37.

BARDEEN, J. M., BOND, J. R., KAISER, N. & SZALAY, A. S. 1986, ApJ, 304, 15.

BARGER, A. J., COWIE, L. L., TRENTHAM, N., FULTON, E., HU, E. M., SONGAILA, A. & HALL, D. 1999, AJ, 117, 102.

BENÍTEZ, N., BROADHURST, T., BOUWENS, R., SILK, J. & ROSATI, P. 1999, ApJ, 515, L65.

BEST, P. N. 2000, MNRAS, 317, 720.

BLANTON, M. R. 2001, AJ, 121, 2358.

BLANTON, E. L., GREGG, M. D., HELFAND, D. J., BECKER, R. H. & WHITE, R. L. 2000, ApJ, 531, 118.

BOUGHN, S. P., SAULSON, P. R. & USON, J. M. 1986, ApJ, 301, 17.

BRANDT, W. N., HORNSCHEMEIER, A. E., SCHNEIDER, D. P., ALEXANDER, D. M., BAUER, F. E., GARMIRE, G. P. & VIGNALI, C. 2001, ApJ, 558, L5.

BRINCHMANN, J. & ELLIS, R. S. 2000, ApJ, 536, L77.

BRUZUAL A., G. & CHARLOT, S. 1993, ApJ, 405, 538.

BRUZUAL A., G. & CHARLOT, S. 2000, priv. comm.

CALZETTI, D., ARMUS, L., BOHLIN, R. C., KINNEY, A. L., KOORNNEEF, J. & STORCHI-BERGMANN, T. 2000, ApJ, 533, 682.

CASERTANO, S. et al. 1995, ApJ, 453, 599.

CASERTANO, S. et al. 2000, AJ, 120, 2747.

CHAPMAN, S. C., LEWIS, G. F., SCOTT, D., RICHARDS, E., BORYS, C., STEIDEL, C. C., ADELBERGER, K. L. & SHAPLEY, A. E. 2001, ApJ, 548, L17.

CHAPMAN, S. C., MCCARTHY, P. J. & PERSSON, S. E. 2000, AJ, 120, 1612.

CHARTAS, G., BAUTZ, M., GARMIRE, G., JONES, C. & SCHNEIDER, D. P. 2001, ApJ, 550, L163.

CLEMENTS, D. L. 2000, MNRAS, 312, L61.

COHEN, J. G. et al. 1999, ApJS, 120, 171.

COHEN, J. G. et al. 2000, ApJ, 538, 29.

COLE, S. et al. 2001, MNRAS, 326, 255.

COLEMAN, G. D., WU, C.-C. & WEEDMAN, D. W. 1980, ApJS, 43, 393.

COLLINS, C. A. & JOSEPH, R. D. 1988, MNRAS, 235, 209.

COWIE, L. L. & HU, E. M. 1998, AJ, 115, 1319.

COWIE, L.L., HU, E.M. & SONGAILA, A. 1995, AJ, 110, 1576.

COWIE, L. L., LILLY, S. J., GARDNER, J. & MCLEAN, I. S. 1988, ApJ, 332, L29.

CRAMPTON, D. & LILLY, S. 1999, in *Photometric Redshifts and the Detection of High Redshift Galaxies,* eds. R. J. Weymann, L. Storrie–Lombardi, M. Sawicki, & R. Brunner, R., (San Francisco: ASP), 229.

DA COSTA, L. et al. 1998, A&A, submitted (astro–ph/9812105).

DADDI, E. et al. 2000, A&A, 361, 535.

DAWSON, S., STERN, D., BUNKER, A. J., SPINRAD, H. & DEY, A. 2001, AJ, 122, 598.

DELTORN, J. M. 1998, Ph. D. thesis, Observatoire de Paris–Meudon.

DELTORN, J. M., LE FÈVRE, O., CRAMPTON, D. & DICKINSON, M. 1997, ApJ, 483, L21.

DE MELLO, D. F., LEITHERER, C. & HECKMAN, T. M. 2000, ApJ, 530, 251.

DEY, A., GRAHAM, J. R., IVISON, R. J., SMAIL, I., WRIGHT, G. S. & LIU, M. C. 1999, ApJ, 519, 610.

DEY, A., SPINRAD, H., STERN, D., GRAHAM, J. R. & CHAFFEE, F. H. 1998, ApJ, 498, L93.

DICKINSON, M. 1995a, in *Galaxies in the Young Universe,* eds. H. Hippelein, K. Meisenheimer & H.-J. Roser, (Springer: Berlin), 144.

DICKINSON, M. 1995b, in *Fresh Views of Elliptical Galaxies,* eds. A. Buzzoni, A. Renzini & A. Serrano, (ASP, San Francisco), 283.

DICKINSON, M. 1997a, in *The Early Universe with the VLT,* ed. J. Bergeron, (Berlin: Springer), 274.

DICKINSON, M. 1997b, in *HST and the High Redshift Universe,* eds. N. Tanvir, A. Aragon-Salamanca & J. V. Wall, World Scientific, 207.

DICKINSON, M. 1997c, in *Galaxy Scaling Relations: Origins, Evolution and Applications,* eds. L. da Costa & A. Renzini, (Berlin: Springer), 215.

DICKINSON, M. 1998, in *The Hubble Deep Field,* eds. M. Livio, S. M. Fall & P. Madau (Cambridge: Cambridge Univ. Press), 219.

DICKINSON, M. 1999, in *After the Dark Ages: When Galaxies were Young,* eds. S. Holt & E. Smith, AIP, 122.

DICKINSON, M. 2000a, in *Building Galaxies: From the Primordial Universe to the Present,* eds. F. Hammer, T. X. Thuan, V. Cayatte, B. Guiderdoni & J. Tranh Than Van, Ed. Frontières, 257.

DICKINSON, M. 2000b, in *Phil. Trans. Royal Soc. Lond. A,* 358, 2001.

DICKINSON, M. *et al.* 2000, ApJ, 531, 624.

DICKINSON, M., DELTORN, J. M. & BUDAVÁRI, T. 2001, in preparation.

DJORGOVSKI, S. G. 1999, in *The Hy–Redshift Universe,* eds. A. J. Bunker & W. J. M. van Breugel, (San Francisco: ASP), 397.

DJORGOVISKI, S., SPINRAD, H., MCCARTHY, P. & STRAUSS, M. A. 1985, ApJ, 299, L1.

DRIVER, S. P., WINDHORST, R. A. & GRIFFITHS 1995, ApJ, 453, 48.

DOWNES, D. *et al.* 1999, A&A, 347, 809.

EISENHARDT, P. *et al.* 1998, in *The Birth of Galaxies,* eds. B. Guiderdoni *et al.* (Paris: Editions Frontieres).

EISENHARDT, P. R., ARMUS, L., HOGG, D. W., SOIFER, B. T., NEUGEBAUER, G. & WERNER, M. W. 1996, ApJ, 461, 72.

ELLINGSON, E., YEE, H. K. C. & GREEN, R. F. 1991, ApJ, 371, 49.

ELLIS, R. S., COLLESS, M., BROADHURST, T., HEYL, J. & GLAZEBROOK, K. 1996, MNRAS, 280, 235.

ELSTON, R., RIEKE, G. H. & RIEKE, M. J. 1988, ApJ, 331, L77.

FAN, X. *et al.* 2001, AJ, 121, 54.

FERGUSON, H. C., DICKINSON, M. & WILLIAMS, R. 2000, ARA&A, 38, 667.

FERNÁNDEZ–SOTO, A., LANZETTA, K. M. & YAHIL, A. 1999, ApJ, 513, 34.

FOLKES, S. *et al.* 1999, MNRAS, 308, 459.

FRANCESCHINI, A. *et al.* 1998, ApJ, 506, 600.

FRANCIS, P. J. *et al.* 1996, ApJ, 457, 490.

FRANCIS, P. J., WILSON, G. M. & WOODGATE, B. E. 2001, PASA, 18, 64.

FRANCIS, P. J., WOODGATE, B. E. & DANKS, A. C. 1997, ApJ, 482, L25.

FRANCIS, P. J., WOODGATE, B. E. & DANKS, A. C. 1998, in *The Young Universe,* eds. S. D'Odorico, A. Fontana & E. Giallongo, (San Francisco: ASP), 496.

FRANX, M. *et al.* 2000, *The Messenger,* 99, 20.

FRANX, M., ILLINGWORTH, G. D., KELSON, D. D., VAN DOKKUM, P. G. & TRAN, K.-V. 1997, ApJ, 486, L75.

FRUCHTER, A. 1999, ApJ, 512, L1.

FRUCHTER, A. *et al.* 2001, in prep.

GARDNER, J. P. *et al.* 2000, AJ, 119, 486.

GARDNER, J. P., SHARPLES, R. M., CARRASCO, B. E. & FRENK, C. S. 1997, MNRAS, 282, L1.

GIAVALISCO, M., MACCHETTO, F. D. & SPARKS, W. B. 1994, A&A, 288, 103.

GIAVALISCO, M., STEIDEL, C. C., ADELBERGER, K. L., DICKINSON, M., PETTINI, M. & KELLOGG, M. 1998, ApJ, 503, 543.

GIAVALISCO, M., STEIDEL, C. C. & MACCHETTO, F. D. 1996, ApJ, 470, 189.

GILLILAND, R. L., NUGENT, P. E. & PHILLIPS, M. M. 1999, ApJ, 521, 30.

GLAZEBROOK, K., ELLIS, R. S., SANTIAGO, B. & GRIFFITHS, R. E. 1995, MNRAS, 275, L19.

GOVERNATO, F., BAUGH, C. M., FRENK, C. S., COLE, S., LACEY, C. G., QUINN, T. & STADEL, J. 1998, Nature, 392, 359.

GRAHAM, J. R. & DEY, A. 1996, ApJ, 471, 720.

GUHATHAKURTA, P., TYSON, J. A. & MAJEWSKI, S. R. 1990, ApJ, 357, L9.

HAINES, C. P., CLOWES, R. G., CAMPUSANO, L. E. & ADAMSON, A. J. 2001, MNRAS, 323, 688.

HALL, P. B. *et al.* 2001, AJ, 121, 1840.

HALL, P. B. & GREEN, R. F. 1998, ApJ, 507, 558.

HILL, G. J. & LILLY, S. J. 1991, ApJ, 367, 1.

HOGG, D. W. *et al.* 2000a, ApJS, 127, 1.

HOGG, D. W. *et al.* 2000b, AJ, 119, 1519.

HOGG, D. W., NEUGEBAUER, G., ARMUS, L., MATTHEWS, K., PAHRE, M. A., SOIFER, B. T. & WEINBERGER, A. J. 1997, AJ, 113, 2338.

HOPKINS, A. M., CONNOLLY, A. J. & SZALAY, A. S. 2000, AJ, 120, 2843.

HORNSCHEMEIER, A. E. *et al.* 2000, ApJ, 541, 49.

HORNSCHEMEIER, A. E. *et al.* 2001, ApJ, 554, 742.

HU, E. M., COWIE, L. L. & MCMAHON, R. G. 1998, ApJ, 502, L99.

HU, E. M., MCMAHON, R. G. & COWIE, L. L. 1999, ApJ, 522, L9.

HU, E. M. & RIDGWAY, S. E. 1994, AJ, 107, 1303.

HUTCHINGS, J. B., CRAMPTON, D. & JOHNSON, A. 1995, AJ, 109, 73.

HUTCHINGS, J. B., CRAMPTON, D. & PERSRAM, D. 1993, AJ, 106, 1324.

IVISON, R. J., DUNLOP, J. S., SMAIL, I., DEY, A., LIU, M. C. & GRAHAM, J. R. 2000, ApJ, 542, 27.

JARVIS, M. J., RAWLINGS, S., WILLOTT, C. J., BLUNDELL, K. M., EALES, S. & LACY, M. 2001, MNRAS, 327, 907.

KAJISAWA, M. *et al.* 2000a, PASJ, 52, 53.

KAJISAWA, M. *et al.* 2000b, PASJ, 52, 61.

KEEL, W. C., COHEN, S. H., WINDHORST, R. A. & WADDINGTON, I. 1999, AJ, 118, 2547.

KINNEY, A. L., CALZETTI, D., BOHLIN, R. C., MCQUADE, K., STORCHI-BERGMANN, T. & SCHMITT, H. R. 1996, ApJ, 467, 38.

KUCHINSKI, L. E., MADORE, B. F., FREEDMAN, W. L. & TREWHELLA, M. 2001, ApJ, in press.

KULKARNI, S. R. *et al.* 1998, Nature, 393, 35.

KURK, J. D. *et al.* 2000, A&A, 358, L1.

LANZETTA, K. M., YAHIL, A. & FERNÁNDEZ–SOTO, A. 1998, AJ, 116, 1066.

LAWRENCE, C. R. *et al.* 1984, Science, 223, 46.

LE FÈVRE, O., DELTORN, J. M., CRAMPTON, D. & DICKINSON, M. E. 1996, ApJ, 471, L11.

LE FÈVRE, O., DELTORN, J. M., CRAMPTON, D. & DICKINSON, M. E. 2001, ApJ, in press.

LEITHERER, C., LEÃO, J., HECKMAN, T. M., LENNON, D. J., PETTINI, M. & ROBERT, C. 2001, ApJ, 550, 724.

LILLY, S. J., LE FÈVRE, O., CRAMPTON, D., HAMMER, F. & TRESSE, L. 1995, ApJ, 455, 50.

LIU, M. C., DEY, A., GRAHAM, J. R., BUNDY, K. A., STEIDEL, C. C., ADELBERGER, K. & DICKINSON, M. 2000, AJ, 119, 2556.

LOWENTHAL, J. D., HOGAN, C. J., GREEN, R. F., CAULET, A., WOODGATE, B. E., BROWN, L. & FOLTZ, C. B. 1991, ApJ, 377, L73.

MADAU, P., FERGUSON, H. C., DICKINSON, M., GIAVALISCO, M., STEIDEL, C. C. & FRUCHTER, A. 1996, MNRAS, 283, 1388.

MADAU, P., POZZETTI, L. & DICKINSON, M. 1998, ApJ, 498, 106.

MALKAN, M., TEPLITZ, H. I. & MCLEAN, I. S. 1995, ApJ, 448, L5.

MANNING, C., STERN, D., SPINRAD, H. & BUNKER, A. J. 2000, ApJ, 537, 65.

MANNUCCI, F., THOMPSON, D., BECKWITH, S. V. W. & WILLIGER, G. M. 1998, ApJ, 501, L11.

MCCARTHY, P. J. 1993, ARAA, 31, 639.

MCCARTHY, P. J. et al. 1999, ApJ, 520, 548.

MCCARTHY, P. J. et al. 2001, in *Deep Fields,* ed. S. Cristiani (Berlin: Springer), in press (astro–ph/0011499).

MCCARTHY, P. J., DICKINSON, M., FILIPPENKO, A. V., SPINAD, H. & VAN BREUGEL, W. J. M. 1988, ApJ, 328, L29.

MCCRACKEN, H. J., METCALFE, N., SHANKS, T., CAMPOS, A., GARDNER, J. P. & FONG, R. 2000, MNRAS, 311, 707.

MEURER, G. R., HECKMAN, T. M. & CALZETTI, D. 1999 ApJ, 521, 64.

MEURER, G. R., HECKMAN, T. M., LEHNERT, M. D., LEITHERER, C. & LOWENTHAL, J. 1997, AJ, 114, 54.

MØLLER, P. & WARREN, S. J. 1993, A&A, 270, 43.

MORIONDO, G., CIMATTI, A. & DADDI, E. 2000, A&A, 364, 26.

O'CONNELL, R. W. 1997, in *The Ultraviolet Universe at Low and High Redshift,* eds. W. H. Waller et al. (Woodbury, NY: AIP), 11.

PAPOVICH, C., DICKINSON, M. & FERGUSON, H. C. 2001, ApJ, 559, 620.

PASCARELLE, S. M., WINDHORST, R. A., DRIVER, S. P., OSTRANDER, E. J. & KEEL, W. C. 1996a, ApJ, 456, L21.

PASCARELLE, S. M., WINDHORST, R. A. & KEEL, W. C. 1998, AJ, 116, 2659.

PASCARELLE, S. M., WINDHORST, R. A., KEEL, W. C. & ODEWAHN, S. C. 1996b, Nature, 383, 45.

PENTERICCI, L. et al. 2000, A&A, 361, L25.

PENTERICCI, L., RÖTTGERRING, H. J. A., MILEY, G. K., CARILLI, C. L. & MCCARTHY, P., J. 1997, A&A, 326, 580.

PETTINI, M., SHAPLEY, A. E., STEIDEL, C. C., CUBY, J., DICKINSON, M., MOORWOOD, A. F. M., ADELBERGER, K. L. & GIAVALISCO, M. 2001, ApJ, 554, 981.

PETTINI, M., STEIDEL, C. C., ADELBERGER, K. L., DICKINSON, M. & GIAVALISCO, M. 2000, ApJ, 528, 96.

PRITCHET, C. J. 1994, PASP, 106, 1052.

QUASHNOCK, J. M., VANDEN BERK, D. E. & YORK, D. G. 1996, ApJ, 472, L69.

RENZINI, A. 1999, in *The Formation of Galactic Bulges,* eds. M. Carollo, H. C. Ferguson & R. F. G. Wyse, (Cambridge: Cambridge University Press), 9.

RHOADS, J. E., MALHOTRA, S., DEY, A., STERN, D., SPINRAD, H. & JANNUZI, B. T. 2000, ApJ, 545, L85.

RICHARDS, E. A., FOMALONT, E. B., KELLERMANN, K. I., WINDHORST, R. A., PARTRIDGE, R. B., COWIE, L. L. & BARGER, A. J. 1999, ApJ, 526, L73.

RIESS, A. G. et al. 2001, ApJ, 560, 49.

ROSATI, P., STANFORD, S. A., EISENHARDT, P. R., ELSTON, R., SPINRAD, H., STERN, D. &

Dey, A. 1999, AJ, 118, 76.

Roukema, B. F. & Mamon, G. A. 2000, A&A, 358, 395.

Rowan–Robinson, M. *et al.* 1991, Nature, 351, 719.

Sargent, W. L. W., Boksenberg, A. & Steidel, C. C. 1988, ApJS, 68, 539.

Sawicki, M. & Yee, H. K. C. 1998, AJ, 115, 1329.

Seitz, S., Saglia, R. P., Bender, R., Hopp, U., Belloni, P. & Zielgler, B. 1998, MNRAS, 298, 945.

Serjeant, S. B. G. *et al.* 1997, MNRAS, 289, 457.

Shanks, T. & Boyle, B. J. 1994, MNRAS, 271, 753.

Shanks, T., Fong, R., Boyle, B. J. & Peterson, B. A. 1987, MNRAS, 227, 739.

Shapley, A. E., Steidel, C. C., Adelberger, K. L., Dickinson, M., Giavalisco, M. & Pettini, M. 2001, ApJ, submitted.

Shaver, P. A. 1984, A&A, 136, L9.

Smail, I. & Dickinson, M. 1995, ApJ, 455, L99.

Smail, I., Ivison, R. J. & Blain, A. W. 1997, ApJ, 409, L5.

Songaila, A., Cowie, L. L., Hu, E. M. & Gardner, J. P. 1994, ApJS, 94, 461.

Songaila, A., Cowie, L. L. & Lilly, S. J. 1990, ApJ, 348, 371.

Spinrad, H., Stern, D., Bunker, A., Dey, A., Lanzetta, K.M., Yahil, A., Pascarelle, S. & Fernández–Soto, A. 1998, AJ, 116, 2617.

Stanford, S. A., Elston, R., Eisenhardt, P. R., Spinrad, H., Stern, D. & Dey, A. 1997, AJ, 114, 2232.

Steidel, C. C., Adelberger, K. L., Dickinson, M., Giavalisco, M., Pettini, M. & Kellogg, M. 1998, ApJ, 492, 428.

Steidel, C. C., Adelberger, K. L., Giavalisco, M., Dickinson, M. & Pettini, M. 1999, ApJ, 519, 1.

Steidel, C. C., Adelberger, K. L., Shapley, A. E., Pettini, M., Dickinson, M. & Giavalisco, M. 2000, ApJ, 532, 170.

Steidel, C. C., Giavalisco, M., Pettini, M., Dickinson, M. & Adelberger, K. L. 1996, ApJ, 462, 17.

Steidel, C. C. & Hamilton, D. 1992, AJ, 104, 941.

Steidel, C. C., Pettini, M. & Hamilton, D. 1995, AJ, 110, 2519.

Stephens, A. W., Schneider, D. P., Schmidt, M., Gunn, J. E. & Weinberg, D. H. 1997, AJ, 114, 41.

Stern, D. & Spinrad, H. 1999, PASP, 111, 1475.

Stiavelli, M. *et al.* 1999, A&A, 343, L25.

Stiavelli, M., Scarlata, C., Panagia, N., Treu, T. & Bertin, G. 2001, ApJ, 561, L37.

Stockton 2000, in *Toward a New Millenium in Galaxy Morphology*,, eds. D. L. Block, I. Puerari, A. Stockton & D. Ferreira (Dordrecht: Kluwer).

Tanaka, I., Yamada, T., Aragón-Salamanca, A., Kodama, T., Miyaji, T., Ohta, K. & Arimoto, N. 2000, ApJ, 528, 123.

Teplitz, H. I. *et al.* 2000, ApJ, 533, L65.

Thompson, D. & Djorgovski, S. G. 1995, AJ, 110, 982.

Thompson, R. I., Storrie–Lombardi, L. J., Weymann, R. J., Rieke, M., Schneider, G., Stobie, E. & Lytle, D. 1999, AJ, 117, 17.

Thompson, R. I., Weymann, R. J. & Storrie-Lombardi, L. J. 2001, ApJ, 546, 694.

van der Werf, P. P., Knudsen, K. K., Labbé, I. & Franx, M. 2001, New Astron. Rev., in press.

Warren, S. J., Hewett, P. C., Irwin, M. J. & Osmer, P. S. 1987, Nature, 330, 453.

Warren, S. J., Hewett, P. C., Lewis, G. F., Møller, P., Iovino, A. & Shaver, P. 1996,

MNRAS, 278, 139.

WARREN, S. J., LEWIS, G. F., HEWETT, P. C., MØLLER, P., SHAVER, P. & IOVINO, A. 1999, A&A, 343, L35.

WEYMANN, R.J., STERN, D., BUNKER, A., SPINRAD, H., CHAFFEE, F.H., THOMPSON, R.I. & STORRIE–LOMBARDI, L.J. 1998, ApJ, 505, L95.

WEYMANN, R. J., STORRIE–LOMBARDI, L., SAWICKI, M. & BRUNNER, R., EDS. 1999, *Photometric Redshifts and the Detection of High Redshift Galaxies,* ASP Conference Series vol. 191 (San Francisco: ASP).

WILLIAMS, R. E. *et al.* 1996, AJ, 112, 1335.

WILLIAMS, R. E. *et al.* 2000, AJ, 120, 2735.

WOLD, M., LACY, M., LILJE, P. B. & SERJEANT, S. 2000, MNRAS, 316, 267.

YAHATA, N. *et al.* 2000, ApJ, 538, 493.

YAMADA, T., TANAKA, I., ARAGÓN-SALAMANCA, A., KODAMA, T., OHTA, K. & ARIMOTO, N. 1997, ApJ, 487, L125.

YAN, L. *et al.* 1999, ApJ, 519, L47.

YATES, M., MILLER, L. & PEACOCK, J. 1989, MNRAS, 240, 129.

YEE, H. K. C., ELLINGSON, E., BECHTOLD, J., CARLBERG, R. G. & CUILLANDRE, J.-C. 1996, AJ, 111, 1783.

YEE, H. K. C. & GREEN, R. F. 1984, ApJ, 280, 79.

YEE, H. K. C. & GREEN, R. F. 1987, ApJ, 319, 28.

ZEPF, S. 1997, Nature, 390, 377.

High-Redshift Galaxies: The Far-Infrared and Sub-millimeter View

By ALBERTO FRANCESCHINI

Dipartimento di Astronomia, University of Padova, I-35122 Padova, IT

Observations at long wavelengths, in the wide interval from a few to 1000 μm, are essential to study diffuse media in galaxies, including all kinds of atomic, ionic and molecular gases and dust grains. Hence they are particularly suited to investigate the early phases in galaxy evolution, when a very rich interstellar medium (ISM) is present in the forming systems.

During the last few years a variety of observational campaigns in the far-IR/sub-mm, exploiting both ground-based and space instrumentation, have started to provide results of relevant cosmological impact. Most crucial among these have been the discovery of an intense diffuse background in the far-IR/sub-mm of extragalactic origin, and the deep explorations from space in the far-IR and with large millimeter telescopes on ground. These results challenge those obtained from optical-UV observations, by revealing luminous to very luminous phases in galaxy evolution at substantial redshifts, likely corresponding to violent events of star-formation in massive systems. This is bringing to significant refinements of the present schemes of galaxy formation, as far as the history of baryon transformations is concerned.

---

1. Introduction

1.1. *The history of baryon transformations*

Although baryons contribute a negligible fraction of the global mass density of the universe, their transformations and the associated energy releases are key elements of the complex, puzzling history bringing from the primeval undifferentiated plasma to the highly structured present-day universe.

Two main driving mechanisms are able to circulate and transform baryons in astrophysical systems: one is related with stars and thermonuclear processes occurring therein, the other with gravitational contraction of gas – an important aspect of which, able to generate vast amounts of energy and producing spectacular effects in Active Galactic Nuclei and quasars, is gravitational accretion onto supermassive black holes (BH).

Obviously, these two fundamental motors of the baryon cycle produce very different outcomes. While gravitational BH accretion irreversibly destroys baryons to produce energy, gas cycling into stars has (more beneficial) effects originating beautiful stellar systems, producing soft-energy photons, heavy elements, dust, and planetary systems in the proper amounts to bring eventually to life.

A basic aim of the present studies of the distant universe, exploiting the current most powerful astronomical instrumentation, is indeed to clarify the history of baryon circulation, and in particular the paths through which the various different galaxy populations, which we observe in the local universe, have built their stellar content, created their hosted nuclear BH's and accumulated material in them.

While the overall story is driven by the evolving background of dark matter distribution, baryons are the observable traces of the evolving large scale structure.

The history of star formation, in particular, is a fundamental descriptor of cosmic evolution. Different cosmogonic scenarios predict very different timetables for the formation of stars and structures. For example, some models predict substantially different formation epochs for stars among the various morphological classes of galaxies, in particular between early-type and late-type galaxy systems. Some others, notably some specializations of the Cold Dark Matter-dominated models, do not.

1.2. Long-wavelength observations of galaxies: a view on the diffuse media and on the "active" phases in galaxy evolution

The build up of stellar populations in high-redshift galaxies is most usually investigated by looking at the optical/UV/near-IR emission from already formed stars in distant galaxies. The complementary approach, less frequently used, is to look at the diffuse media – atomic and molecular gas and dust – in high-z systems, and their progressive transformation into stars.

While observations of the redshifted starlight emission in the optical/near-IR can exploit large telescopes on ground and very efficient photon detectors, reliable probes of the diffuse media require longer-wavelength observations in the far-IR and sub-millimeter: a large variety of lines from atomic species and molecules in the Inter-Stellar Medium (ISM) at all ionization levels are observable there. Another fundamental component of the ISM, dust grains present in all astrophysical settings ranging from planetary disks to nuclear accretion torii around quasars, have the property to emit at these wavelengths, typically between a few μm to 1000 μm.

Observations at long-λ are then essential to study diffuse media in galaxies and *are particularly suited [and needed] to study the early phases in galaxy evolution, when a very rich ISM is present in the forming system*.

Under the generic definition of *galaxy activity* we indicate transient phases in the secular evolution of a galaxy during which the various transformations of the baryons undergo a significant enhancement with respect to the average rate, for reasons to be ascertained. These phenomena concern both enhanced rates of conversion of the ISM gas into stars (the *starburst* phenomenon), and phases of increased activity of the nuclear emission following an event of fast accretion of gas into the super-massive BH (the so-called AGN phase, reaching parossistic levels of photon production of up to $\sim 10^{50} erg/s$ in some high-z quasars).

As we will describe in this paper, IR and sub-mm wavelengths provide a privileged viewpoint to investigate galaxy "activity" in general, for two main reasons: (a) in many cases this λ-interval includes a dominant fraction of the whole bolometric output of active objects; (b) at long wavelengths the screening effect of diffuse dust, present in large amounts in "active" galaxies, is no more effective and an unimpeded access to even the most extreme column-density regions is possible.

1.3. Observational issues

Unfortunately, the IR and sub-millimeter constitute a very difficult domain to access for astronomy: from ground this is possible only in a few narrow bands from 2.5 to 30 μm and at $\lambda > 300$ μm. From 30 to 300 μm observations are only possible from space platforms, the atmosphere being completely opaque.

In any case, however, infrared observations even from space are seriously limited by several factors. The most fundamental limitation is intrinsic in the energies ϵ of photons we are looking at: the quantum-mechanics uncertainty principle sets a boundary to the best achievable angular resolution θ due to diffraction of photons in the primary mirror of a telescope of size D: $\theta[FWHM]/[arcsec] \geq 1.4 \times 57.3 \times 3600 \lambda/D$, ($\lambda = ch/\epsilon$). For a typical cooled space telescope of 1 meter diameter working at $\lambda = 100$ μm this corresponds to $\theta \sim 30$ $arcsec$. For deep surveys of high-redshift IR galaxies this limited spatial resolution implies a limiting flux detectable above the noise due to confusion of several faint sources in the same elementary sky pixel. This confusion limit sets in at flux levels corresponding to ~ 0.04 *sources/area element*, or 0.16 *sources/arcmin*^{2} = 570 *sources/degree*^{2} in the above example (see eq. 8.26 and further details below). On this regard, recent surveys (see §10 and §11) have revealed that the far-IR sky is very much populated by luminous extragalactic sources, which implies that confusion starts to manifest already at relatively bright fluxes for even large space observatories.

Other limiting factors for IR observations come from the difficulty to reduce the instrumental background of (even space) telescopes due to photons generated by the optics. This adds to the ambient photon backgrounds, due to Zodiacal light from interplanetary dust, dust emission from the Milky Way, and the terrestrial atmospheric emission.

The instrumental backgrounds are reduced by cooling the instrumentation, in particular for space IR observatories, but this requires either inserting the whole telescope in large dewars (ISO, SIRTF), or by passively cooling the telescope with a very efficient Sun-shielding (*Herschel*, NGST). All this is technologically very much demanding and tends to limit the duration of space IR missions (because of the finite reservoir of coolant) and the size of the primary photon collector.

Finally, photon detection is not as easy in the IR as it is in the optical, and limited performances are offered by bolometers in the sub-mm and by photo-conductors in the mid- and far-IR. Furthermore, the need to cool detectors to fundamental temperatures entails problems of response hysteresis and detector instabilities due to slow reaction of the electrons to the incoming signal.

1.4. *These lectures*

In spite of the mentioned difficulties to observe at long wavelengths, it was clear since the IRAS survey in 1984 that very important phenomena can be investigated here. Only recently, however, pioneering explorations of the high-redshift universe at these long-wavelengths have been made possible by new space and ground-based facilities, and a new important chapter of observational cosmology has been opened.

These lectures are dedicated to a preliminary assessment of some results in the field. Because of the very complex, often still elusive, nature of many of the discovered sources, and because of the complicated astrophysical processes involved, we dedicate a significant fraction of this paper to review properties of diffuse media (particularly dust) in local galaxies, and of their relation with stars (§2, §3, §4 and §5). We also devote a substantial chapter (§6) to the description of local IR starbursts and ultra-luminous IR galaxies, to improve our chances of understanding their high-redshift counterparts.

Then after a brief mention of historical (IRAS) results in the field (§7), we come to discuss in §8 the discovery and recent findings about the Cosmic Infrared Background (CIRB), in §9 the deep IR surveys by the Infrared Space Observatory (ISO), and in §10 the pioneering observations by millimetric telescopes (JCMT/SCUBA, IRAM). Interpretations of the deep counts are given in §11, and the question of the nature of the fast-evolving IR source populations is addressed in §12. §13 is dedicated to discuss the global properties of the population and some constraints set by the CIRB observations. A concise summary is given in §14. A Hubble constant $H_0 = 50$ $km/s/Mpc$ will be adopted unless otherwise stated.

2. Dust in Galaxies

2.1. *Generalities*

Dust is one of the most important components of the ISM, including roughly half of the heavy elements synthesized by stars. The presence of dust is relevant in many astrophysical environments and has a crucial role in shaping the spectra of many cosmic bodies. However, its existence has been inferred from very indirect evidences up untill recently. The first evidence came from the discovery of a tenuous screen of small particles around the Earth producing the *zodiacal light*. Other evidences came from observations of dust trails of comets, circumstellar dust envelopes around evolved stars, diffuse dust in the Milky Way (MW) producing the interstellar extinction, the discovery of IR emission by galaxies and ultra-luminous IR galaxies in the IRAS era, circumnuclear dust in AGNs (essential ingredient of the unified model for AGNs), the cosmological IR background (COBE, 1996-1998), and eventually the discovery of sites of extremely active star formation at high redshifts (SCUBA and ISO, 1998-2000).

Accounting for the effects of dust is essential not only to understand the extinction, but, even more importantly, to evaluate the energy re-emitted by dust at longer wavelengths, typically at $\lambda \sim 5$ to $1000\,\mu$. This is crucial for estimating all basic properties of distant galaxies: the *Star Formation Rate* (SFR) from various optical and IR indicators, the *ages of stellar populations*, which, based as they are on optical colours, have to distinguish the reddening of the spectrum due to aging from that due to dust extinction, and finally to constrain the stellar *initial mass function* (IMF).

2.2. *Dust grains in the ISM*

Rather than by stars, the available volume in a galaxy is occupied by the ISM, which in local late-type systems amounts to $\sim 10\%$ of the baryonic mass. The ISM includes gas mixed with tiny solid particles, the *dust*, with sizes ranging from a few Å (the PAH molecules) up to $\sim 10\,\mu$m. The mass in dust is typically 0.5 to 1% of the ISM mass.

2.2.1. *Grain production*

The mechanisms of birth, growth and destruction of grains are very complex and poorly understood. It is believed that condensation nuclei for dust grains mostly form in dense regions of the ISM, which are better shielded from UV photons. Main dust production sites are hereby listed.

Envelopes of protostars: during the process leading to the birth of a star a solar nebula is produced, where silicate grains can be formed and then blown away by a Pre-Main Sequence wind (T Tauri phase).

Cold evolved stars: in the cold atmospheres of evolved giants, dust grains can form and drive a strong stellar wind, in particular graphite grains from carbon stars and silicate grains in OH-IR stars. Stars with $M < 8M_\odot$ are important dust producers; higher-M stars, like Wolf-Rayets with high mass-loss rates, are too rare.

Type-II supernovae are probably the most important contributors, as revealed by a variety of tests, like those provided by the IR excesses in the light-curve and the extinction of background stars in SN ejecta. Direct evidences of dust production came from the case-study of SN 1987a (CO and SiO molecules found in the ejecta), the dark spots observed in the synchrotron nebula of Crab, the IR mapping by ISO of Cas-A which resolved clumpy emission associated with the fast moving knots (Lagage et al. 1996).

Type-I supernovae have an uncertain role, with no evidence yet for dust formation (which would be otherwise relevant to solve the problem of the Fe depletion).

The general interstellar medium is also the site of a slow growth around pre-existing condensation nuclei (refractory cores); it is in this way that dirty icy grains are produced.

2.2.2. *Grain destruction*

Grain survival is another, uncertain, chapter of the complex story of dust enrichment of the ISM. Grain destruction is not likely a problem in stellar winds, the grain should survive the injection into the ISM, while it is more a problem for SN ejecta (which have typical velocities in excess of 1000 km/s).

Even after the ejection phase, the ISM is in any case a difficult environment for grain survival: grains can be destroyed there by evaporation, thermal sublimation in intense radiation fields, evaporation in grain-grain collisions, and by radiative SN shocks.

2.2.3. The evolution of the dust content in a galaxy

Modelling the complex balance between grain production and destruction is also guided by observations of isotopic anomalies in meteorites and of the elemental depletion pattern. A detailed account of most plausible intervening processes in the dust life cycle can be found in Dwek (1998). The author also discusses evolution paths of the elemental abundances in the gas and dust phases in a typical spiral galaxy, based on standard assumptions for the infall of primordial gas and chemical evolution. Type-II SN are found to be the main producers of silicate dust in a galaxy, while carbon dust is due to lower mass (2-5 $M_\odot$) stars. The different lifetimes of the two imply likely anomalous abundance ratios between the various dust grain types during the course of galaxy evolution, naturally evolving from an excess of silicate to an excess of carbon grains with galactic time.

Altogether, the dust mass is found to be linearly proportional to the gas metallicity and equal to 40% of the total mass in heavy elements in a present-day galaxy. Although the details can depend to some extent on the evolution of the SFR with time (e.g. in the case of elliptical galaxies this evolution could have been more rapid, see Mazzei, de Zotti & Xu 1994), these general results are not believed to be much affected.

2.3. Interactions between dust and radiation

Dust particles interact with photons emitted by astrophysical sources by absorbing, scattering, and polarizing the light (the combined effect of absorption and scattering takes the name of *extinction*). They also emit photons at wavelengths typically much greater than those of the absorbed photons. The total intensity radiation field $I_\nu(\vec{r}, t)$ (defined as usual by $dE \equiv I_\nu d\nu d\Omega dA dt$, dE being the differential amount of radiant energy) is related to the field sources by the *transfer equation*:

$$\frac{dI_\nu}{d\tau} = -I_\nu + S_\nu, \qquad (2.1)$$

where $d\tau_\nu \equiv \alpha_\nu ds$ is the differential optical depth corresponding to a spatial path ds, $S_\nu \equiv j_\nu/\alpha_\nu$ is the *source function*, α_ν and j_ν being the *extinction* (true absorption + scattering) and emission (true emission + scattering) coefficients. A medium is said *optically thin* or *thick* if τ_ν along a typical path through the medium is $<< 1$ or $>> 1$. Absorption includes those processes in which the energy of photons is turned into other forms (may be internal energy of matter or fields), true emission is the opposite process, whilst in scattering the energy of photons is simply *deviated* into other directions. Dust scattering is usually elastic. A formal solution to eq.(2.1) [e.g. Rybicki & Lightman 1979] is given by:

$$I_\nu(\tau_\nu) = I_\nu(0)\exp(-\tau_\nu) + \int_0^{\tau_\nu} \exp(-\tau_\nu + \tau'_\nu)\, S_\nu(\tau'_\nu)\, d\tau'_\nu \qquad (2.2)$$

If each dust grain has a λ-dependent effective *cross section* σ_ν and spatial density n, then $\alpha_\nu = n\sigma_\nu$ or $\tau_\nu = N\sigma_\nu$, where N is the *column density*. For dust grains it is common to write

$$\sigma_\nu = Q_{\nu,e}\,\sigma_g = (Q_{\nu,a} + Q_{\nu,s})\,\sigma_g$$

where σ_g is the geometrical cross section (πa^2 for spheres) and $Q_{\nu,e}$ is the extinction efficiency (true absorption + scattering). At short-λ (UV), diffraction effects in the photon-grain interaction become negligible, and the effective cross-section coincides with the geometric one, $Q_{\nu,e} \sim 1$. Altogether: $\alpha_\nu = Q_{\nu,e} \sigma_g n$.

The albedo $a_\nu = Q_{\nu,s}/Q_{\nu,e}$ is the fraction of extinguished light being scattered by the grain rather than absorbed.

The emission coefficient j_ν includes a *true* emission $j_{\nu,e}$ and an elastic scattering component, $j_{\nu,s}$, given by:

$$j_{\nu,s}(\hat{\omega}) = Q_{\nu,s} \sigma_g n_d \frac{1}{4\pi} \int_{4\pi} I_\nu(\hat{\omega}') f_\nu(\hat{\omega} - \hat{\omega}') d\Omega$$

where f_ν is the *phase function*, depending on the incidence–scattering angle.

The true emission of dust grains is thermal. From Kirchoff's law $[j_\nu = \alpha_\nu B_\nu(T)]$:

$$j_{\nu,e} = n_d Q_{\nu,a} \sigma_g B_\nu(T_d). \tag{2.3}$$

It is clear that both terms of the emission coefficient depend on the radiation field I_ν. In particular $j_{\nu,e}$ depends on it through the dust grain temperature T: grain heating is almost always dominated by the radiation field. Thus a primary task is to compute T. Two situations apply.

(a) *Grains sufficiently large and massive* don't cool in the time interval between absorption of two photons: they are in thermal equilibrium with the radiation field. Their temperature can be determined by solving for T an energy conservation equation *absorbed energy = emitted energy*:

$$\int Q_{\nu,a} J_\nu d\nu = \int Q_{\nu,a} B_\nu(T) d\nu \tag{2.4}$$

where $J_\nu = 1/4\pi \int I(\nu,\omega) d\Omega$ is the angle-averaged I_ν.

(b) *Small grains fluctuate in temperature* at any acquired photon. They never reach thermodynamic equilibrium (the cooling time is shorter than that between two photons arrivals). A probability distribution $P(T)dT$ to find a grain between T and $T+dT$ can then be computed based on a statistical approach (Puget et al. 1985; Guhathakurta & Draine 1989, Siebenmorgen & Kruegel 1992). Basic ingredients for this computation are:

- the specific heat $C(T,a)$ per C-atom of PAH's of size a and the number N_c of C atoms in the grain;
- the maximum T a PAH can attain after absorption of a photon $h\nu$, and given by the relation:

$$h\nu' = \int_{T_{min}}^{T_{max}} N_c(\nu',a) C(T,a) dT;$$

- the cooling rate of a PAH after being heated to T_{max} is

$$dT/dt = \frac{4\pi a^2 F(T,a)}{C(T,a)}$$

where $F(T,a) = \int Q_{abs}(\nu,a) \pi B(\nu,T) d\nu$ is the power radiated per unit grain surface. The total IR spectrum radiated during the cooling down is:

$$S(\nu',\nu,a) = \int_0^t \pi B(\nu,T) Q_a(\nu,a) 4\pi a^2 dt = \int_{T_{min}}^{T_{max}} \pi B(\nu,T) Q_a(\nu,a) \frac{N_c C(T,a)}{F(T,a)} dT .$$

In any case, dust grains are destroyed by radiation-induced temperatures above $\sim 1000-2000$ K (depending mainly on composition). This is the reason why their emission is relevant only longwards a few μm.

For mixtures of different species of particles the equations must be summed over all the species. For spherical grains of different compositions and sizes a and density $n_i(a)$:

$$\alpha_\nu = \sum_i \int n_i(a)\, Q_{i,\nu,e}\, \pi a^2\, da.$$

The interaction of a dusty medium with the radiation field then requires the knowledge of the quantities $Q_{\nu,a}$, $Q_{\nu,s}$ and f_ν. The Mie (1908) theory provides analytic solutions for homogeneous spheres and infinite cilinders. Otherwise, for irregularly shaped and inhomogeneous grains good approximations can be obtained by simple generalizations of the exact solutions for spheres and cylinders (e.g. Hoyle & Wickramasinghe 1991; Bohren & Huffman 1983).

As a source of scattering (like the e^-), another important effect of dust is to induce polarization in the emitted light. Two ways for dust to produce this are through (a) light transmission in a dusty medium including oriented bipolar components; or (b) dust reflection (e.g. in AGNs). Should we be interested in modelling these effects of dust on polarization, then solutions of the transfer equation (2.1) for all four Stokes parameters would be required.

2.4. Alternative heating mechanisms for dust

Two other heating mechanisms for dust grains can operate (Xu 1997).

(a) *Collisional heating* for dust mixed with thermal gases. In the HI component of the solar neighbourhood the ratio of collisional heating G_{coll} to radiative heating G_{rad} turns out to be

$$G_{coll}/G_{rad}|_{ISRF} \simeq v_{HI,thermal}/c;$$

i.e. the collisional is 5 orders of magnitude less than radiative heating! Only in very hot plasmas (IC plasmas at $T > 10^7$) the two can get comparable.

(b) *Chemical heating*, a process occurring typically in the cold gas component of the ISM, e.g. when an H_2 molecule is formed on the surface of a grain from the combination of 2 H atoms:

$$H + H \to H_2 + 4.48 eV.$$

Most of this chemical energy is absorbed by the grain (the rest is taken by the molecule). The released energy turns out comparable with the collisional one (hence negligible).

2.5. The interstellar extinction curve

Before IRAS, the properties of interstellar dust were mainly inferred from the dimming of optical light of stars inside the Galaxy. If we observe the light from a source through a dust screen, dust emission is negligible in the optical (dust emits significantly only in the IR), offline scattering is unimportant, and the formal solution (eq. 2.1) simplifies to $I_\nu = I_\nu(0) e^{-\tau_\nu}$. Given a source with unextinguished magnitude $m_\lambda(0)$, the extinction in magnitudes is:

$$A_\lambda \equiv m_\lambda - m_\lambda(0) = \frac{2.5}{\ln 10} \tau_\nu \simeq 1.08\, \tau_\nu.$$

The knowledge of the intrinsic colors for a source population allows to determine the wavelength dependence of the extinction curve. The mean extinction curve along most line-of-sights in the Milky Way has been studied by many authors (see references in Hoyle & Wickramasinghe 1991; and see Figure 1). Its main properties are: (a) a growth in the optical–near UV, steeper with frequency, $\tau \propto \nu^{1.6} \propto \lambda^{-1.6}$ $(0.6 - 5~\mu m)$; (b) a bump around 2175 Å; (c) a steeper rise in the far–UV; (d) two features in the mid–IR at 9.7 and 18 μm.

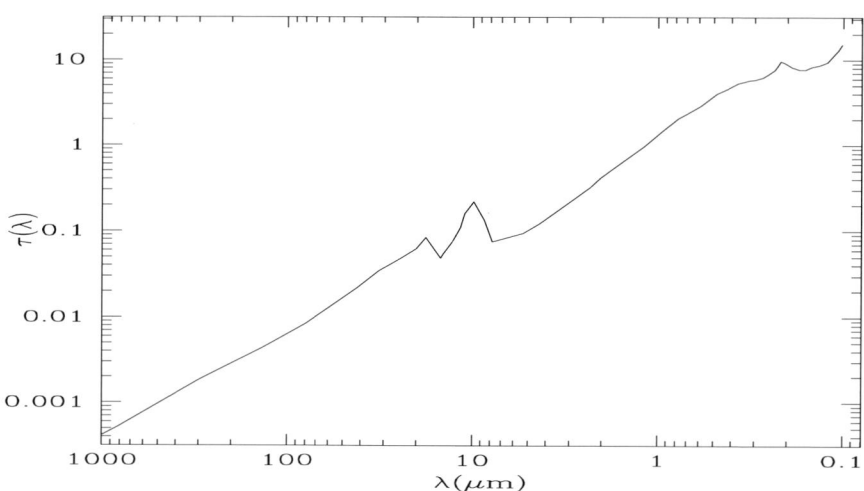

FIGURE 1. The galactic extinction curve, in optical depth per unit value of E(B-V). The two silicate features at 10 and 18 μm and that of carbonaceous grains at 2175 Å can be recognised.

The extinction curve is not universal: in the Milky Way it depends on the line of sight. Data on other stellar systems (LMC and SMC for example) suggest a variable behaviour, in particular in UV.

More recently it has been possible to evaluate indirectly the extinction curve in distant galaxies, by means of accurate photometric observations in narrow-band filters. Gordon et al. (1997) (see also Calzetti 1997) analyze colour-colour plots for 30 starburst galaxies, inferring starburst ages and extinction properties. The 2175 Å bump is absent and the rise in the far-UV slower than observed for the Milky Way. The authors suggest that the starburst has modified the grain distribution, in particular suppressing the 2175 Å feature observed in the MW. Alternatively, Granato et al. (2000) reproduce the observed extinction law in starbursts as a purely geometrical effect, by using the same dust grain mixture than for the MW and accounting for differential extinction for young and old stars (see §4 below).

2.6. Models of the interstellar dust

The extinction curve, whose main features are reported in Fig.1, can be explained by a mixture of grains with different sizes and compositions. The curve in the optical is reproduced by grains with $a \sim 0.1$ μm, while the fast growth of the extinction curve in the UV requires smaller particles with $a \sim 0.01$ μm. Silicate grains explain the 9.7 μm and 18 μm emission features, whose large widths suggest the presence of many impurities (*dirty* or *astronomical* silicates).

On the contrary, silicates cannot explain the optical extinction, because of their excessive albedo. Here carbonaceous grains (graphite or amorphous carbon) are proposed as main absorbers, their resonance at 2175 Å nicely fitting the observed UV bump. The non-linear growth in the FUV is probably due to very small grains and PAH (polycyclic aromatic hydrocarbons) molecules, required also to explain the interstellar IR emission bands, e.g. Puget & Leger 1989.

Unfortunately, the extinction curve does not constrain enough the properties of interstellar dust. For this reason, a variety of models, all with the above basic ingredients, have been proposed to reproduce it.

Draine & Lee (1984) adopt a power law size distribution of silicate and graphite grains $dn/da \propto n_H a^{-3.5}$ for 0.005 μm $< a < 0.25$ μm. A quite more complex model by Siebenmorgen & Krugel (1992) includes five classes of grains (amorphous carbon, silicates, very small grains, small PAH and PAH clusters), providing an impressive fit to the extinction curve. The one by Rowan–Robinson (1992) with a discrete set of nine kinds of grain (amorphous carbon size $a = 30$ μm and $a = 0.1$ μm; graphites with $a =$0.03, 0.01, 0.002 and 0.0005 μm; amorphous silicate $a = 0.1$ μm and silicates with $a = 0.03$ and 0.01 μm) explains also the FIR emission from circumstellar envelopes. The population of very big grains is assumed here to explain the sub-mm emission of carbon stars.

The most relevant recent improvement with respect to the classical models by Draine & Lee is the addition to the grain mixture of very small particles and macro-molecules reaching temperatures higher than equilibrium because of their small size, as described above. Two regions of the extinction curve are particularly sensitive to the presence of these small particles: the mid-IR spectrum (including the emission bands at 3.28, 6.2, 7.7, 8.6 and 11.3 μm, and appreciable continuum), and the fast far-UV rise.

The mid-IR emission bands, in particular, are most commonly interpreted as due to a family of very stable planar molecules, the PAH's, whose vibrational spectra closely resemble, according to laboratory tests, those of emission bands. PAH emission features originate mainly in the so-called *photo-dissociation regions*, i.e. in the interfaces between molecular clouds and the HII regions, where the cloud surfaces are illuminated by the high energy field of the young stars. There are evidences that in denser environments and stronger UV field intensities the PAHs (and the associated mid-IR bands) could be depleted. In the circum-nuclear dusty regions around AGNs PAH emission is not observed.

PAH emission features have been observed by ISO to display Lorentian profiles, whose broad overlapping wings may mimic a kind of continuum (Boulanger et al. 1998). This may possibly explain the observed underlying mid-IR continuum in many astrophysical objects.

3. Evaluating the Dust Emission Spectra

Knowing, or guessing, the optical properties of dust, one can predict the spectra of dusty systems. From a computational point of view, we have to distinguish two cases.

- If the IR dust emission is not self-absorbed ($\tau_{IR} << 1$), the emitted spectrum is simply the volume integral of the local emissivity. An example is the diffuse dust in the IR galactic *cirrus*. Solution of the energy balance equation (2.4) provides the T distribution for the various grain species. Since, in particular, $Q_{\nu,(a,s)} \sim 1$ in UV and $Q_{\nu,(a,s)} \propto a/\lambda^{1.5-2}$ in the far-IR, and considering that the left-hand side is dominated by absorption of UV photons while the right hand by emission at long wavelengths, eq. (2.4) can be re-written for a given grain specie as:

$$I_{bol} = \int J_\nu d\nu \simeq \int B_\nu[T_g(a)] Q_{\nu,(a)} d\nu. \qquad (3.5)$$

Since $\int B_\nu[T_g(a)] d\nu = aT_g^4$ and because of the additional dependence implied by $Q_{\nu,(a)} \propto \nu^{1.5-2} \propto T^{1.5-2}$, the grain equilibrium temperature T_g is found to depend very weakly on the intensity of the local radiation field:

$$T_g \propto (I_{bol}/a)^{1/6}. \qquad (3.6)$$

This implies that dust emission spectra in a variety of galactic environments (from quiescent to actively starbursting galaxies and AGNs) are quite stable and robust, with peak emission mostly confined to the wavelength interval $\lambda_{peak} \simeq 100$ to 30 μm. Longward of λ_{peak} and after eq. (2.3), dust spectra converge according to the Rayleigh Jeans law as

$$I_\nu \propto B_\nu[T_g(a)]Q_{\nu,(a)} \propto \nu^{3.5-4},$$

in agreement with mm observations of local IRAS galaxies by Andreani & Franceschini (1996) and Chini et al. (1995).

- Otherwise, in the presence of IR-thick media (e.g. dense molecular clouds and dusty torii in AGNs), one is faced by the difficult task to solve the transfer equation. We expect that in thick media the IR spectrum will be completely oscured at the short wavelengths (typically in the near- and mid-IR, but sometimes even in the far-IR) by self-absorption.

3.1. Radiative transfer in thick dusty media

In most practical cases, the radiative transfer equation can be solved only with numerical techniques. We mention in this Section a couple of such approaches quite often used.

3.1.1. Numerical solutions based on iterative schemes

A first class of solutions adopt an iterative numerical scheme based on applications of the formal solution of the transfer equation (eq. 2.1). This was originally developed for interpreting AGN spectra (Granato & Danese 1994; Pier & Krolik 1992; Granato, Danese & Franceschini 1997), but is useful to treat more generally radiative transfer in thick media. Although the source function can be any kind in principle, we discuss here an application by Granato & Danese for a central point-source and for a planar and azimuthal symmetry of the dust distribution within a minimum r_m and maximum r_M radii. A condition is set on r_m because of dust sublimation: it cannot be lower than $r_m = L_{46}^{0.5} T_{1500}^{-2.8}$ (pc) to avoid exceeding an equilibrium grain temperature of $T_{gr} = 1500$ for graphite and $T_s = 1000$ for silicates.

The two fields to solve for are the radiation field intensity $I_\nu(r,\Theta,\theta,\phi)$ and the grain temperature distribution $T(r,\Theta)$. The solution is found by representing the field intensity as the contribution of two terms

$$I_\nu = I_\nu^1 + I_\nu^2, \tag{3.7}$$

the first term being the radiation field emitted by the central source and extinguished by the dust, with trivial solution from eq.(2.1):

$$I_\nu^1 = \frac{1}{4\pi}\frac{L_\nu(\Theta)}{4\pi r^2}\exp[-\tau_\nu(r,\Theta)], \tag{3.8}$$

$L_\nu(\Theta)$ becoming dependent on direction because of differential extinction, τ being the optical depth to the point (r, Θ, Φ). The second term originates from thermal emission by dust, and may be expressed at the zero-th order as the formal solution (eq. 2.2) of the transfer equation:

$$I_\nu^2(r,\Theta,\theta,\phi) = \int_{(r,\Theta)}^{(\infty,\infty)} S_\nu(r',\Theta')\exp[-\tau_\nu(r',\Theta')]d\tau_\nu \tag{3.9}$$

The quantity S_ν is the source function j_ν/α_ν which, if the scattering is isotropic, can be expressed as a weigthed average of the scattering and absorption (Rybicki & Lightman 1979) summed over all grain species:

$$S_\nu = \frac{\sum_i \sigma n[Q_{\nu,a}B_\nu(T) + Q_{\nu,s}J_\nu(r,\Theta)]}{\sum_i \sigma n Q_{\nu,s}} \tag{3.10}$$

The function J_ν is the direction-averaged radiation field intensity $\int I_\nu d\Omega$: this integral obviously includes both contributions to the total intensity in eq.(3.7). Finally, assuming radiative equilibrium for the dust grains, the grain temperature distribution is found from eq.(2.4). The following iterative scheme is used to obtain a solution for I_ν:

(1) the zero-th order approximation for I_ν^1 in eq.(3.7) is obtained from eq.(3.8) given L_ν and the adopted dust distribution;

(2) then a zero-th value for the T-field is found from eq.(2.4);

(3) the source function S_ν is then computed from eq.(3.10) including the contribution from thermal dust emission;

(4) after eq.(3.9) the second term I_ν^2 of the radiation field is computed and the total field intensity in eq.(3.7) is updated;

(5) convergence is achieved when e.g. dT from one step to the other is less than a small fixed amount. Suitable scaling rules are usually adopted to accelerate the convergence.

3.1.2. *Monte Carlo solutions*

The advantage of brute-force solutions like a Monte Carlo simulation is that it is better suited to treat complex situations for the geometries of the source function and of the spatial distribution of the absorber. Also velocity fields can be naturally considered in the code to map the kinematical structure of the emission lines (e.g. Jimenez et al. 2000).

The usual approach is to assume a given geometrical distribution for the absorber, possibly including a velocity field, and to generate inside (or outside) it photons according to a given source function (plus a background photon distribution). All these fields are usually discretized into appropriate spatial grids. Each photons are then followed through the distribution of the absorber, their interaction being ruled by the optical depth, albedo and scattering phase functions at that point. The simplest geometrical distributions adopted are (e.g. Disney et al. 1989; Gordon et al. 1997): *the mixed*, in which the source and absorber are homogeneously distributed; *the shell*, where the source and absorber are separated, typically the former inside and the latter outside acting as a screen. However, much more complex situations can be described this way, up to fully 3D distributions without any symmetries (Jimenez et al. 2000).

4. Generalized Spectro-Photometric Models of Galaxies

Twenty years after the first serious models of stellar population synthesis (Tinsley 1977; Bruzual 1983), the most relevant recent progresses have been the attempts to provide a self-consistent description of the effects of dust (and gas) in galaxy spectra and spectral evolution. We review in this Section some recent efforts of generalized spectral synthesis of galaxies from the UV to the sub-mm, including dust effects (as for both the extinction of the primary optical spectrum, and dust re-radiation at longer λ) in the various galactic environments.

Dust plays an important role in all relevant galactic sites: *(1)* the neutral interstellar medium, whose associated dust is heated by the general radiation field (infrared cirrus, prominent in the 100 μm IRAS band); *(2)* the dense cores of molecular clouds, where dust optical-depth is very high and prevents light from very young stars to be observed; *(3)* dust in the external layers of molecular clouds (PRD regions), heated by the interstellar radiation field and OB associations formed in the clouds; *(4)* dust around protostars; *(5)* dust around evolved giants and young planetary nebulae; *(6)* hot dust associated with HII regions.

The inclusion of dust means a dramatic complication of spectro-photometric models: the usual assumption of population-synthesis codes – that the global emission of a whichever complex stellar system is simply the addition of the integrated flux of all components independently on the geometry of the system – is no longer valid: not only the extinction process depends in a complex way on the relative distributions of stars and dust, but also dust emission itself, at high dust column densities and according to geometry, may be optically thick.

In principle, accounting for dust effects in detail may require a very complex description of: (1) the physical-chemical-geometrical properties of grains, determining their interactions with the radiation field (e.g. amorphous, porous low-albedo grains vs. highly reflective grains); (2) the chemical composition of the ISM where grains have condensed (which affects the dust composition), given by the integrated contribution of all previously active stellar populations in the galaxy; (3) the modifications that grains and molecules undergo during the course of evolution, i.e. sublimation in strong UV radiation fields, sputtering, etc. (see §2.2).

These complications of the classical purely stellar evolutionary codes cannot be avoided, if we want a complete and reliable description of physical processes inside galaxies. As we will discuss in later Sections, this turns out to be particularly critical when describing what we called the *active phases* during galaxy evolution: *neglecting dust effects in such cases would bring to entirely wrong conclusions.*

On the other hand, the uncertainties introduced by the large number of new parameters are largely reduced by adopting a multi-wavelength (UV through mm) approach, which balances the unknowns with the number of constraints coming from a wide-band observed spectrum.

4.1. *Semi-empirical approaches*

A phenomenological approach to a global spectrophotometric description of galaxy evolution was recently discussed by Devriendt, Guiderdoni & Sadat (1999). This paper elaborates separately the code for stellar population synthesis from that of dust emission. The former is treated with the most recent prescriptions. The dust emission is schematically represented as the contribution of four different components: the PAH emission features, very small grains, big grains illuminated by the general galactic radiation field (cold dust), and big grains illuminated by young stars in starburst regions. These four components are modelled using typical parameter values for the dust composition, radiation field intensity, mass, etc. Relative normalizations of the four components are finally calibrated using the observed relationship between the IRAS colours of galaxies and the bolometric luminosity.

This approach is quite fast as for computation time (in particular it overcomes the problem of solving the radiative transfer equation), and is particularly useful for statistical analyses of large galaxy databases.

4.2. *Detailed self-consistent spectro-photometric models*

More physically detailed descriptions of the galactic dust emission are discussed by several teams. These models interface two logical procedures:

• (1) the first is to describe, given a prescription for the IMF, the history of star-formation in the galaxy as a function $\psi(t)$ detailing the mass in stars formed per unit time t, the actual gas metallicity $Z(t)$, the abundances of various elements produced by stars during evolution, and the residual gas fraction $g(t)$ as a function of time;

• (2) the second step is to sum up, at any galactic age t, the fluxes from all populations of stars, by solving the radiative transfer equation taking into account how stars and the residual gas and dust are geometrically distributed.

4.2.1. Chemical evolution of the ISM

While point (1) above is addressed in detail by other contributions to this Book (Bruzual), we remind here a few basic concepts.

A galaxy is usually modelled from the chemical point of view as a single environment where primordial gas flows in according to an exponential law

$$\dot{M}(t) \propto \exp(-t/t_{inf}). \tag{4.11}$$

The SFR follows a general Schmidt law

$$\psi(t) = \nu M_g(t)^k \propto g(t)^k \tag{4.12}$$

with the addition of one or more bursts of star-formation to describe starburst episodes possibly triggered by galaxy interactions or mergers. The typically adopted value for k is 1. For the initial mass function (IMF) the usual assumption is a Salpeter law

$$d\phi(M) \propto M^{-x}dM, \quad x = 2.35, \quad M_{min} < M < M_{max} = 100\ M_\odot \tag{4.13}$$

with typically $M_{min} = 0.1\ M_\odot$ (but higher values may apply for example in the case of starbursts). The observed photometric properties of galaxies of various types and morphologies are reproduced by varying in particular t_{inf} and ν.

Given the above parameters, the solution of the equations of chemical evolution allow to compute at any given galactic time all basic quantities, in particular the functions $g(t)$ and $Z(t)$, and then, after eq.(4.12), the number of stars generated at that time with metallicity $Z(t)$. The integrated spectrum of each stellar generation (Single Stellar Population, SSP) then evolves according to the prescriptions of stellar evolution, defining a 2D sequence (spectral intensity $L[\nu, t]$ vs. frequency ν as a function of time, t).

4.2.2. Geometrical distributions of gas and stars

In the model by Silva et al. (1998) three different stellar and ISM components are considered in the generic galaxy: *(a)* star-forming regions, comprising molecular clouds (MC), with young stars, gas and dust in a dense phase, and HII regions; *(b)* young stars escaped from the MC complexes; *(c)* diffuse dust ("cirrus") illuminated by the general interstellar radiation field.

For disk galaxies the adopted geometry is a flattened system with azimuthal simmetry and a density distribution for the 3 above components described by *double exponentials*:

$$\rho = \rho_0 \exp(-r/r_d) \exp(-|z|/z_d).$$

For spheroidal galaxies, spherical symmetry is adopted with King profiles:

$$\rho = \rho_0((1 + [r/r_c]^2)^{-\gamma} - (1 + [r_t/r_c]^2)^{-\gamma}),$$

with $\gamma = 3/2$, $[r_t/r_c] \sim 200$, $r_c \sim 300$ pc as typical values.

4.2.3. Models of the molecular clouds (MC)

High-resolution CO and radio observations show that MCs are highly structured objects containing very dense cores where stars are actually formed. Typical values for the MCs are: size ~ 10 pc, mass $\sim 10^6\ M_\odot$.

All star-formation in the Galaxy happens in dusty MCs, the early evolution phases of young star clusters occurr inside such dusty regions, hence are optically hidden. Later, on the lifetime of OB stars ($10^6 - 10^7\ yrs$), the radiation power of young stars, stellar winds and the first SNs destroy the parent MCs and allow the young stellar population to appear in the optical.

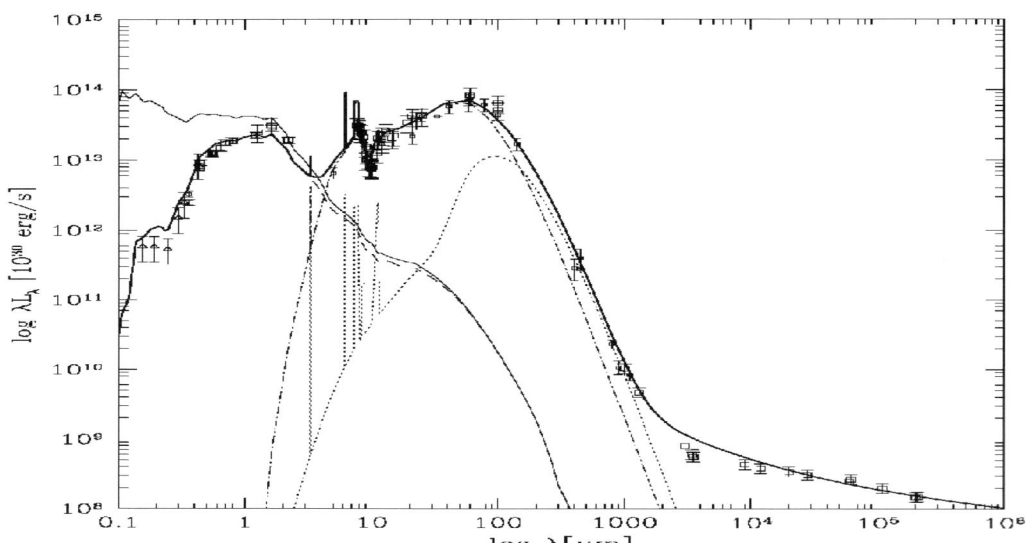

FIGURE 2. The broad-band (UV through radio) data on the prototype nearby starburst galaxy M82. The ordinate axis is normalized to 10^{30} erg/s. [Courtesy of G.L. Granato].

Note that, because of the clumpiness of MCs, this is in any case a *statistical process*: in some clouds even the emission of the youngest OB stars is already visible, while in others all young stars are completely embedded in dust. Silva et al. (1998) describe schematically this transition of the MC from a dust-embedded phase to the optically dominated phase, as a process in which the fraction f of the light from the SSP generated within the cloud still embedded in dust decreases linearly with time as $f(t) = 2 - t/t_0$, t_0 being the time interval during which the SSP is entirely extinguished.

The spectrum emitted by the MC and filtered by dust is computed by solving the *transfer equation*, e.g. by assuming that the primary SSP spectrum comes from a point source in the center of the cloud (this rather crude assumption allows substantial simplifications in the numerical code, see above).

A more detailed description of molecular cloud structure and emission is provided by Jimenez et al. (2000). Their model is based on fully three-dimensional simulations of the *density* and *velocity fields* obtained by solving 3D compressible magneto-hydrodynamical (MHD) equations in supersonic turbulent flows, as typical of the motions in Galactic molecular clouds (Padoan et al. 1998). The MHD turbulence generates a large density contrast, with the density field spanning a range of 4 to 5 orders of magnitude. This brings to a *highly filamentary and clumpy morphology*. All this is consistent with observed properties of the clouds.

Young stars with $M > 15 - 20$ $M_\odot$ in this model are heavily extinguished for virtually all their live. A detailed Monte Carlo approach is required to solve the radiative transfer equation. The simultaneous knowledge of the density and velocity fields allows also to estimate in great detail the molecular emission lines (CO).

4.2.4. Models of diffuse dust (cirrus)

Diffuse dust in the galaxy is responsible for a general attenuation of the light emitted by all stars and MC complexes. In this case the dust column density is not so high to require a detailed solution of the transfer eq. (τ_ν for IR photons is small). One can express an effective optical depth to account for combined absorption and scattering (Rybicky and Lightman 1979): $\tau_{eff} = \sqrt{\tau_a(\tau_a + \tau_s)}$. The galaxy is divided into small volume elements V_i, such that the local radiation field in this elementary volume is

$$J(\lambda)_i = \sum_k \frac{V_k [j(\lambda)_k^{mc} + j(\lambda)_k^{star}] e^{-\tau_{eff}(i,k)}}{r_{i,k}^2},$$

$r_{i,k}^2$ being the distance between the i-th and k-th volumes. This determines the temperature of the local diffuse dust, whose integrated flux seen by an observer in a direction θ is a simple sum over all volume elements of the diffuse dust emissivity:

$$S(\lambda, \theta) = 4\pi \sum_k V_k j(\lambda)_k e^{-\tau_{eff}(k,\theta)}$$

τ being the optical depth from the V-element to the outskirts in that direction and $j(\lambda)_k = j(\lambda)_k^{mc} + j(\lambda)_k^{star} + j(\lambda)_k^{cirrus}$.

4.2.5. Modelling the SEDs of normal and starburst galaxies

Figure 2 shows the broad-band (UV through radio) spectrum of the prototype starburst galaxy M82, a closeby well studied object at 3.2 Mpc. The lines in the figure come from the fit obtained by Silva et al. (1998). The thin continuous line peaking at 0.1 μm corresponds to the unextinguished integrated spectrum of all stellar population, while the long-dashed line is the reddened stellar continuum. The dot-dashed line is the contribution of dust in molecular clouds, while the dotted line comes from diffuse dust in the "cirrus". In this model, the optical-NIR spectrum of the galaxy is contributed mostly by old stellar populations unrelated to the ongoing starburst, whereas the starburst emission is mostly observable at $\lambda > 4$ μm in the form of dust re-radiation, radio SN and free-free emissions.

Equal areas in the $\lambda L(\lambda)$ plot of Fig. (2) subtend equal amounts of radiant energy: it is then clear from the figure that in this moderate starburst $\sim 80\%$ of the bolometric flux emerges as dust re-radiation above 5 μm. In higher luminosity starbursts and in Ultra-Luminous IR Galaxies (ULIRGs, e.g. Arp 220) this fraction gets close to 100%. On the contrary, for local normal galaxies the average fraction is only $\sim 30\%$, as found from comparison of the far-IR with the optical luminosity functions of galaxies (Saunders et al. 1990).

5. Infrared and Sub-mm Line Spectra

IR/sub-mm spectroscopy offers unique opportunities to probe the physical conditions ($n[atoms]$, P, T, extinction, ionization state) in the various components of the ISM, because:
- the widespread presence of dust makes optical-UV-NIR line diagnostics completely unreliable;
- most of the line emission by MCs is extinguished and does not appear in the optical;
- lines from molecular phases (including most of the ISM mass) appear in the FIR-mm;
- several fundamental cooling lines of gas happen in the FIR;
- lines from an extremely wide range of ionization states are observable in the IR.

TABLE 1. Relevant components and line tracers of the ISM

Component	Temperature	Density	Tracers and IR lines
Cold gas	10–100 K	1–1000 cm^{-3}	H_2, CO, PAH's
Diffuse HI	100–1000 K	1 cm^{-3}	HI 21cm, [CII], [OI]
HII regions	1000–10000 K	3-300 cm^{-3}	$H\alpha$, [OII], [OIII]

Table 1 summarizes IR tracers of the various ISM components. Clearly, IR spectroscopy is essential for studies of galaxy activity, though it requires a continuous coverage of the IR spectrum, possible only from space. While ISO allowed to investigate spectroscopically nearby IR active galaxies, future missions (SIRTF, NGST, *Herschel*) will make possible similar studies for galaxies at any redshifts.

5.1. *The cold molecular gas*

Looking at the mm/sub-mm spectral lines is the usual way to study the cold molecular gas, which typically includes the largest mass fraction of the ISM. The lines come from *rotational* and *vibrational* transitions of diatomic and polyatomic molecules.

The very many molecules observable allow to accurately sample the various regimes of ρ, T and elemental abundance. Unfortunately, the most abundant molecule (H_2) is not easily observed directly. It is seen in absorption in UV, or in the NIR roto-vibrational transitions at 2.121 and 2.247 μm. Only with mid-IR spectroscopy by ISO it was possible to observe the fundamental rotational lines at 17 μm (S[1]), 28.2 μm (S[0]), and 12.3 μm (S[2]) in NGC6946, Arp220, Circinus, NGC3256, NGC4038/39). These observations indicate very cool gas to be present with very high column densities (the transition probabilities of the lines are very low).

Because of the difficulty of a direct measure, the amount of molecular gas (H_2) is often inferred from easier measurement of CO emission lines, assumed an H_2/CO conversion. CO rotational transitions allow excellent probes of cold ISM in galaxies: the CO brightness temperature ($\propto$ line intensity) is almost independent on z at $z = 1$ to 5, due to the additional $(1 + z)^2$ factor with respect to the usual scaling with the luminosity distance (Scoville et al. 1996). CO line measurements have been performed for all IRAS sources in the Bright Galaxy Sample, the majority have been detected with single-dish telescopes. In the most luminous objects the molecular mass is $0.2 - 5 \; 10^{10} \; M_\odot$, i.e. 1 to 20 times the content of the Milky Way. Typically 50% or more of this mass is found within the inner kpc from the nucleus, the molecular mass substantially contributing to the total dynamical mass ($> 50\%$ of M_{dyn}). Unfortunately, detecting CO emission by high-z galaxies has proven to be difficult (see below).

5.2. *The cold neutral gas*

The diffuse neutral ISM is commonly traced by the HI 21 cm line from ground-based observations. HI cooling, which is essential to achieve temperatures and densities needed to trigger SF, depends mainly on emission by the 158 μm [CII] line, the 21 cm line and the 63 μm [OI] line.

The 158 μm [CII] line is a major coolant for the diffuse neutral gas and a fundamental cooling channel for the photo-dissociation regions (PDR's), the dense phase interfacing cold molecular clouds with the HII or HI lower-density gas. Carbon is the most abundant element with ionization potential (11.3 eV) below the H limit (13.6 eV): CII atoms are then present in massive amounts in neutral atomic clouds. The two levels in the ground

TABLE 2. The most important IR fine-structure lines. [a] Line intensity compared with the observed [CII]158 μm for the prototypical starburst M82, when available, or predicted by Spinoglio & Malkan (1992) from a model reproducing the physical conditions in M82.

Species	Excitation potential	λ (μm)	n_{crit} cm^{-3}	F/F[CII][a]
OI	-	63.18	5 10^5	1.4
OI	-	145.5	5 10^5	0.06
FeII	7.87	25.99	2 10^6	
SiII	8.15	34.81	3 10^5	2.6
CII	11.26	157.7	3 10^2	1
NII	14.53	121.9	3 10^2	0.37
NII	14.53	203.5	5 10^1	0.11
ArII	15.76	6.99	2 10^5	0.11
NeII	21.56	12.81	5 10^5	2.1
SIII	23.33	18.71	2 10^4	0.68
SIII	23.33	33.48	2 10^3	1.1
ArIII	27.63	8.99	3 10^5	0.23
NIII	29.60	57.32	3 10^3	0.31
OIII	35.12	51.82	5 10^2	0.74
OIII	35.12	88.36	4 10^3	0.66
NeIII	40.96	15.55	3 10^5	0.16
OIV	54.93	25.87	10^4	–

state of CII responsible for the $\lambda = 158$ μm transition correspond to a relatively low critical density $n_{crit} \simeq 300$ cm^{-3} [the density at which collisional excitation balances radiative de-excitation]: CII is excited by electrons and protons and cools down by emitting a FIR photon. The CII line intensity is also weakly dependent on T, hence a good measure for P. The [OI] 145 μm and 63 μm lines are also coolants, though less efficient.

5.3. The ionized component of the ISM

Again, a number of lines from atomic species, covering an extremely wide range of ionization conditions, are observable in the far-IR. Their observations allow extensive analyses of the physical state of the gas. This, coupled with the modest sensitivity to dust extinction, provides the ideal tool to probe even the most compact, extinguished sites, e.g. in the inner galactic nuclei.

For a detailed physical investigation, line ratios sensitive to either gas temperature T or density n are used. To estimate electron density n one can use the strong dependence of the fine-structure line intensities for doublets of the same ion on n: one example are the [OIII] lines at 5007 Å, 52 μm and 88 μm. Similarly one can estimate T and the shape of the ionizing continuum.

Particularly relevant to test the spectral shape of the ionizing continuum are the *fine-structure lines from photo-ionized gas*, which allow to discriminate spectra of stellar and quasar origin. Low-ionization transitions typically strong in starbursts are [OIII] 52 μm and 88 μm, [SiII] 34 μm, [NeII] 12.8 μm, [NeIII] 15.6 μm, [SIII] 18.7 μm and 33.4 μm, while higher ionization lines in AGNs are [OIV] 25.9 μm and [NeV] 24 μm. Table 2 reports a few of the most important IR ionic lines.

One important application of IR spectroscopy was by Genzel et al. (1998), to investigate the nature of the primary energy source in IR luminous galaxies (see §6.8).

6. IR Starburst and Ultra-Luminous Galaxies in the Local Universe

For a variety of reasons it is unlikely that star-formation (SF) in galaxies has proceeded quietly during the Hubble time. 'A posteriori' evidence has accumulated that a fraction of stars in stellar systems was produced during short-lived events (see e.g. the excellent review in Moorwood 1996). These SF events are expected to be very luminous, either in the optical or in the IR, and are expected to contribute substantially to the global energetics from baryon thermonuclear reactions, to the synthesis of metals, and the generation of background radiations in the optical, IR and sub-mm. Also the origin of Es, S0s and of the bulges of spirals may have some relationship with luminous and ultra-luminous starburst events at high-z.

If the study of star-formation in high-redshift sources is a primary task for modern cosmology, it is obvious that relevant information for the interpretation of distant objects comes from a close up on local galaxies with enhanced SF. For this reason we consider in this Section a sub-class of local galaxies, *the starburst galaxies and the IR luminous and ultra-luminous galaxies*, including a small fraction (few %) of all local objects, but accounting for a large percentage of the present-day star formation in galaxies.

The discovery of the starburst phenomenon dates back to the 1970's and came almost simultaneously from two quite independent lines of investigation: from objective-plate (Markarian) surveys of UV-excess galaxies, and from the first pioneering IR observations of galaxies in the local universe. IR observations, in particular, revealed the existence of galaxies with IR luminosities and L/M ratios appearently too high to be sustained over their lifetimes (Harwit & Pacini 1975). This brought to the idea that some galaxies undergo a sudden burst of massive star formation, with dust reprocessing of UV photons emitted by the young stars interpreted as the source of the IR light.

From these observations it was clear that SF has a twofold appearance, a UV excess and an IR excess, which may be explained by the *stocastical nature of the interaction between photons and dust* in star-forming regions of galaxies (see above).

However, the abilities of UV and IR surveys to sample the starburst phenomenon are very different: while at low bolometric luminosities UV and IR surveys sample roughly the same kind of objects, at high luminosities the UV flux is no more a good tracer of the SF, which is better sampled by the IR emission. This effect is due to dust extinction of the UV-light by young stars becoming more and more relevant at the higher bolometric luminosities ($L_{bol} > 10^{11}$ $L_\odot$, Sanders & Mirabel 1996). At the highest values of L_{bol} ($> 10^{12}$ $L_\odot$) most ($> 95\%$) of the flux comes out in the IR.

L_{bol} is also tightly correlated with the optical morphology: while at low-L there is a "natural" mix of various (mostly late) types, at the higher-L nearly all objects appear to be interacting galaxies, and at the highest-L they look as advanced mergers. Also, the correlation is in the sense that while in low-L objects the SF activity is spread over the galactic disk (enhanced in the spiral arms), at increasing luminosity the SF gets more and more concentrated in the nuclear regions.

In the higher-L objects in particular, it is often observed a concomitant stellar and nuclear non-thermal (AGN) activity, usually the latter occurring in the dynamical center of the galaxy and the former in a circum-nuclear ring (at ~ 1 kpc).

A basic difficulty encountered in studies of active galaxies is to disentangle between starburst-dominated and AGN-dominated energy sources of the IR-luminosity. In fact, the two astrophysical processes are quite often associated in the same object. Optical line ratios (high vs. low excitation, e.g. [OIII]5007/Hβ vs. [NII]6583/Hα, the Osterbrock diagram) and line widths (few hundreds km/s for starbursts, larger for AGNs) are sometimes useful indicators, even in the presence of dust.

Useful near-IR lines, accessible from ground, are the Hydrogen Brγ 2.166 μm, HeI 2.058 μm, H$_2$, but also higher atomic number species, [FeII] among others. The Brγ 2.166 μm and HeI 2.058 μm, in particular, so close in λ that differential extinction is negligible, constrain the underline ionization spectrum.

However, the most reliable information is provided by mid- and far-IR spectroscopy by space observatories. Extremely promising in this field, in addition to ISO and SIRTF in the next few years, are the planned large space telescopes: NGST in the mid-IR and *Herschel* in the far-IR.

6.1. *The infrared-radio correlation*

While there is no direct proof for the basic interpretation of the IR starburst phenomenon (i.e. being due to UV light from newly formed stars absorbed by dust and re-emitted in the IR), an indirect support comes from the well-known radio to far-IR luminosity correlation (de Jong et al. 1985, Helou et al. 1985). This, which is the tightest correlation involving global properties of galaxies, provides an important constraint on the physics ruling starbursts of any luminosities. It not only involves luminous active starbursting galaxies, but also many other galaxies, like quiescent spirals.

The correlation is parametrized by the ratio of the bolometric far-IR flux F_{FIR} (in $erg/s/cm^2$) to the radio flux S_ν (in $erg/s/cm^2/Hz$):

$$q = \log[F_{FIR}/(3.75\ 10^{12}\ \text{Hz})/S_\nu(1.4\ \text{GHz})] \simeq 2.35, \qquad \sigma(q) \simeq 0.2 \qquad (6.14)$$

which is observed to keep remarkably constant with L_{bol} ranging over many orders of magnitude, from low-luminosity spirals up to ultraluminous objects (Arp 220) [small departures from linearity appearing at the low- and high-luminosity ends].

The relation is interpreted as an effect of the ongoing star formation: the far-IR emission comes from dust heated by UV photons from young stars, which also heat the ISM producing free-free emission. Supernovae from young stars produce high-energy e$^-$ which results in synchrotron flux mostly by interaction with the general galactic magnetic field. This same scheme explains the departures from linearity: e.g. q slightly increases at the low-luminosity end because L_{FIR} is also contributed by the flux by old stars heating the dust. The radio emission tends to be less concentrated than the far-IR, because of fast e$^-$ diffusion.

6.2. *Estimates of the star formation rate (SFR)*

As the bolometric luminosity increases, the optical indicators of the SFR (e.g. the UV flux, or the EW of Hα) become increasingly uncertain, as a larger and larger fraction of short-λ photons are extinguished. In such a situation, the IR luminosity (proportional to the luminosity by young stars) becomes the most reliable indicator of the SFR. A slight complication here is that older stars illuminating the diffuse cirrus dust in galaxies also contribute to the far-IR flux, particularly in low-luminosity inactive systems.

The SFR is estimated by Telesco (1988) from the energy released by the CNO cycle and assuming a Salpeter IMF (eq. 4.13):

$$SFR(OBA) = 2.1\ 10^{-10}\ L_{FIR}/L_\odot\ [M_\odot/yr], \qquad SFR(All) = 6.5\ 10^{-10}\ L_{FIR}/L_\odot\ [M_\odot/yr]$$

the former relation referring to the OBA star formation. A refined calibration is given by Rowan-Robinson et al. (1997):

$$SFR(All) = 2.6\ 10^{-10}\ \phi\ \epsilon\ L_{60\mu m}/L_\odot\ [M_\odot/yr]$$

where ϕ incorporates the correction from a Salpeter IMF to the true IMF ($\phi \sim 3.3$ going to a Miller-Scalo) and includes corrections for the cut in the IMF (e.g. $\phi \sim 1/3$ if only OBA stars are formed), ϵ being the fraction of photons re-radiated in the IR.

Another mean of estimating the ongoing SFR exploits the radio flux (Condon 1992), by relating the SN rate to the rate of SF and using observations of the radio luminosity of the Milky Way to calibrate the relation. Since the synchrotron emission (proportional to the rate of SN remnant production) and thermal radiation (from HII regions heated by young OB stars) dissipate in $10^7 - 10^8$ yrs, the radio flux provides a good measure of the instantaneous SFR. Operatively, one needs to estimate the fraction of stars with masses $M > 8\ M_\odot$, progenitors of type-II SN, formed per unit time. The problem with faint radio-source observations is that the radio emission of stellar origin gets easily confused with non-thermal emission by a radio-loud AGN.

Finally, ISO observations indicate that also the mid-IR flux [dominated by hot dust and PDR emission] traces very well the SFR (see §12.1.2 below).

All these long-wavelength methods provide obvious advantages, in terms of robustness with respect to dust-extinction, compared with the optical ones, namely the relation of SFR with the UV continuum flux by Madau et al. (1996):

$$SFR\ (all\ stars) = 5.3\ 10^{-10}\ L_{2800\text{\AA}}/L_\odot\ [M_\odot/yr];$$

$$SFR\ (metals) = 1/42\ SFR\ (stars),$$

and that between the $H\alpha$ line flux and the SFR (Kennicut 1998):

$$SFR\ (all) = 7\ 10^{-42}\ L_{H\alpha}\ [erg/s].$$

Poggianti, Bressan & Franceschini (2000) and Franceschini et al. (2000) have shown that even after correcting for extinction the $H\alpha$ flux using measurements of the Balmer decrement, the $H\alpha$-based SFR is typically a factor ~ 3 lower than the appropriate value inferred from the bolometric flux in IR-luminous galaxies.

Altogether, with these calibrations, moderate luminosity IR starbursts have SFR$\sim$3–30 $[M_\odot/yr]$, (corresponding to $\sim 10^5$ O stars present during a typical burst). The most luminous objects, if indeed powered by SF, have SFR up to 1000 $[M_\odot/yr]$. Bolometric flux and SFR are correlated with the broad-band IR to optical luminosity ratio: $L_{IR}/L_B \sim 0.1$ in inactive galaxies (M31, M33), $L_{IR}/L_B \sim 3-10$ in luminous ($L \sim 10^{11}\ L_\odot$) SBs, $L_{IR}/L_B \sim 100$ in ultra-luminous objects ($L > 10^{12}\ L_\odot$, e.g. Arp 220).

6.3. Gas reservoirs, depletion times, starburst duration

The duration of the starburst is critically related with the mass fraction of stars produced during the event and to the available gas reservoir. Assuming that the SB dominates the spectrum on top of the old stellar population emission, an estimator of the SB duration is the EW of the Brγ line, which is a measure of the ratio of the OB stellar flux (the excitation flux) to the red supergiant star flux (evolved OBA stars). The EW is then expected to evolve monotonically with time. Also, the comparison of the Brγ line with the CO NIR absorption lines is an age indicator (Rieke et al. 1988). Moorwood et al. (1996) find in this way ages of 10^7 to $\sim 10^8$ yrs.

However, *the most direct way to estimate at least an upper limit to the burst duration is the comparison of the total mass of molecular material in the galaxy nucleus with the estimated SFR*, which is also a measure of the efficiency of SF. The gas mass is usually estimated from mm-wave CO line emission and from mm continuum observations of dust emission (assuming suitable conversion factors for H_2/CO and dust/gas). Chini et al. (1995) have found that the two independent evaluations of the molecular mass provide consistent results, showing that luminous IR galaxies are very rich in gas ($2\ 10^9$ to $2\ 10^{10}\ M_\odot$). The ratio L_{IR}/M_{gas} takes enormously different values in different stages of galaxy activity: in normal inactive spirals $L_{IR}/M_{gas} \sim 5$ ($L_\odot/M_\odot$) (e.g. M31), in moderate starbursts $L_{IR}/M_{gas} \sim 20$ (M82, NGC253), in ultra-luminous IR galaxies $L_{IR}/M_{gas} \sim 200$ (Arp 220), in quasars $L_{IR}/M_{gas} \sim 500$.

A limit to the SB duration is then given by $t_{depletion} = 10^{10}\ M_{gas}/L_{IR}$ yr, ranging from typically several Gyr for inactive spirals down to a few 10^7 yr for the more active SBs.

6.4. Starburst-driven super-winds

There are several evidences that extremely energetic outfows of gas are taking place in starbursts: (a) from optical spectroscopy, evidence for Wolf-Rayet lines indicative of very young SBs ($< 10^7$ yrs) and outflow of ionized gas, with velocities up to 1000 km/s (Heckman, Armus & Miley 1990; Lehnert & Heckman 1996); (b) from optical imaging there are evidences of bubbles and cavities left over by large, galactic-scale explosions; (c) from X-ray spectroscopy, evidence for plasmas at very high temperatures (up to few keV), far in excess of what the gravitational field could explain (Cappi et al. 1999).

These highly energetic processes are interpreted as due to *radiative pressure by massive stars, stellar winds and supernovae explosions* occurring in a small volume in the galaxy core, able to efficiently energize the gas and to produce a dynamical unbalance followed by a large scale outflow of the remaining gas.

This phenomenon has relevant implications. It is likely at the origin of the huge amounts of metals observed in the Intracluster Plasma (ICP) in local galaxy clusters and groups. It should be noted that the estimated Fe metallicity of the outflowing plasma ($\sim 0.2 - 0.3$ solar, Cappi et al. 1999) is similar to the one observed in the ICP. Also the higher abundances (1.5-2 solar) observed in the 2 archetypal starbursts for α elements (Si, O, Mg) may indicate that type-II SN (those produced by very massive stars, $M > 8\ M_\odot$) are mostly responsible (Gibson, Loewenstein & Mushotzky 1997). Similar properties are observed in the hot halo plasmas around elliptical galaxies, also rising the question of a possible relationship of the hyper-luminous IR galaxy phenomenon with the formation of early-type galaxies. Therefore, *the enriched plasmas found in local clusters and groups may represent the fossil records of ancient starbursts of the kind we see in local luminous IR starbursts.*

6.5. Starburst models

More precise quantifications of the basic parameters describing the SB phenomenon require detailed modelling. The first successful attempt accounting in some detail for the observed IR and radio data was by Rieke et al. (1980), who demonstrated that the remarkable properties of M82 and N253 are consistent with SB activity.

Since then, a number of groups elaborated sophisticated models of SBs. These successfully reproduce SB properties assuming exponentially decaying SFRs with burst durations of 10^7 to 10^8 yrs, whereas both instantaneous and long duration bursts are excluded.

An important issue addressed by these models is about the stellar IMF during the burst: Rieke et al. found that assuming for M82 a Salpeter IMF with standard low-M cutoff at 0.1 $M_\odot$ resulted in a stellar mass exceeding the limit implied by dynamical mass evaluations. The problem was resolved by assuming that formation of stars with masses less than a few $M_\odot$ is strongly suppressed. This result however is not univocally supported by more recent studies of M82: e.g. Leitherer & Heckman (1995) solution is for a 1 to 30 $M_\odot$ IMF.

Interesting constraints on the IMF come in particular from the analysis of CO line kinematics in Arp 220: Scoville et al. (1996) indicate that the dynamical mass, the Lyman continuum, the SFR and the burst timescale can be reconciled by assuming a IMF truncated outside 5 to 23 $M_\odot$, with a SFR $\sim$ 90 $M_\odot/yr$ for stars within this mass range. Altogether, there seem to be fairly clear indications for a "top-heavy" mass function in the more luminous SBs, as compared with quiescent SF in the Milky Way and in spirals. This has relevant implications for the SFR history in galaxies, the cosmic production of light and of heavy elements.

A very detailed modellistic study of starbursts was given by Leitherer & Heckman (1995) and Leitherer et al. (1999), incorporating all up-to-date improvements in the treatment of stellar evolution and non-LTE stellar atmospheric models. The model successfully explains most basic properties of starbursts, as observed in the optical. Model predictions for a continuous SF over 10^8 yrs and a 1-30 $M_\odot$ Salpeter IMF, normalized to a SFR = 1 $M_\odot/yr$ are: bolometric luminosity = $1.3\ 10^{10}$ $L_\odot$; number of O stars = $2\ 10^4$; ionizing photon flux ($\lambda < 912$ Å) = $1.5\ 10^{53}$ $photons/s$; SN rate = 0.02 yr^{-1}; K = –20.5 mag; mass deposition rate = 0.25 $M_\odot/yr$; mechanical energy deposition rate = $6\ 10^{41}$ erg/s. Important outcomes of these papers are predictions for the EW of most important line tracers of the SF ($H\alpha$, Paβ, Brγ), as a function of the time after the onset of SF and of IMF shape.

Models of dusty starbursts have been discussed by Silva et al. (1998), Jimenez et al. (2000), Siebenmorgen, Rowan-Robinson & Efstathiou (2000), Poggianti & Wu (2000), Poggianti, Bressan & Franceschini (2000). The latter two, in particular, address the question of the classification and interpretation of optical spectra of luminous IR starbursts: they find that the elusive class of e(a) (emission+absorption) spectra, representative of a large fraction ($>$ 50%) of all IR SBs, are better understandable as ongoing active and dusty starbursts, in which the amount of extinction is anti-correlated with the age of the population (the youngest stars are the more extinguished, see also §4.2.3), rather than post-starburst galaxies as sometimes have been interpreted.

6.6. Statistical properties of active galaxy populations

Statistical properties of SB galaxies provide guidelines to understand the origin and triggering mechanisms of the phenomenon. A fundamental descriptor of the population properties is provided by the Local Luminosity Function (LLF), detailing the distribution of space density as a function of galaxy luminosity in a given waveband.

While the faint luminosity end is important for cosmogonic purposes (providing constraints on the formation models, its flattish shape being roughly similar at all wavelengths), SBs and their complex physics dominate at the bright end of the LLF. Indeed, the latter is observed to undergo substantial changes as a function of λ: if the optical/near-IR LLF's display the classical "Schechter" exponential convergence at high-luminosities (essentially tracing the galaxy mass function), LLF's for galaxies selected at longer wavelengths show flatter and flatter slopes (see Fig. 8 below). This flattening is progressive with λ going from the optical up to 60 μm. When expressed in differential form ($Mpc^{-3}L^{-1}$), the bright-end slope of the 60 μm LLF is $\propto L_{60\mu m}^{-2}$, according to the extensive sampling by IRAS (Saunders et al. 1990). Note that this flattening is not due to the contribution of AGNs at 60 μm, which is modest here and quite more important instead at 12 μm. What is progressively increasing with λ up to λ = 60 μm is the incidence of the starburst contribution to the luminosity: it is the starbursting nature of 60 μm selected galaxies responsible for the shape of the LLF.

It is interesting to consider that almost the same slopes $\propto L^{-2}$ are found for all known classes of AGNs, from the luminous radio-galaxies (Auriemma et al. 1977; Toffolatti et al. 1987), to the optical and X-ray quasars (Miyaji, Hasinger & Schmidt 2000; Franceschini et al. 1994a). Also to note is the evidence that the L^{-2} slope for AGNs keeps almost exactly the same at any redshifts, in spite of the drastic increase of the source number-density and luminosity with z due to evolution.

There should be a ruling process originating the same functional law in a wide variety of categories of active galaxies and remarkably invariant with cosmic time, in spite of the dramatic differences in the environmental and physical conditions of the sources. This remarkable behaviour may be simply understood as an effect of the triggering mechanism for galaxy (AGN and starburst) activity: *the galaxy-galaxy interactions (either violent mergers between gas rich objects or encounters triggering a slight increase of the activity).*

The physical mechanism ruling the process is the variation in the angular momentum $\Delta J/J$ of the gas induced by the interaction, and the consequent gas accretion $\Delta m/m$ in the inner galaxy regions (Cavaliere & Vittorini 2000). Starting for example from a $\delta-$function-shaped LLF, the starburst triggered by the interaction produces a transient increase of L which translates into a distortion of the LF towards the high-L's. Assumed $\Delta m/m$ is ruled by the probability distribution of the impact parameter b, it is simple to reproduce in this way the LLF's observed asymptotic shape at the high luminosities.

All this points at the interactions as ruling the probability to observe a galaxy during the active phase.

6.7. *Starburst triggering*

In normal inactive spirals the disk SFR is enhanced in spiral arms in correspondence with density waves compressing the gas. This favours the growth and collapse of molecular clouds and eventually the formation of stars. This process is, however, slow and inefficient in making stars (also because of the feedback reaction to gas compression produced by young stars). This implies that very long timescales (several Gyr) are needed to convert the ISM into a significant stellar component.

On the contrary, because of the extremely high compression of molecular gas inferred from CO observations in the central regions of luminous starburst galaxies, SF can proceed there much more efficiently. Both on theoretical and observational grounds, it is now well established that the trigger of a powerful nuclear starburst is due to a galaxy-galaxy interaction or merger, driving a sustained inflow of gas in the nuclear region. This gas has a completely different behaviour with respect to stars: it is extremely dissipative (gas clouds have a much larger cross-section and in cloud collisions gas efficiently radiates thermal energy generated by shocks). A strong dynamical interaction breaks the rotational symmetry and centrifugal support for gas, induces violent tydal forces producing very extended tails and bridges and triggers central bars, which produce shocks in the leading front, and efficiently disperse the ordered motions and the gas angular momentum. *The gas is then efficiently compressed in the nuclear region and allowed to form stars.*

These concepts are confirmed by numerical simulations of galaxy encounters. Toomre (1977) was the first to suggest that ellipticals may be formed by the interaction and merging of spirals. This suggestion is supported by various kinds of morphological features (e.g. tidal tails, rings) observed in the real objects and predicted by his pioneering numerical simulations.

Much more physically and numerically detailed elaborations have more recently been published by Barnes & Hernquist (1992), who model the dynamics of the encounters between 2 gas-rich spirals including disk/halo components, using a combined N-body and gas-dynamical code based on the Smooth Particle Hydrodynamics (SPH). Violent tidal forces act on the disk producing extended tails and triggering central bars, who sweep the inner half of each disk and concentrate the gas into a single giant cloud. The final half-mass radii of gas are much less than those of stars: for an M^* galaxy of 10^{11} $M_\odot$, $\sim 10^9$ $M_\odot$ of gas are compressed within 100-200 pc, with a density of 10^3 $M_\odot/pc^3$ (Barnes & Hernquist 1996).

Various other simulations confirm these findings. SPH/N-body codes show in particular that the dynamical interaction in a merger has effects not only on the gas component, but also on the stellar one, where the stars re-distribute following the merging and violent relaxation of the potential.

6.8. *Ultra-luminous IR galaxies (ULIRGs)*

Defined as objects with bolometric luminosity $L_{bol} \simeq L_{IR} > 10^{12}$ $L_\odot$, they are at the upper rank of the galaxy luminosity function. A fundamental interpretative problem for this population is to understand the primary energy source, either an extinguished massive nuclear starburst, or a deeply buried AGN.

A systematic study of this class of sources was published by Genzel et al. (1998), based on ISO spectroscopy of low-excitation and high-excitation IR lines, as well as of the general shape of the mid-IR SED (the intensity of PAH features vs. continuum emission; see also Lutz et al. 1998). While the general conclusion of these analyses is that star-formation is the process dominating the energetics in the majority of ultraluminous IR galaxies, they have also proven that AGN and starburst activity are often concomitant in the same source. This fact is also proven by the evidence (e.g. Risaliti et al. 2000; Bassani et al. 2001) that many of the ULIRGs classified by Genzel et al. as starburst-dominated also show a hidden, strongly photoelectrically absorbed, hard X-ray spectrum of AGN origin. Soifer et al. (2000) have also found that several ULIRGs show very compact (100-300 pc) structures dominating the mid-IR flux, a fact they interpret as favouring AGN-dominated emission. The relative role of SF and AGN in ULIRGs is still to be quantified, hard X-ray (spectroscopic and imaging) observations by CHANDRA and XMM, as well as IR spectroscopy by space observatories (SIRTF, *Herschel*) will provide further crucial information.

6.9. *Origin of elliptical galaxies and galaxy spheroids*

As pointed out for the first time by Kormendy & Sanders (1992), the typical gas densities found by interferometric imaging of CO emission in ultra-luminous IR galaxies turn out to be very close to the high values of stellar densities in the cores of E/S0 galaxies. This is suggestive of the fact that ULIRG's have some relationships with the long-standing problem of the origin of early-type galaxies and spheroids.

Originally suggested by Toomre (1977), the concept that E/S0 could form in mergers of disk galaxies immediately faced the problem to explain the dramatic difference in phase-space densities between the cores of E/S0 and those of spirals. Some efficient dissipation is required during the merger, which can be provided by the gas. Indeed, the CO line observations in ULIRG's, also combined with those of the stellar nuclear velocity dispersions and effective radii, show them to share the same region of the "cooling diagram" occupied by ellipticals.

Detailed analyses of the H_2 NIR vibrational lines in NGC 6240 and Arp 220 (van der Werf 1996) have provided interesting information about the mass, kinematics, and thermodynamics of the molecular gas. The conclusion is that shocks, the fundamental drivers for dissipation, can fully explain the origin of the H_2 excitation. The evidence that the H_2 emission is more peaked than stars, and located in between the two merging nuclei, is consistent with the fact that gas dissipates and concentrates more rapidly, while stars are expected to relax violently and follow on a longer timescale the new gravitational potential ensuing the merger.

A detailed study of Arp 220 by van der Werf (1996) has shown that most of the H$_2$ line emission, corresponding to $\sim 2\ 10^{10} M_\odot$ of molecular gas, comes from a region of 460 pc diameter, the gas mass is shocked at a rate of $\sim 40\ M_\odot/yr$, not inconsistent with a SFR $\sim 50-100\ M_\odot/yr$ as discussed in §6.5. Compared with the bolometric luminosity of Arp 220, this requires a IMF during this bursting phase strongly at variance with respect to the Salpeter's one (eq. 4.13) and either cut at $M_{min} >> 0.1\ M_\odot$ or displaying a much flatter shape.

In support of the idea that ellipticals may form through merging processes there is evidence coming from high-resolution K-band imaging that the starlight distribution in hyper-luminous IR galaxies follows a de Vaucouleurs $r^{-1/4}$ law typical of E/S0 (Clements & Baker 1997).

Also proven by simulations, after the formation of massive nuclear star clusters from the amount of gas (up to $10^{10}\ M_\odot$) collapsed in the inner kpc, part of the stellar recycled gas has low momentum and further contracts into the dynamical center, eventually producing a super-massive Black Hole with the associated AGN or quasar activity (Norman & Scoville 1988, Sanders et al. 1988).

7. IR Galaxies in the Distant Universe: Pre-ISO/SCUBA Results

We have summarized in previous paragraphs the main properties of local galaxies when observed at long wavelengths, and emphasized the unique capability of these observations to unveil classes of sources, unnoticeable at other wavelengths, but extremely luminous in the IR. It was clear from this that the most luminous objects in the universe and the most violent starbursters can be reliably studied only at these wavelengths.

Our previous discussion has also illustrated the complexity and difficulty of modelling the long-wavelength spectra of galaxies, heavily dependent on the relative geometries of stars and dust.

Now, assumed we have a decent understanding of the local universe and its IR galaxy populations, we dedicate the next Sections to illustrate and discuss new emerging facts about their distant counterparts, which entail important discoveries for cosmology.

The IRAS survey in 1983, allowing the first sensitive all-sky view of the universe at long wavelengths, is considered as the birth date of IR astronomy. Most of our knowledge about local IR galaxies, as previously discussed, comes from the IRAS database. The fair sensitivity of the IRAS surveys, coupled with the prominent emission of IR galaxies at 60-100 μm, have also allowed to sample and study galaxies at cosmological distances and to derive first tentative indications for evolution.

Counts of IRAS galaxies (mostly at 60 μm, where the S/N was optimum, S including the source signal and N the instrumental and sky [cirrus] noise) have been obtained by Hacking & Houck (1987), Rowan-Robinson et al. (1991), Gregorich et al. (1995), Bertin, Dennefeld & Moshir (1997). Samples at 60 μm with optical identifications and radial velocities have been published by Saunders et al. (1990, 1997), Lonsdale et al. (1990), and Oliver et al. (1996).

Early evidence in favour of evolution for IRAS-selected galaxies have been discussed by Hacking et al. (1987), Franceschini et al. (1988) and Lonsdale et al. (1990), among others. In the models by Franceschini et al. and Pearson & Rowan-Robinson (1996), a sub-population of starburst galaxies including a substantial fraction (30%) of all galaxies in the local universe evolves as $L(z) = L(0)(1+z)^{3.1}$ (Pearson & Rowan-Robinson) or $L(z) = L(0)e^{4.3\tau(z)}$ (Franceschini et al.), roughly reproducing counts and redshift distributions.

However, the IRAS sensitivity was not enough to detect galaxies at substantial redshifts, apart from a handful of exceptions (essentially due to gravitational lensing amplifying the flux): the most distant were found at $z \simeq 0.2 - 0.3$. Any conclusions based on IRAS data are to be considered as preliminary, large-scale inhomogeneities badly affecting these shallow samples.

Another problem for the IRAS surveys was the uncertain identification with faint optical counterparts, because of the large [$\sim 1\ arcmin^2$] IRAS error-box: this implied a systematic bias towards associating IRAS sources with the brightest galaxy falling inside it, which may systematically miss the fainter higher-redshift correct identification.

8. The Breakthrough: Discovery of the CIRB

Cosmological background radiations are a fundamental channel of information about cosmic high-redshift sources, particularly if, for technological limitations, observations of faint sources in a given waveband are not possible. This was clearly the case for the IR/sub-mm domain. The present Section is dedicated to a review on a recently discovered new cosmic component, the cosmological background at IR and sub-mm wavelengths (CIRB), an important achievement made possible by the NASA's Cosmic Background Explorer (COBE) mission.

To appreciate the relevance of this discovery (anticipated by a detailed modellistic prediction by Franceschini et al. 1994b), consider that extragalactic backgrounds at other wavelengths contain only modest (undiscernible) contributions by distant galaxies. The Radio background is clearly dominated by radio-loud AGNs; the Cosmic Microwave Background includes photons generated at $z \sim 1500$; the X-ray and γ-ray backgrounds are dominated by distant quasars and AGNs. Also, diffuse light in the optical-UV (and partly the near-IR) will be hardly depurated of the foreground contaminations (in particular, Galactic starlight reflected by high latitude "cirrus" dust, and Zodiacal-reflected Sun-light).

On the other hand, the recently completed third experiment (DIRBE) of the COBE mission has brought to the *first detection ever (with surprisingly small uncertainties) of the integrated emission of distant galaxies* in the form of an isotropic signal in the far-IR and sub-mm (Puget et al. 1996, Guiderdoni et al. 1997, Hauser et al. 1998, Fixsen et al. 1998).

8.1. Observational status about the CIRB

In spite of the presence of very bright foregrounds (Zodiacal and Interplanetary dust emission, Galactic Starlight, high-latitude "cirrus" emission), relatively clean spectral windows exist in the IR suitable for extragalactic research: the near-IR cosmological window (2-4 μm) and the sub-mm window (100-500 μm). At these wavelengths the Zodiacal, Starlight, and emission by high galactic latitude dust produce two minima in the total foreground intensity, which is much lower here than it is in the optical-UV.

These spectral windows occur where we would expect to observe the redshifted photons from the two most prominent galaxy emission features: the stellar photospheric peak at $\lambda \sim 1$ μm and the one at $\lambda \sim 100$ μm due to dust re-radiation. The best chances to detect the integrated emission of distant and primeval galaxies are here.

For a curious coincidence, the (expected) integrated emission of distant galaxies turns out to be comparable by orders of magnitude to the Galaxy emission at the Pole and to the Zodiacal light in the near-IR window. This implies that a delicate subtraction of the foreground emissions is required to access the extragalactic domain.

Three main observational routes have been followed to measure the CIRB:

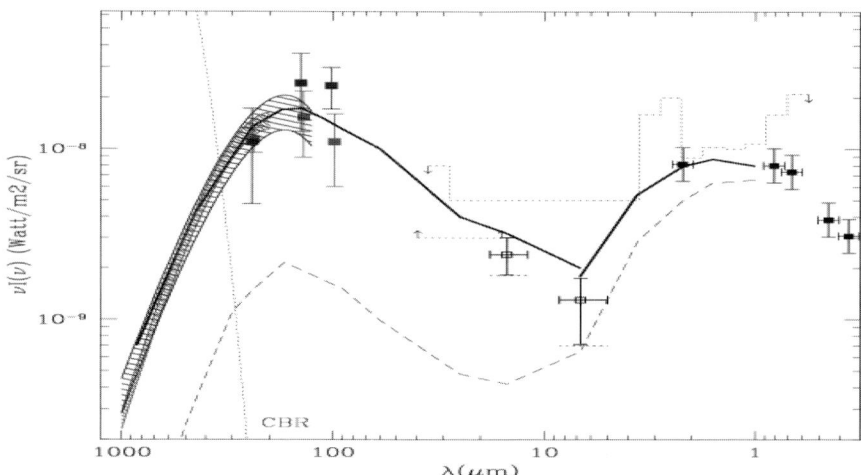

FIGURE 3. The Cosmic Infrared Background (CIRB) spectrum as measured by independent groups in the all-sky COBE maps (e.g. Hauser et al. 1998), compared with estimates of the optical extragalactic background based on ultradeep optical integrations by the HST in the HDF (Madau & Pozzetti 2000). The three lower datapoints in the far-IR are from a re-analysis of the DIRBE data by Lagache et al. (1999), the shaded areas from Fixsen et al. (1998) and Lagache et al. The two mid-IR points are the resolved fraction of the CIRB by the deep ISO surveys IGTES (Elbaz 2001), while the dashed histograms are limits set by TeV cosmic opacity measurements (§8.2). The lower dashed line is the expected intensity based on the assumption that the IR emissivity of galaxies does not change with cosmic time. The thick line is the predicted CIRB spectrum by the model discussed in §11 The dotted line marked CBR corresponds to the Cosmic Microwave Background spectrum.

• rocket flights of dedicated instrumentation (1970-1990), now only of historical interest;
• all-sky surveys by space telescopes (IRAS and COBE, 1984-1996);
• indirect estimates based on very high-energy spectral observations of extragalactic γ-ray sources (Stecker, de Jager & Salamon 1992; Stanev & Franceschini 1998).

In sky directions outside obvious Galactic sources, like star-forming and low-galactic latitude regions, the total far-IR background is due to the contribution of various dust components in the ISM: galactic dust associated with neutral and ionized hydrogen, the interplanetary dust emission, all adding to the isotropic diffuse flux, the CIRB. The way to subtract these various foregrounds when estimating the CIRB intensity is to exploit the different spatial dependencies of the various components, using the correlations with appropriate dust tracers like the HI 21 cm or H$_\alpha$ lines.

To subtract the most important foreground in the far-IR, the galactic dust emission, the simplest procedure is to determine the parameters of the correlation between the background intensity I_ν and the dust tracers expressed in terms of equivalent hydrogen column density N_H, and then to evaluate the CIRB as the intercept of the total flux at $N_H = 0$.

Another method is to perform an all-sky best-fit analysis of a relation like $I_\nu = C_1 N_{HI} + C_2 N_{HII} + CIRB$, N_{HI} and N_{HII} being the column densities of the dust components associated with the neutral and ionized H, respectively, $CIRB$ being the extragalactic background intensity at the working wavelength (e.g, Lagache et al. 1999). The best-fitting determines the constant C_1 and C_2 and allows to estimate a value for the parameter $CIRB$.

Puget et al. (1996) first recognized in the all-sky FIRAS/COBE maps an isotropic signal (independent of Galactic coordinates) with an intensity that can be represented by the law $\nu B_\nu \simeq 3.4 \times 10^{-9} (\lambda/400 \ \mu m)^{-3} \ W \ m^{-2} \ sr^{-1}$ in the 400–1000 μm interval.

This tentative detection has been later confirmed with independent analyses by various other groups (e.g. by Fixsen et al. 1998, who find significant isotropic signal from 200 and 1000 μm), as well as by analyses of data from the DIRBE experiment on COBE in two broad-band channels at $\lambda = 140$ and 240 μm (Hauser et al. 1998). Finkbeiner, Davies & Schlegel (2000), after a very delicate subtraction of the far dominant Galactic and IPD foregrounds, found an isotropic signal at 60 and 100 μm with intensity at the level of $\sim 30 \ 10^{-9} \ W \ m^{-2} \ sr^{-1}$. This latter result is presently under discussion, but appears to conflict with independent estimates (see §8.2).

Recent analyses by Dwek & Arendt (1998) and Gorjian, Wright & Chary (2000) have tentatively found also a signal in the near-IR cosmological window at 3.5 μm and in the J, H and K DIRBE bands, however with large uncertainties because of the problematic evaluation of the Zodiacal (scattered) light. Because of this, CIRB estimates particularly in J, H and K are to be taken more reliably as upper limits.

To avoid overcrowding, we report in Figure 3 only the most recent results from DIRBE (Lagache et al. 1999; Finkbeiner, et al. 2000) and FIRAS (Fixsen et al. 1998).

No isotropic signals are significantly detected at 4 μm $< \lambda <$ 60 μm, any cosmological flux being far dominated here by the Zodiacal light, the Interplanetary dust (IPD) emission and by Galactic dust emission (only missions to the outer Solar System would have chances to reduce the dominant IPD flux to achieve detection of the CIRB here). The constraints we report at these wavelengths come from indirect estimates based on the cosmic high-energy opacity (§8.2 below).

Altogether, after four years of very active debate among various teams working on the COBE data, first about the existence and later on the intensity and spectral shape of CIRB, there is now ample consensus even on details of CIRB's spectral intensity, at least from 140 to 300 μm where it is most reliably measured and where two completely independent datasets (FIRAS and DIRBE, with independent absolute calibrations) are available. The CIRB flux has in particular stabilized at values $\nu I_\nu \simeq 24 \pm 5$ and $\nu I_\nu \simeq 15 \pm 5 \ 10^{-9} \ Watt/m^2/sr$ at $\lambda = 140$ and 240 μm, respectively. Modest differences in the absolute calibration of FIRAS and DIRBE around 100 μm have been reported (Hauser et al. 1998), but these do not seem to affect the overall result.

This was a fundamental achievement for observational cosmology, providing the global energy density radiated by cosmic sources at any redshifts. Two concomitant facts, the very strong K-correction for galaxies in the far-IR/sub-mm implied by the very steep and featureless dust spectra, and their relative robustness due to the modest dependence of dust equilibrium temperature T on the field intensity (eq.[3.6]) have suggested to use the CIRB spectrum to infer the evolution of the galaxy long-wavelength emissivity as a function of redshift (Gispert, Lagache & Puget 2000). Indeed, while the peak intensity at $\lambda = 100$ to 200 μm constrains the galaxy emissivity at $z = 0$ to $z = 1$, the quality of the FIRAS intensity maps and the low foreground contamination at $\lambda > 200$ μm allow to set important constraints on the universal emissivity at $z > 1$.

Between 100 and 1000 μm the integrated CIRB intensity turns out to be $\sim 30 \pm 5 \ 10^{-9} \ Watt/m^2/sr$. In addition to this measured part of the CIRB, one has to consider the presently un-measurable fraction resident in the frequency decade between 100 and 10 μm. This flux is larger than the integrated "optical background" ($\sim 17 \ nWatt/m^2/sr$, see Fig.3), obtained by counting all galaxies detected between 0.3 and 3 μm by HST down to the faintest detectable sources. This procedure to estimate the "optical background" relies on the fact that optical counts show a clear convergence at magnitudes $m_{AB} \geq 22$ (Madau & Pozzetti 2000), such that the expected contribution by sources fainter than HST limiting fluxes appears negligible (a significant upwards revision of this optical background suggested by Bernstein (1998) to account for low surface brightness emission by galaxies is not confirmed).

Already the directly measured part of the CIRB sets a relevant constraint on the evolution of cosmic sources, when compared with the fact mentioned in §4.2.5 that for local galaxies only 30% of the bolometric flux is absorbed by dust and re-emitted in the far-IR. *The CIRB's intensity matching or even exceeding the optical background tells unequivocally that galaxies in the past should have been much more "active" in the far-IR than in the optical, and very luminous in an absolute sense. A substantial fraction of the whole energy emitted by high-redshift galaxies should have been reprocessed by dust at long wavelengths.*

8.2. *Constraints from observations of the cosmic high-energy opacity*

As originally suggested by F. Stecker soon after the discovery of high-energy photon emissions from distant blazars, high-energy spectral observations may provide a suitable alternative to the direct detection of the CIRB at wavelengths where it is currently impossible. The idea is to infer the CIRB intensity from combined GeV and TeV observations of a set of Blazars by exploiting the $\gamma \to \gamma$ interaction of their emitted high energy photons with those of the CIRB.

The absorption cross-section of γ-rays of energy E_γ [TeV] has a maximum for IR photons with energies obeing the condition (Stecker, de Jager & Salomon 1992):

$$\epsilon_{max} = 2(m_e c^2)^2/E_\gamma,$$

which implies

$$\lambda_{peak} \simeq 1.24 \pm 0.6 (E_\gamma [TeV]) \; \mu\text{m}. \tag{8.15}$$

The optical depth for a high-energy photon E_0 travelling through a cosmic medium filled of low-energy photons with density $\rho(z)$ from z_e to the present time is

$$\tau(E_0, z_e) = c \int_0^{z_e} dz \frac{dt}{dz} \int_0^2 dx \frac{x}{2} \int_0^\infty d\nu (1+z)^3 \frac{\rho_\nu(z)}{h\nu} \sigma_{\gamma\gamma}(s) \tag{8.16}$$

$$\sigma_{\gamma\gamma}(s) = \frac{3\sigma_T}{16}(1-\beta^2)[2\beta(\beta^2-2) + (3-\beta^4)\ln(\frac{1+\beta}{1-\beta})]$$

$$s \equiv 2E_0 h\nu x (1+z); \quad \beta \equiv (1 - 4m_e^2 c^4/s)^{1/2}.$$

Coppi & Aharonian (1999) report the following analytical approximation, good to better than 40%, to eq.(8.16):

$$\tau(E_0, z_e) \simeq 0.24 \frac{E_\gamma}{TeV} \frac{\rho(z=0)}{10^{-3} eV/cm^3} \frac{z_e}{0.1} h_{60}^{-1} \simeq 0.063 \frac{E_\gamma}{TeV} \frac{\nu I_\nu}{nW/m^2/sr} \frac{z_e}{0.1} h_{60}^{-1} \tag{8.17}$$

Interesting applications of this concept have been possible when data from the Compton Gamma Ray Observatory and from hard X-ray space telescopes have been combined with observations at TeV energies by the Whipple and other Cherenkov observatories on the Earth.

Stanev & Franceschini (1998) have obtained model-independent upper limits on the CIRB with no a-priori guess about the CIRB spectrum, using HEGRA data for the Blazar MKN 501 (z=0.034) during an outburst in 1997, on the assumption that the high-energy source spectrum is the flattest allowed by the data. These limits (see Fig. 3) get quite close to the CIRB background already resolved by the ISO mid-IR deep surveys (see §9).

More recently, Krawczynski et al. (2000) have combined the observations of MKN501 during the 1997 outburst with X-ray data from RossiXTE and BeppoSAX, providing a simultaneous high-quality description of the whole high-energy spectrum. These data are very well fitted by a Synchrotron Self Compton (SSC) model in which the spectrum at $\nu = 10^{27} Hz$ is produced by Inverse Compton of the hard X-ray spectrum at $\nu = 10^{18} Hz$: the combination of the two provides solid constraints on the shape of the "primary" (i.e. before cosmic attenuation) spectrum at TeV energies. This is used to derive $\tau_{\gamma\gamma}$ as a function of energy and, after eqs. 8.16 and 8.17, a constraint on the spectral intensity of the CIRB. The result is compatible with the limits by Stanev & Franceschini (1998) and allows to get a tentative estimate of the CIRB intensity in the interval from $\lambda = 10$ to 40 μm (see Fig.[3]), which is formally dependent, however, on the SSC model adopted for the intrinsic source spectrum.

Less model dependent is the constraint set by the observations of purely power-law Blazar spectra around $E_\gamma \simeq 1\ TeV$, which translates into the upper limit of about $10\ nWatt/m^2/sr$ at $\lambda \simeq 1\ \mu m$ shown in Fig. 3. Substantially exceeding that, as suggested by some authors (Bernstein 1998, Gorjian et al. 2000), would imply either very "ad hoc" γ-ray source spectra or new physics (Harwit, Protheroe & Biermann 1999).

8.3. Contribution of cosmic sources to the CIRB: the formalism

A simple formalism relates background intensity and cell-to-cell anisotropies to the statistical properties (luminosity functions and number counts) of the contributing sources.

8.3.1. Source contribution to the background intensity

The differential number counts (sources/unit flux interval/unit solid angle) at a given flux S write:

$$\frac{dN}{dS} = \int_{z_l}^{z_h} dz \frac{dV}{dz} \frac{d \log L(S; z)}{dS} \rho[L(S,z), z] \qquad (8.18)$$

where $\rho[L(S,z), z]$ is the epoch-dependent luminosity function and dV/dz is the differential volume element. Flux S and rest-frame luminosity L are related by

$$S_{\Delta\nu} = \frac{L_{\Delta\nu} K(L, z)}{4\pi d_L^2}, \qquad (8.19)$$

where d_L is the luminosity distance and $K(L,z) = (1+z)\frac{L[\nu(1+z)]}{L(\nu)}$ the K-correction. The contribution of unresolved sources (sources fainter than the detection limit S_d) to the background intensity is given by:

$$I = \int_0^{S_d} \frac{dN}{dS} S\, dS = \frac{1}{4\pi} \frac{c}{H_0} \int_{z(S_d, L_{min})}^{z_{max}} \frac{dz}{(1+z)^6 (1+\Omega z)^{1/2}} j_{\text{eff}}(z), \qquad (8.20)$$

having defined the volume emissivity $j_{\text{eff}}(z)$ as

$$j_{\text{eff}}(z) = \int_{L_{min}}^{\min[L_{max}, L(S_d, z)]} d\log L\ L\ n_c(L, z) K(L, z), \qquad (8.21)$$

where L_{min} and L_{max} are the minimum and the maximum source luminosities. From eq.(8.20) we can note that, when the counts converge like $dN/dS \propto S^{-2}$ or flatter, the contribution by faint sources to the background intensity becomes almost insensitive to the source minimum flux $[I \propto \ln(S_{min})$ or less$]$. This property has been used by Madau

& Pozzetti (2000) to estimate the optical background intensity (see Fig. 3) from ultra-deep HST counts of galaxies, by exploiting the convergence of the optical counts fainter than $m_{AB} \sim 22$. A similar property of faint IR sources is used in §9.4 to estimate the contribution of IR galaxies to the CIRB.

8.3.2. Small scale intensity fluctuations

In addition to the average integrated flux by all sources in a sky area, the background radiation contains also spatial information (the cell-to-cell fluctuations) which can be used to further constrain the source flux distribution and spatial correlation properties (e.g. De Zotti et al. 1996). The usually most important contribution to the cell-to-cell intensity fluctuations comes from the stochastic nature of the spatial distribution of sources among elementary cells with an effective solid angle $\omega_{\text{eff,P}}$ (Poisson fluctuations). They can be expressed as

$$(\delta I)^2 \equiv C(0) = \frac{\omega_{\text{eff,P}}(0)}{4\pi} \int_0^{S_d} S^2 \frac{dN}{dS} dS. \tag{8.22}$$

What is really measured, however, is not the flux S but the detector's response $x = f(\vartheta,\varphi)S$, $f(\vartheta,\varphi)$ being the angular power pattern of the detector. Let $R(x) = \int dN\, [x/f(\vartheta,\varphi)]/dS \cdot d\omega/f(\vartheta,\varphi)$ be the *mean number of source responses of intensity x*. For a Poisson distribution of the number of sources producing a response x, its variance equals the mean $R(x)dx$. Adding the variances of all responses up to the cutoff value x_c (brighter sources are considered to be individually detected) gives the contribution of unresolved sources to fluctuations:

$$(\delta I)^2 = \int_0^{x_c} x^2 R(x)\, dx. \tag{8.23}$$

The cutoff x_c is chosen to be some factor q times $(\delta I)^2$; usually $q = 3 - 5$. The rms background fluctuations (δI) imply a sky noise $\sigma_{conf} = \langle(\delta I)^2\rangle^{1/2}$ for observations with spatial resolution ω_{eff}.

The integrated signal D recorded by the detector is the sum of the responses x due to all sources in the angular resolution element. Its probability distribution function $P(D)$ is informative on the amplitude and slope of counts of unresolved sources. Scheuer (1957) has shown that its Fourier transform, $p(\omega)$, is a simple function of the FT $r(\omega)$ of $R(x)$: $p(\omega) = \exp[r(\omega) - r(0)]$. It follows:

$$P(D) = \int_{-\infty}^{\infty} p(\omega) \exp(-2\pi i\omega D)\, d\omega = \int_{-\infty}^{\infty} \exp\left[r(\omega) - r(0) - 2\pi i\omega D\right] d\omega =$$
$$2\int_0^{\infty} \exp\left\{-\int_0^{\infty} R(x)\left[1 - \cos(2\pi\omega x)\right] dx\right\} \cdot \cos\left[\int_0^{\infty} R(x)\sin(2\pi\omega x)\, dx - 2\pi\omega D\right] d\omega. \tag{8.24}$$

This synthetic $P(D)$ has to be convolved with the noise distribution to be compared with the observations. Assumed that the number count distribution below the detection limit can be represented as a power-law, $N(>S) = K(S/S_k)^{-\beta}$, then eq. [8.24] can be integrated to get (Condon 1974):

$$\sigma_{conf} = \left[\frac{q^{2-\beta}}{2-\beta}\right]^{1/\beta} (\omega_{eff}\beta K)^{1/\beta} S_k, \qquad \omega_{eff} = \int f(\vartheta,\varphi)^{\beta} d\Omega \tag{8.25}$$

which allows to estimate the slope of the counts (β) below the detection limit from a given measured value of the cell-to-cell fluctuations σ_{conf}. This constraint on $N(S)$ applies down to a flux limit corresponding to ~ 1 source/beam. Assumed that S_k represents the confusion limit ($S_k = q \times \sigma_{conf}$) of a survey having an areal resolution ω_{eff}, then eq. 8.25 further simplifies to a relation between the number of sources K resolved by the survey (and brighter than S_k) and the parameters q and β:

$$K = \frac{2-\beta}{\beta q^2} \frac{1}{\omega_{eff}}, \qquad (8.26)$$

this implies the confusion limit to occur at the flux corresponding to an areal density of $(\beta q^2/[2-\beta])^{-1}$ sources per unit beam area ω_{eff}. For Euclidean counts and $q=3$, this corresponds to 1 source/27 beams. Confusion limits based on this criterion for various IR observatories are indicated in Figs. 6 and 7 below.

9. Deep Sky Surveys with the Infrared Space Observatory (ISO)

ISO has been the most important IR astronomical mission of the 1990s. Launched by ESA, it consisted of a 60 cm telescope operative in a highly eccentric 70000 km orbit. It included two instruments of cosmological interest (in addition to two spectrographs): a mid-IR 32×32 camera (ISOCAM, 4 to 18 μm), and a far-IR imaging photometer (ISOPHOT, with small 3×3 and 2×2 detector arrays from 60 to 200 μm). The whole payload was cooled to 2 K by a He^3 cooling system so performant to allow ISO to operate for 30 months (Nov 1995 to Apr 1998), instead of the nominal 18 months. An excellent review of the extragalactic results from ISO can be found in Genzel & Cesarsky (2000).

9.1. Motivations for deep ISO surveys

While designed as an observatory-type mission, the vastly improved sensitivity offered by ISO with respect to the previous IRAS surveys motivated to spend a relevant fraction of the observing time to perform a set of deep sky explorations at mid- and far-IR wavelengths. The basic argument for this was to parallel optical searches of the deep sky with complementary observations at wavelengths where, in particular, the effect of dust is far less effective in extinguishing optical light. This could have been particularly relevant for investigations of the distant universe, given the large uncertainties implied by the (probably large) extinction corrections in optical spectra of high redshift galaxies (e.g. Meurer et al. 1997).

Observations in the mid- and far-IR also sample the portion of the electromagnetic spectrum dominated by dust re-processed light, and are then ideally complementary to optical surveys to evaluate the global energy output by stellar populations and active nuclei.

Organized in parallel with the discovery of the CIRB, a major intent of the deep ISO surveys was to start to physically characterize the distant sources of the background and to single out the fraction contributed by nuclear non-thermal activity in AGNs.

Finally, exploring the sky to unprecedented sensitivity limits should have provided an obvious potential for discoveries of new unexpected phenomena from our local environment up to the most distant universe.

9.2. Overview of the main ISO surveys

Deep surveys with ISO have been performed in two wide mid-IR (LW2: 5-8.5 μm and LW3: 12-18 μm) and two far-IR (λ_{eff} = 90 and 170 μm) bands. The diffraction-limited spatial resolutions were ~ 5 arcsec at 10 μm and ~ 50 arcsec at 100 μm. Mostly because of the better imaging quality, ISO sensitivity limits in the mid-IR are 1000 times better than at the long wavelengths (0.1 mJy versus 100 mJy). At some level the confusion problem will remain a fundamental limitation also for future space missions (SIRTF, *Herschel*,

ASTRO-F). A kind of compensation to these different performances as a function of λ derives from the typical FIR spectra of galaxies and AGNs, which are almost typically one order of magnitude more luminous at 100 μm than at 10 μm. We detail in the following the most relevant programs of ISO surveys.

9.2.1. The ISOCAM Guaranteed Time (GT) Extragalactic Surveys

Five extragalactic surveys with the LW2 and LW3 filters have been performed in the ISOCAM GT (GITES, P.I. C. Cesarsky), including large-area shallow surveys and small-area deep integrations. A total area of 1.5 square degree in the Lockman Hole and the "Marano" southern field have been surveyed, where more than one thousand sources have been detected (Elbaz et al. 1999). These two areas were selected for their low zodiacal and cirrus emissions and because of the existence of data at other wavelengths (optical, radio, X).

9.2.2. The European Large Area ISO Survey (ELAIS)

ELAIS is the most important program in the ISO Open Time (377 hours, P.I. M. Rowan-Robinson, see Oliver et al. 2000a). A total of 12 square degree have been surveyed at 15 μm with ISOCAM and at 90 μm with ISOPHOT, 6 and 1 square degree have been covered with the two instruments at 6.7 and 170 μm. To reduce the effects of cosmic variance, ELAIS was split into 4 fields of comparable size, 3 in the north, one in the south, plus six smaller areas. While data analysis is still in progress, a source list of over 1000 (mostly 15 μm) sources is being published, including starburst galaxies and AGNs (type-1 and type-2), typically at $z < 0.5$, with several quasars (including various BAL QSOs) found up to the highest z.

9.2.3. The ISOCAM observations of the two Hubble Deep Fields

Very successful programs by the Hubble Space Telescope have been the two ultra-deep exposures in black fields areas, one in the North and the other in the South, called the Hubble Deep Fields (HDF). These surveys promoted a substantial effort of multi-wavelength studies aimed at characterizing the SEDs of distant and high-z galaxies. These areas, including the Flanking Fields for a total of ~ 50 sq. arcmin, have been observed by ISOCAM (P.I. M. Rowan-Robinson) at 6.7 and 15 μm, achieving completeness to a limiting flux of 100 μJy at 15 μm.

These have been among the most sensitive surveys of ISO and have allowed to discover luminous starburst galaxies over a wide redshift interval up to $z = 1.5$ (Rowan-Robinson et al. 1997; Aussel et al. 1999). In the inner 10 square arcmin, the exceptional images of HST provided a detailled morphological information for ISO galaxies at any redshifts (see Figure 4). Furthermore, these two fields benefit by an almost complete redshift information (Cohen et al. 2000), allowing a very detailed characterization of the faint distant IR sources.

9.2.4. ISOCAM survey of two CFRS fields

Two fields from the Canada-France Redshift Survey (CFRS) have been observed with ISOCAM to intermediate depths: the '14 + 52' field (observed at 6.7 and 15 μm) and the '03+00' field (with only 15 μm data, but twice as deep). The CFRS is, with the HDFs, one of the best studied fields with multi-wavelength data. Studies of the galaxies detected in both fields have provided the first tentative interpretation of the nature of the galaxies detected in ISOCAM surveys (Flores et al. 1999).

9.2.5. *The ISOPHOT FIRBACK survey program*

FIRBACK is a set of deep cosmological surveys in the far-IR, specifically aimed at detecting at 170 μm the sources of the far-IR background (P.I. J.L. Puget, see Puget et al. 1999). Part of this survey was carried out in the Marano area, and part in collaboration with the ELAIS team in ELAIS N1 and N2, for a total of 4 square degrees. This survey is limited by extragalactic source confusion in the large ISOPHOT beam (90 arcsec) to $S_{170} \geq 100$ mJy. Some constraints on the counts below the confusion limit obtained from a fluctuation analysis of one Marano/FIRBACK field are discussed by Lagache & Puget (2000) (§9.4). The roughly 300 sources detected are presently targets of follow-up observations, especially using deep radio exposures of the same area to help reducing the large ISO errorbox and to identify the optical counterparts. Also an effort is being made to follow-up these sources with sub-mm telescopes (IRAM, SCUBA): this can provide significant constraints on the redshift of sources which would be otherwise very difficult to measure in the optical (§12.2).

9.2.6. *The Lensing Cluster Surveys*

Three lensing galaxy clusters, Abell 2390, Abell 370 and Abell 2218, have received very long integrations by ISOCAM (Altieri et al. 1999). The lensing has been exploited to achieve even better sensitivities with respect to ultra-deep blank-field surveys (e.g. the HDFs), and allowed detection of sources between 30 and 100 μJy at 15 μm. However this was obviously at the expense of distorting the areal projection and ultimately making uncertain the source count estimate.

9.2.7. *The Japanese Guaranteed Time Surveys*

An ultra-deep survey of the Lockman Hole in the 7 μm ISOCAM band was performed by Taniguchi et al. (1997; the survey field is different from that of the GITES Lockman survey). Another field, SSA13, was covered to a similar depth (P.I. Y. Taniguchi). The Lockman region was also surveyed with ISOPHOT by the same team: constraints on the source counts at 90 and 175 μm were derived by Matsuhara et al. (2000) based on a fluctuation analysis.

9.3. *Data reduction*

ISOCAM data need particular care to remove the effects of glitches induced by the frequent impacts of cosmic rays on the detectors (the 960 pixels registered on average 4.5 events/sec). This badly conspired with the need to keep them cryogenically cooled to reduce the instrumental noise, which implied a slow electron reaction time and longterm memory effects. For the deep surveys this implied a problem to disentangle faint sources from trace signals by cosmic ray impacts.

To correct for that, tools have been developed by various groups for the two main instruments (CAM and PHOT), essentially based on identifying patterns in the time history of the response of single pixels, which are specific to either astrophysical sources (a jump above the average background flux when a source falls on the pixel) or cosmic ray glitches (transient spikes followed by a slow recovery to the nominal background). The most performant algorithm for CAM data reduction is PRETI (Starck et al. 1999), a tool exploiting multi-resolution wavelet transforms (in the 2D observable plane of the position on the detector vs. time sequence). An independent method limited to brighter flux sources, developed by Desert et al. (1999), has been found to provide consistent results with PRETI, in the flux range in common. Other methods have been used by Oliver et al. (2000a) and Lari et al. (2001). These various detection schemes and photometry algorithms have been tested by means of very sophisticated Monte Carlo simulations, including all possible artifacts introduced by the analyses.

With simulations it has been possible to control as a function of the flux threshold: the detection reliability, the completeness, the Eddington bias and photometric accuracy (∼10% where enough redundancy was available, as for CAM HDFs and Ultradeep surveys). Also the astrometric accuracy is good (of order of 1-2 arcsec for deep highly-redundant images), allowing straightforward identification of the sources (Aussel et al. 1999, see Fig. 4). The quality of the results for the CAM surveys is proven by the very good consistency of the counts from independent surveys (see Fig. 5 below).

Longer wavelength ISOPHOT observations also suffered from similar problems. The 175 μm counts from PHOT C200 surveys are reliable above the confusion limit $S_{170} \sim 100$ mJy, and required only relatively standard procedures of baseline corrections and "de-glitching". More severe are the noise problems for the C100 90μm channel, which would otherwise benefit by a better spatial resolution than C200. The C100 PHOT survey dataset is still presently under analysis.

9.4. Mid-IR and far-IR source counts from ISO surveys

IR-selected galaxies have typically red colors, because of the dust responsible for the excess IR emission. The most distant are also quite faint in the optical. For this reason the redshift information is available only for very limited subsamples (e.g. in the HDF North and CFRS areas). In this situation, the source number counts, compared with predictions based on the local luminosity function, provide important constraints on the evolution properties.

Particularly relevant information comes from the mid-IR samples selected from the CAM GITES and HDF surveys in the LW3 (12-18 μm) filter, because they include the faintest, most distant and most numerous ISO-detected sources. They are also easier to identify because of the small ISO error box for redundant sampling at these wavelengths.

Surveys of different sizes and depths are necessary to cover a wide dynamic range in flux with enough source statistics, which justified performing a variety of independent surveys at different flux limits. The differential counts based on these data, shown in Fig. 5, reveal an impressive agreement between so many independent samples. Including ELAIS and IRAS survey data, the range in fluxes would reach four orders of magnitude. The combined 15 μm differential counts display various remarkable features (Elbaz et al. 1999): a roughly Euclidean slope from the brightest IRAS observed fluxes down to $S_{15} \sim 5$ mJy, a sudden upturn at $S_{15} < 3$ mJy, with the counts increasing as $dN \propto S^{-3.1} dS$ to $S_{15} \sim 0.4$ mJy, and evidence for a flattening below $S_{15} \sim 0.3$ mJy (where the slope becomes quickly sub-Euclidean, $N \propto S^{-2}$).

The areal density of ISOCAM 15 μm sources at the limit of ~ 50-80 μJy is ~ 5 arcmin^{-2}. This is nominally the ISO confusion limit at 15 μm, if we consider that the diffraction-limited size of a point-source is ~ 50 arcsec^2: from eq. (8.26) and for $\beta = -2$, confusion sets in at a source areal density of 0.1/resolution element, or $7/arcmin^2$ in our case. The IR sky is so populated at these wavelengths that ISO was confusion limited longwards of $\lambda = 15$ μm. This will also be the case for NASA's SIRTF (due to launch in mid 2002), in spite of the moderately larger primary collector (85 cm).

Obviously, far-IR selected samples are even more seriously affected by confusion. The datapoints on the 175 μm integral counts reported in Fig. 6 come from the FIRBACK survey. Similarly deep observations at 90, 150 and 175 μm are reported by Juvela, Mattila & Lemke (2000). Given the moderate depth of these direct counts, background fluctuation analyses were used to constrain their continuation below the survey detection limit. The analysis of small-scale fluctuations in one FIRBACK field by Lagache & Puget

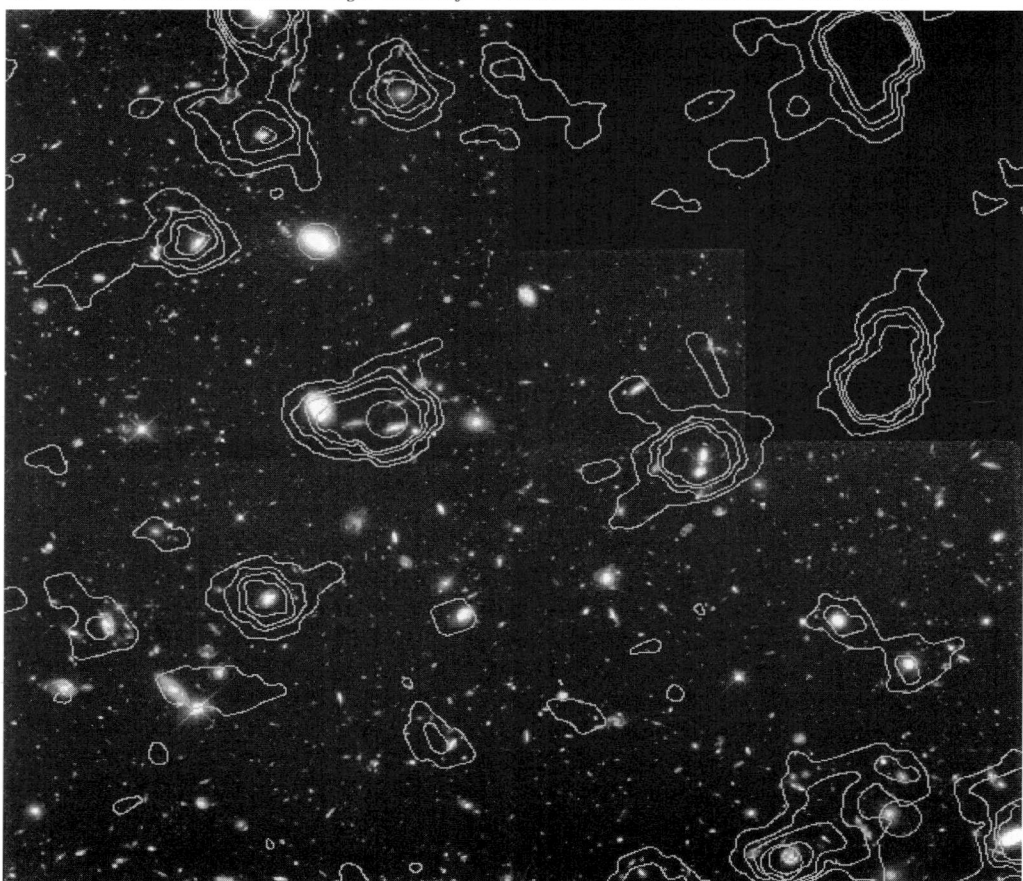

FIGURE 4. ISOCAM LW3 map ($\lambda_{eff} = 15$ μm, contours) of the Hubble Deep Field North by Aussel et al. (1999), overimposed on the HST image. The figure illustrates the spatial accuracy of the ISO deep images with LW3, allowing a reliable identification of the IR sources.

(1999) produced $\sigma_{conf} \simeq 0.07\ MJy/sr$ with a beam of size $\omega \simeq 6\ 10^{-4}\ sr$. From eq. 8.25, this may be used to constrain the continuation of the counts in Fig. 6 fainter than 100 mJy.

The 15 μm counts in Fig. 5 display a remarkable convergence below $S_{15} \sim 0.2$ mJy, proven by at least three independent surveys. The asymptotic slope flatter than -1 in integral count units implies a modest contribution to the integrated CIRB flux by sources fainter than this limit, unless a sharp upturn of the counts would happen at much fainter fluxes with very steep number count distributions, a rather unplausible situation. A meaningful estimate of the CIRB flux can then be obtained from direct integration of the observed mid-IR counts (the two datapoints at 15 and 7 μm in Fig.3). If we further consider how close these are to the upper limits set by the observed TeV cosmic opacity (Fig. 3), the ISOCAM surveys appear to have resolved a significant (50 - 70%) fraction of the CIRB in the mid-IR. On the other hand, the depth of the ISO far-IR surveys (FIRBACK) is not enough to resolve more than ten percent of the CIRB at its peak wavelength.

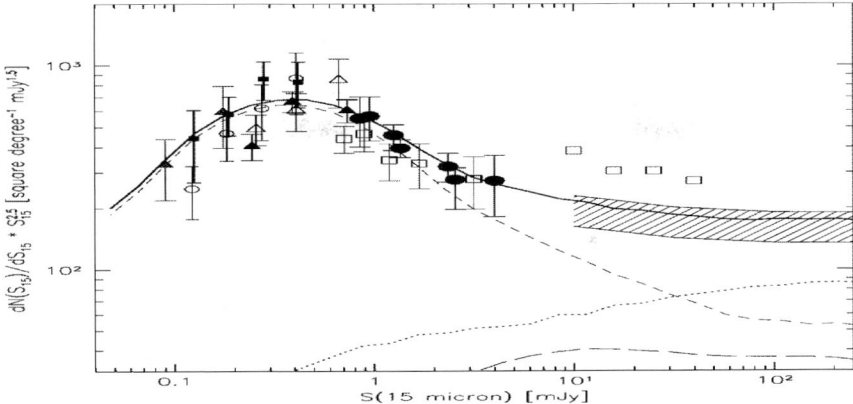

FIGURE 5. Differential counts at $\lambda_{\text{eff}} = 15$ μm normalized to the Euclidean law ($N[S] \propto S^{-2.5}$; the differential form is preferred here because all data points are statistically independent). The data come from an analysis of the IGTES surveys by Elbaz et al. (1999). The dotted line corresponds to the expected counts for a population of non-evolving spirals. The short dashed line comes from our modelled population of strongly evolving starburst galaxies, the long-dashed one are type-I AGNs. The shaded region at $S_{15} > 10$ mJy comes from an extrapolation of the faint 60 μm IRAS counts (Mazzei et al. 2001).

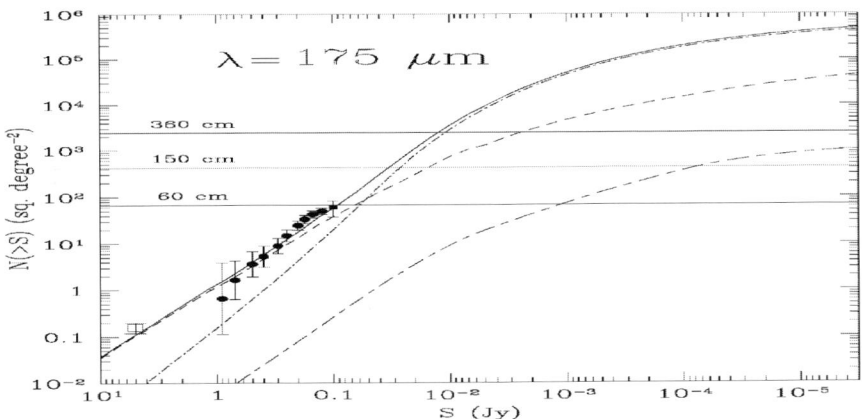

FIGURE 6. Integral counts based on the ISOPHOT FIRBACK survey (§10.2.5) at $\lambda_{eff} = 175$ μm (filled circles, from Dole et al. 2000) and on the ISOPHOT Serendipitous survey (open squares, Stickel et al. 1998). The dashed and dot-dashed lines correspond to the non-evolving and the strongly evolving populations as in Fig. 5. The lowest curve is the expected (negligible) contribution of type-I AGNs. The horizontal lines mark the confusion limits for three telescope sizes (based on eq. 8.26): the lines marked "60cm" and "360cm" correspond to the ISO and Herschel/*Herschel* limits for faint source detection.

10. Explorations of the Deep Universe by Large Millimeter Telescopes

Galaxy surveys in the sub-millimeter waveband offer a unique advantage for the exploration of the distant universe: the capability to naturally generate volume-limited samples from a flux-limited survey. This property is due to the peculiar shape of galaxy spectra in the sub-mm, with an extremely steep slope from 1 mm to 100 μm, as illustrated in Figure 2 for the prototype dusty starburst galaxy M82.

While above a few mm the luminosity is dominated by synchrotron and free-free radio emission, from 100 μm to 1 mm dust continuum emission dominates, with slopes as steep as $L(\nu) \propto \nu^{3.5}$ (see §3). Then, as we observe at sub-mm wavelengths galaxies at larger and larger redshifts, the rest-frame flux density moves to higher and higher frequencies along a steeply increasing spectrum, and the corresponding K-correction almost completely counter-balances the cosmic dimming of the observed flux, for a source of given luminosity at $z \geq 1$. The source flux keeps roughly constant with redshift up to $z \sim 10$, assuming cosmic sources were already present and dusty so early.

A further related advantage of sub-mm surveys is that local galaxies emit very modestly at these wavelengths. Together with the very favorable K-correction, this implies that a sensitive sub-mm survey will avoid local objects (stars and nearby galaxies) and will select preferentially sources at high and very high redshifts: a kind of direct picture of the high-redshift universe, impossible to obtain at other frequencies, where surveys are dominated by galaxies at modest redshifts if not by galactic stars. Finally, and similarly to the ISO surveys, observing in the sub-mm has the advantage of producing samples completely unaffected by intergalactic opacity and dust extinction.

The third breakthrough event after 1996 for IR/sub-mm cosmology has come from operation of a powerful array of bolometers (SCUBA) at the focal plane of the sub-mm telescope JCMT on Mauna Kea. The success of SCUBA on JCMT was due to a combination of three crucial factors: a sensitive detector array with good multiplexing capability (37 bolometers on a field of 2 arcmin diameter, with a diffraction-limited spatial resolution of 15 arcsec), put at the focal plane of a powerful sub-mm telescope (15m dish), on a site allowing to operate at short enough wavelengths (850 μm) to exploit the very steep shape of sub-mm SED's of galaxies. For comparison, in spite of the larger collecting area, the competing bolometer array camera on the IRAM 30m telescope at Pico Veleta (Spain) is limited to work at wavelengths > 1.2 mm by the poorer, lower-altitude site, which means by itself a factor 5 penalty in the detectable source flux with respect to SCUBA/JCMT.

The latter had a long development phase (almost like a space project!), partly because of the difficulty to keep the microphonic noise within acceptable limits. But eventually, its long-sought results have come, and the instrument is providing new very exciting facts to observational cosmology.

Basically, SCUBA/JCMT has allowed to partly resolve the long-λ (850 μm) CIRB background into a population of faint distant, mostly high-z sources, as discussed in §12.3 below. During three years of activity, largely dedicated to deep surveys, SCUBA has discovered several tens of sub-mm sources, mostly at 850 μm.

Four main groups have used SCUBA for a variety of deep integrations. Smail et al. (1997, 1999) have undertaken an ingenious program exploiting distant galaxy clusters as cosmic lenses to amplify the flux of background sub-mm sources and to improve the spatial resolution at the source. Their sample includes now 17 sources brighter than $S_{850} = 6$ mJy. Hughes et al. (1998) published a single very deep image of the HDF North containing 5 sources at $S_{850}(4\sigma) \geq 2$ mJy.

Barger et al. (1998), while detecting only 2 sources down to 3 mJy, have carried out a very successful program of follow-up of SCUBA sources with optical telescopes on Mauna Kea. Eales et al. (1999) and Lilly et al. (1999) have published 12 sources to 3 mJy [a richer sample of 20 more sources is being published].

All these deep integrations are requiring many tens of hours each of especially good weather, which meant a substantial fraction of the JCMT observatory time. In spite

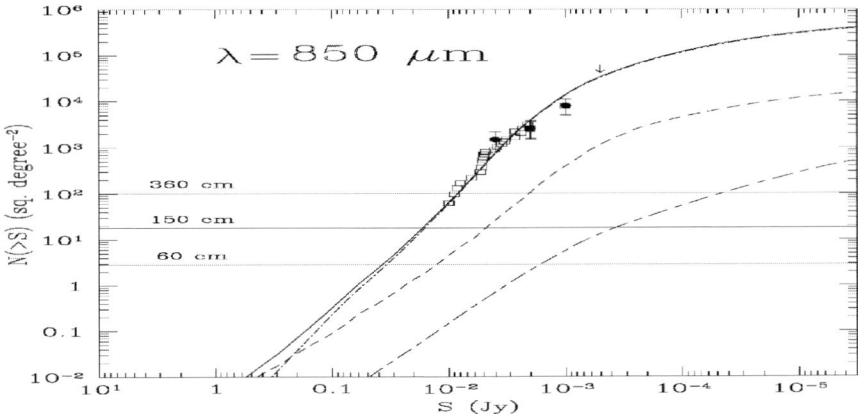

FIGURE 7. Integral counts at $\lambda_{eff} = 850$ μm (see also caption to Fig. 6).

of this effort, the surveyed areas (few tens of $arcmin^2$) and number of detected sources are quite modest, which illustrates the difficulty to work from the ground at these wavelengths.

The extragalactic source counts at 850 μm, reported in Figure 7, show a dramatic departure from the Euclidean law $[N(>S) \propto S^{-2}$ in the crucial flux-density interval from 1 to 10 mJy], a clear signature of the strong evolution and high redshift of SCUBA-selected sources. Only 4 of them have been detected also at 450 μm, the sky transmission at Mauna Kea in this atmospheric channel is usually poor.

More recently, a new powerful bolometer array (MAMBO) has been put in operation on IRAM. Bertoldi et al. (2000) report the first results of observations at $\lambda_{eff} = 1.2$ mm from a survey of 3 fields with a total area of over 300 $arcmin^2$ to a flux limit of few mJy.

11. Interpretation of Faint IR/mm Galaxy Counts

11.1. *Predictions for non-evolving source populations in the mid-IR*

A zero-th order approach to interpret the deep count observations is to compare them with the expectations of models assuming no-evolution for cosmic sources. Any such calculations have to account for the effects of the very complex spectrum of galaxies in the mid-IR (including strong PAH emission and silicate absorption features, see Fig.9) in the K-correction factor appearing in eq.(8.19), which in terms of the system transmission function $T(\lambda)$ is more appropriately written as:

$$K(L,z) = \frac{(1+z) \int_{\lambda_1}^{\lambda_2} d\lambda \left(\frac{\lambda_0}{\lambda}\right) T(\lambda) L[\nu(1+z)]}{\int_{\lambda_1}^{\lambda_2} d\lambda \left(\frac{\lambda_0}{\lambda}\right) T(\lambda) L[\nu, z=0]}.$$

The effect on the source flux and on the counts [eq. 8.18] may be particularly important in the wide LW3 (12-18 μm) filter. The prominent mid-IR features imply a complication when interpreting the counts, but at the same time they imply an enhanced sensitivity of the LW3 source selection to the details of the evolution of sources in the redshift interval $0.5 < z < 1.3$, which is known to be so critical for the formation of structures in the universe.

Local mid-IR luminosity functions (LLF) have been published by Rush et al. (1993), Xu et al. (1998) and Fang et al. (1998) based on the 12 μm all-sky IRAS survey, see

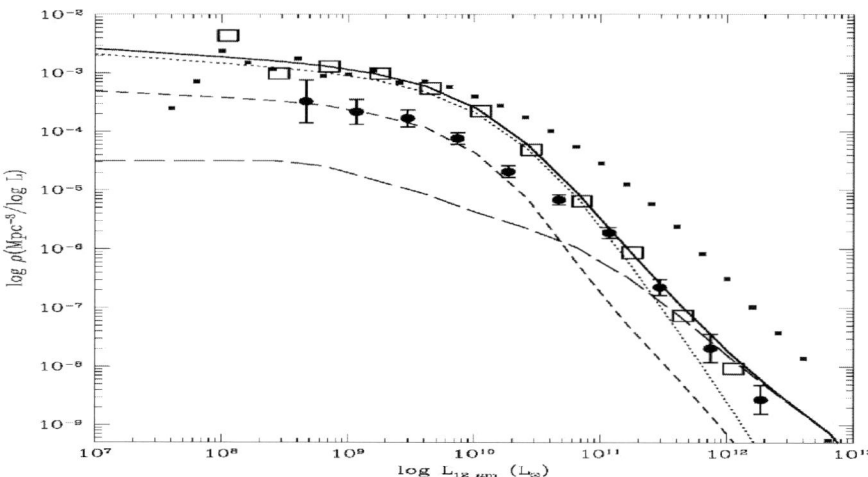

FIGURE 8. Galaxy LLF's at 12 μm from Fang et al. (1998, open squares) and adapted from Xu et al. (1998) in the low-luminosity regime. A comparison is made with the IRAS 60 μm LLF by Saunders et al. (1990, small filled squares). Circles are an estimate, based on the Rush et al. (1993) catalogue, of the contribution to the 12 μm LLF by active galaxies (including type-I AGNs [long-dashed line] and type-II AGNs plus starbursts [short-dashed line]). Type-II AGNs and starbursts are included in the same population on the assumption that in both classes the IR spectrum is dominated by starburst emission. The dotted line is the separate contribution of normal spirals, while the continous line is the total LLF. Active galaxies clearly dominate the LLF at high luminosities.

Figure 8. Unfortunately, in spite of the proximity of the CAM LW3 and IRAS 12 micron bands, at the moment we do not have a reliable LLF at 15 μm because of: a) uncertainties in the IRAS 12 μm photometry, b) the effects of local inhomogeneities, particularly the local Virgo super–cluster; and c) the flux conversion between the IRAS and CAM-LW3 bands (Elbaz et al. 1999).

The dotted line in Fig. 5 corresponds to the present best estimate of the contribution from a non-evolving population with a luminosity function consistent with that in the IRAS 12 μm band derived by Xu et al. and Fang et al. The correction to the CAM LW3 band is made assuming a 12 to 15 μm flux ratio which is a function of the 12 μm luminosity: for the less luminous objects the ratio is based on the observed mid-IR spectrum of quiescent spirals, while for the highest luminosity galaxies the ratio is the one expected for ultraluminous IR galaxies, and for intermediate objects it is close to a typical starburst spectrum like the one of M82 (see continuous line in Fig. 9). The 15 to 12 μm flux ratio increases continuously with luminosity, the flux at long-wavelength being increasingly dominated by the starburst emission.

It is clear that the no-evolution prediction, even taking into account the effects of the PAH features on the K-corrections, falls very short of the observed counts at fluxes fainter than a few mJy. Also the observed slope in the 0.4 to 4 mJy flux range ($N[S] \propto S^{-3\pm0.1}$) is very significantly different from the no-evolution predicted dependence $N(S) \propto S^{-2}$. The extrapolation to the bright fluxes is instead consistent, within the uncertainties, with the IRAS 12 μm counts with a slope close to Euclidean.

11.2. Evidence for a strongly evolving population of mid-IR galaxies

The shape of the differential counts shown in Fig. 5 contains relevant indications about the properties of the contributing source populations. In particular the almost flat (Euclidean) normalized counts extending from the bright IRAS fluxes down to a few mJy, followed by the sudden upturn below, suggests that is not likely the whole population of IR galaxies that evolve: in this case and for the observed IR galaxy LLF, the super-Euclidean increase in the counts would appear at brighter fluxes and not be as abrupt. This behaviour is better consistent with a locally small fraction of IR galaxies to evolve.

The IR counts in Fig. 5 are reproduced with the contribution of two source populations, one evolving, the other with constant properties as a function of time. The local fraction of the evolving starburst population is only several percent of the total, consistent with the observed fraction of interacting galaxies ($\sim 5\%$ locally), the quick upturn in the counts then requiring quite a strong evolution to match the peak in the normalized counts around $S_{15} \simeq 0.5$ mJy. The details of the fit depend on the assumed values for the geometrical parameters of the universe. For a zero-Λ open universe (in our case $H_0 = 50\ km/s/Mpc$, $\Omega_m = 0.3$), a physically credible solution would require a redshift increase of the comoving density of the starburst sub-population and at the same time an increase of the luminosities respectively as

$$n(L[z], z) = n_0(L_0) \times (1+z)^6 \qquad L(z) = L_0 \times (1+z)^3. \tag{11.27}$$

These are quite extreme evolution rates, if compared with those observed in optical samples for the merging and interacting galaxies (e.g. Le Fèvre et al. 2000). The inclusion of a non-zero cosmological constant, and the corresponding increase of the cosmic timescale from z = 0 to 1, tend to make the best-fitting evolution rates less extreme. For $H_0 = 50\ km/s/Mpc$, $\Omega_m = 0.2$, $\Omega_\Lambda = 0.8$, a best-fit to the counts requires:

$$n(L[z], z) = n_0(L_0, z) \times (1+z)^{4.5}, \qquad L(z) = L_0 \times (1+z)^2 \tag{11.28}$$

To be consistent with data on the $z-$distributions from the ISO source samples in the HDF (Aussel et al. 1999, 2001, see Fig. 10) and with the observed CIRB intensity, this fast evolution should turn over at $z \simeq 1$ and the IR emissivity should keep roughly constant at higher z. An accurate probe, however, of hidden SF in the interval $1 \leq z \leq 2$ will only be possible with the longer-wavelength broad-band channel of SIRTF at $\lambda_{eff} = 24\ \mu m$.

In our scheme, *any single galaxy would be expected to spend most of its life in the quiescent (non-evolving) phase, being occasionally put by interactions in a short-lived (few 10^7 yr) starbursting state. The evolution for the latter may simply be due to an increased probability in the past to find a galaxy in such an excited mode. Then the density evolution in eq. (11.28) scales with redshift as the rate of interactions due to a simple geometric effect following the increased source volume density. The luminosity evolution may be interpreted as an effect of the larger gas mass available to the starbursts at higher z.*

Note, however, that the above evolutionary scheme is by no means the only one able to fit the data, other solutions may be devised (e.g. the one by Xu [2000] allowing the whole local population to evolve with cosmic time).

11.3. A panchromatic view of IR galaxy evolution

Deep surveys at various IR/sub-mm wavelengths can be exploited to simultaneously constrain the evolution properties and broad-band spectra of faint IR sources. Franceschini et al. (2001) have compared the 15 μm survey data with those coming from the IRAS 60

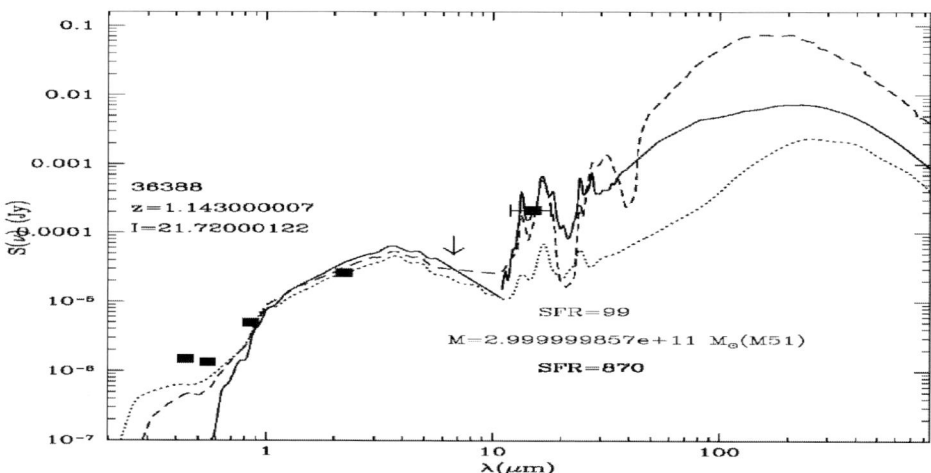

FIGURE 9. Broad-band spectrum of a mid-IR source selected by ISOCAM LW3 in the Hubble Deep Field North (Aussel et al. 1999), compared with the SED's of M82 (thick continuous line), Arp 220 (dashed line), and M51 (dotted line). Estimates of the SF rate [based on the M82 and Arp 220 templates] and of the stellar mass [based on the M51 template] are indicated.

μm, the FIRBACK 175 μm, the ELAIS 90 μm, and the SCUBA 850 μm surveys, which are the deepest, most reliable available at the moment. Information on both number counts and the source redshift distributions were used in these comparisons.

Further essential constraints, providing the local boundary conditions on the evolution histories, are given by the multi-wavelength local luminosity functions. In addition to the 12 and 15 μm LLF's, as discussed in §11.1, the galaxy LLF is particularly well known at 60 μm after the IRAS all-sky surveys and their extensive spectroscopic follow-up (Saunders et al. 1990). Dunne et al. (2000, see also Franceschini, Andreani & Danese 1998) attempted to constrain the galaxy LLF in the millimeter, based on mm observations of complete samples of IRAS 60 μm galaxies.

As previously mentioned, the properties of LLF's observed at various IR/sub-mm wavelengths can be explained only assuming that the galaxy IR SED's depend on bolometric luminosity. Fig. 8 shows that the 60 μm LLF has a flatter (power-law) shape at high-L compared with the mid-IR LLF's (a fact explained in §6.6 as an effect of spectra for luminous active galaxies showing excess 60 μm emission compared to inactive galaxies [see also the L-dependence of the IRAS colours]).

Franceschini et al. (2001) have modelled in some detail the redshift-dependent multi-wavelength LLF's of galaxies by assuming for both non-evolving spirals and active starburst galaxies spectral energy distributions dependent on luminosity, with spectra ranging from those typical of inactive spirals for low-luminosities, to the 60 μm–peaked spectra of luminous and ultra-luminous IR galaxies as previously described. For the SED's of intermediate luminosity objects, linear interpolations between the two as a function of bolometric luminosity were assumed. This allows to simultaneously fit the LLF's at the various wavelengths. For comparison, solutions with single spectral energy distributions for the evolving populations were also tried.

Altogether, the observed long-wavelength counts and CIRB intensity, when compared with typical galaxy SED's and the multi-wavelength LLFs, require a substantial increase of the IR volume emissivity of galaxies with redshift (see Figs. 3, 5, 7).

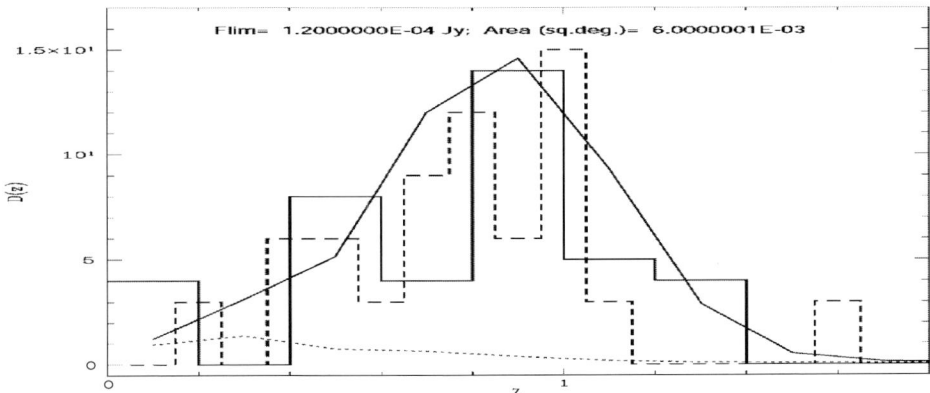

FIGURE 10. Redshift distribution of 15 μm sources with $S_{15} > 120$ μJy in the HDF North (continuous histogram; Cohen et al. 2000; Aussel et al. 2001), compared with our best-fit evolutionary model (continuous line). The dashed histogram is the z-distribution of sources in the CFRS field reported by Flores et al. (1999). The dotted line at the bottom corresponds to the no-evolution prediction.

Should one assume that the IR SED of the ultra-luminous galaxy Arp 220 is representative of the average spectrum of the evolving population detected by ISOCAM LW3, then the consequence would be that the observed far-IR counts and the CIRB intensity are far exceeded. On the contrary, if we assume for the IR evolving sources a more typical starburst spectrum (like the one of M82, by all means similar to those of other luminous starbursts observed by ISO), then most of the observed properties of far-IR galaxy samples (number counts, redshift distributions, luminosity functions) are appropriately reproduced. Best-fits to the counts based on the M82 template are given in Figs. 6 and 7.

The good match to the multi-wavelength counts obtained by assuming a typical starburst spectrum for the evolving population already indicates that the faint IR-selected source population is likely dominated by processes of star-formation in distant galaxies more than by AGN emissions. This seems indeed the result of the first spectroscopic studies of faint ISO sources (§12.1), although a more substantial effort is required to confirm it. Considering the different shapes of the IR SEDs for SBs and AGNs, this would imply that the population detected by ISO in the mid-IR not only contributes a major fraction of CIRB at 15 μm, but is also responsible for a majority contribution of the CIRB at any wavelengths.

12. Nature of the Fast Evolving Source Population

12.1. Tests of the evolving IR population in the HDFs and CFRS fields

The ISO observatory has deeply surveyed with CAM LW3 some of the best investigated sky areas, in particular the two Hubble Deep Fields (North & South, Rowan-Robinson et al. 1997, Oliver et al. 2000b) and the area CFRS 1415+52 (Flores et al. 1999). Given the variety of multi-wavelength data and the almost complete spectroscopic follow-up, the surveys in these areas have allowed to achieve important tests of the evolving population responsible for the upturn of the ISO mid-IR counts and for a substantial fraction of the CIRB.

Aussel et al. (1999 and 2001) report reliably tested (see §9.3) complete samples of 49 and 63 sources to $S_{15} \geq 100$ μJy in the HDF North and South respectively, covering similar areas of 25 square arcmin each. Flores et al. (1999) analyse a sample of 41 sources brighter than $S_{15} \sim 300$ μJy ($S/N > 4$) over an area of $10'$x$10'$ in CFRS 1415+52. The vast majority (90%) of the ISO sources in the HDF surveys have spectroscopic redshifts, and for the remaining objects photometric redshifts are easily estimated. The redshift distributions $d(z)$ for the HDF and CFR1415 surveys are reported in Figure 10, and compared with the model fitting the multi-wavelength counts mentioned in §11.2. Although the two surveys cover individually small sky areas, the fair match between them gives some confidence about the overall reliability of the result. These data set a stringent limit on the rate of cosmological evolution for IR galaxies above $z \sim 1$, which needs to level off to avoid exceeding the observed $d(z)$ on the high-z tail. Note however that the observed high-z convergence of $d(z)$ is also partly an effect of the strong K-correction in the LW3 flux for dust-rich galaxies (see an example in Fig. 9): disentangling K- from evolutionary-corrections at $z > 1$ will require SIRTF and *Herschel*.

HST imaging data on these fields provide detailed morphological information on ISO sources. Elbaz et al. (1999) and Aussel et al. (1999) find that 30 to 50% of them show clear evidence of peculiarities and multiple structures, in keeping with the local evidence that galaxy interactions are the primary trigger of luminous IR starbursts. From their Caltech redshift survey in the HDF North, Cohen et al. (2000) report that over 90% of the faint LW3 ISO sources are members of galaxy concentrations and groups, which they identify as peaks in their redshift distributions. Indeed, it is in these dense galaxy environments with low velocity dispersion that interactions produce resonant perturbation effects on galaxy dynamics.

12.1.1. *Optical and NIR spectral properties: nature of the IR sources*

Flores et al. (1999) report a preliminary analysis of the spectra of IR sources in CFRS 1415+52, noting that a majority fraction of these display both weak emission (OII 3727 Å) and absorption ($H\delta$) lines, as typical of the *e(a)* galaxy spectral class: the latter is mentioned in the literature as a post-starbursting population, one in which a vast population of A-type absorption-line stars from a ~ 1 Gyr old massive starburst combine with a small residual of ongoing SF evidenced by the weak [OII] emission. Given the far-IR selection of the faint ISO sources, which is expected to preferentially detect dusty star-forming galaxies, this result would be difficult to understand, as it lets open the question of "why the ongoing active starbursts are not detected".

Rigopoulou et al. (2000) and Franceschini et al. (2000) have observed with ISAAC on VLT a sample of 13 high-z ($0.2 < z < 1.4$) galaxies selected in the HDF South to $S_{15} > 100$ μJy: *the Hα line is detected in virtually all of the sources, and found quite prominent (EW > 50 Å), indicating substantial rates of SF after de-reddening corrections, and demonstrating that these optically faint but IR luminous sources are indeed powered by an ongoing massive dusty starburst.*

The *e(a)* spectral appearance is interpreted by Poggianti & Wu (2000) and Poggianti, Bressan, Franceschini (2000) as due to selective dust attenuation, extinguishing more the newly-formed stars than the older ones which have already disrupted their parent molecular cloud.

These papers independently found that $\sim 70 - 80\%$ of the energy emitted by young stars and re-processed in the far-IR leaves no traces in the optical spectrum, hence can only be accounted for with long-wavelength observations.

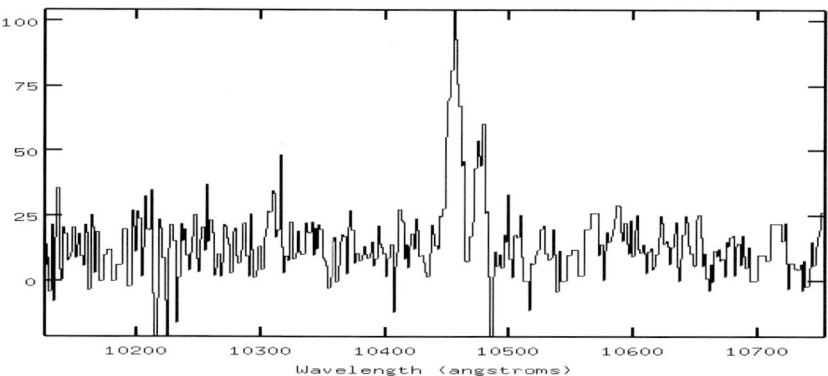

FIGURE 11. ISAAC/VLT spectrum of HDFS source # 53 at z=0.58. The $H\alpha$ and [NII] redshifted lines are clearly visible [from Rigopoulou et al. 2000].

12.1.2. *Evaluating baryonic masses and the SFR of the IR population*

Further efforts of optical-NIR spectroscopic follow-up of faint IR sources are planned for the next years, including attempts to address the source kinematics and dynamics based on line studies with the next-generation of IR spectrographs (e.g. SINFONI on VLT). The latter would be particularly relevant in consideration of the typically complex dynamical structure of luminous IR starbursts. At the moment, for an evaluation of the main properties of the IR population we have to rely on indirect estimates exploiting the near-IR and far-IR fluxes. One important parameter is the baryonic mass in stars, for measure of which fits of local template SEDs to the near-IR broad-band spectrum can be used. Our estimated values of the baryonic mass ($\sim 10^{11}$ $M_\odot$, with 1 dex typical spread, see Figure 12) indicate that already evolved and massive galaxies host the powerful starbursts.

As a measure of the rate of star-formation (SFR), the other fundamental parameter describing the physical and evolutionary status of the sources, we have exploited the mid-IR flux as an alternative to the (heavily extinguished) optical emissions, since it is much more directly related to the bolometric (mostly far-IR) flux, which is the most robust indicator of the number of massive reddened newly-formed stars. Vigroux et al. (1998) find that the ISOCAM mid-IR fluxes (from both LW3 and LW2 ISOCAM observations) are tightly and linearly related with the bolometric emission in local galaxies, evidence contradicted only in very extinguished peculiar sources (e.g. Arp 220), for which the mid-IR spectrum is self-absorbed. Using several HDF North sources having both the mid-IR and radio flux, Aussel et al. (2001) find that the two SFR estimators, both largely unaffected by dust extinction, provide consistent results on the SFR. However, the mid-IR flux has the advantage over the radio to be less affected by AGN emission, providing a more reliable SF measure (Cohen et al. 2000; Aussel et al. 2001; Franceschini et al. 2000). Also the fact that only 7 of the 49 IR SBs in the HDFN are detected in radio to a flux limit of few tens of μJy tells that the mid-IR flux is a more sensitive indicator of SF. This state of affairs will change when dedicated space missions (in particular the 3.6m *Herschel* observatory) will measure the peak of dust emission at $\lambda \sim 100$ μm in high-redshift galaxies with high accuracy.

Altogether, the galaxy population dominating the faint mid-IR counts and substantially contributing to the bolometric CIRB intensity (assumed typical SB SEDs) appears

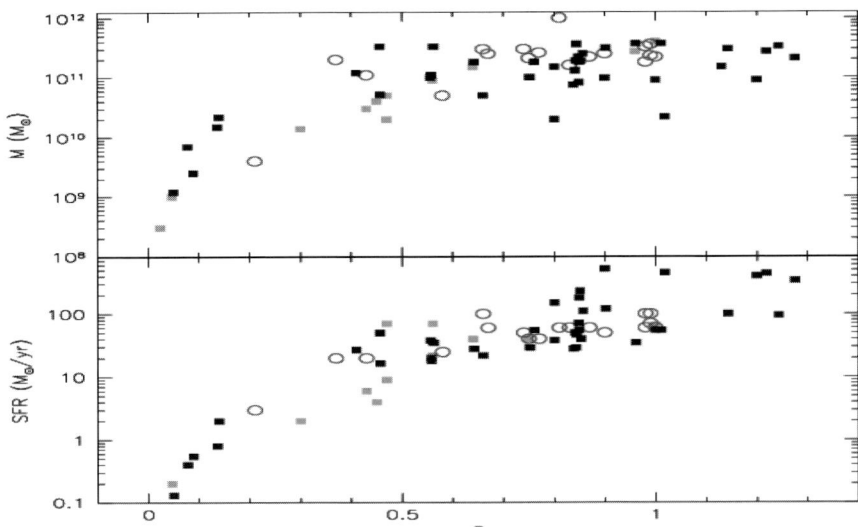

FIGURE 12. Evaluations of the star formation rates [from an estimator based on the mid-IR flux] and baryonic masses [from fits of the NIR SED] as a function of redshift for galaxies selected by ISOCAM LW3 at 15 μm in the HDFN (filled squares) and CFRS 1415+52 (open circles).

to be composed of luminous ($L_{bol} \sim 10^{11}$–10^{12} $L_\odot$) starbursts in massive ($M \sim 10^{11}$ $M_\odot$) galaxies at $z \sim 0.5 - 1$, observed during a phase of active stellar formation. The typically red colors of these systems suggest that they are mostly unrelated to the faint blue galaxy population dominating the optical counts (Ellis 1997), and should be considered as an independent manifestation of (optically hidden) star formation (Elbaz 1999; Aussel 1998).

12.2. What are the FIRBACK 175 μm sources?

The nature of the 175 μm sources discovered by FIRBACK/ISO, and contributing $\sim 10\%$ of the CIRB intensity, is presently the target of intense observational and modelling investigations, although no conclusions are possible at the moment. Because of the missing knowledge of the LLF, the interpretation of the 175μm counts themselves is subject to some uncertainties: is there strong or marginal evidence for evolution at the survey limit of 100 mJy (Fig.6)? Dole et al. (2000) argue in favour of the former, while Fig. 6 reports a solution in which a moderate-redshift ($z \sim 0.5$) population still dominates there.

The basic limitation comes from the difficulty to identify the optical counterparts, due to the large (40 arcsec) ISOPHOT error-box. Progress is being achieved by cross-correlating with deep radio surveys available in the FIRBACK fields (exploiting the good radio/FIR correlation, eq. 6.14) and by means of some limited SCUBA follow-up. Scott et al. (2000) have obtained data at 450 and 850 μm for 10 FIRBACK sources: the FIR-mm SEDs tentatively indicate, for plausible far-IR spectra, redshifts in the range from 0 to 0.4 for the majority of the sources, while a few may be at $z > 1$.

Mid-IR 15 μm fluxes from an ISOCAM map are available in the "FIRBACK Marano" area, which indicate that the 15 μm counterparts of the 175 μm sources are rather faint (Elbaz 1999). Three interpretations have been suggested: (a) FIRBACK sources are typically very high-luminosity Arp220-like at low redshift ($z \sim 0.1 - 0.4$); (b) they are

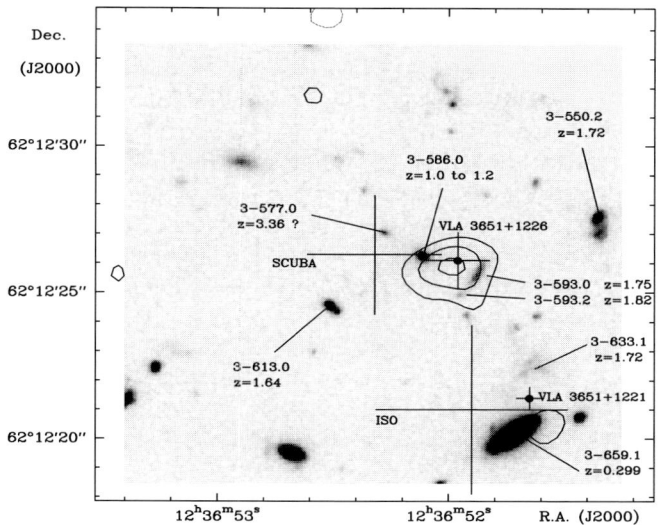

FIGURE 13. Map of the 1.3 mm continuum obtained with the IRAM interferometer in the field of the source HDF 850.1 by Downes et al. (1999). HDF 850.1 is the brightest source discovered at 850 μm by SCUBA (Hughes et al. 1998), and has a flux density of 2.2 mJy at 1.3 mm. The field center coincides with the center position of the SCUBA error-box, whose size is however comparable to the whole image area. The image is a composite of BVI data from HDF. Positions of VLA and ISO sources, as well as photometric redshift data, are also indicated. IRAM and VLA position clearly point to a faint optical counterpart of HDF 850.1 (3-593.0), possibly influenced by gravitational lensing by the elliptical 3-586.0, in a similar configuration to the prototypical primeval galaxy IRAS F10214 [courtesy of D. Downes].

more standard starbursts at $z > 1$; (c) they are low-activity spirals at moderate z with significant amounts of cold-dust and excess emission at $\lambda > 100$ μm.

Although the results of the SCUBA observations might indicate that the last interpretation could be more probable, the nature of the FIRBACK source population is far from proven, further multi-wavelength data being required to address it. Deeper far-IR observations will be possible with SIRTF, but a more final solution will probably require the *Herschel*'s better spatial resolution.

12.3. *The nature of the high-z galaxies detected in the millimeter*

Thanks to the unique advantage for deep sub-mm observations offered by the very peculiar $K-$ correction, sub-mm surveys with sensitivities of few mJy at 850 μm, have been able to detect high-redshift (very luminous) sources in flux-limited samples. The observed 850 μm counts, far in excess of the no-evolution prediction, already tell incontrovertibly about the cosmological distance and evolutionary status of the SCUBA-selected source population.

Unfortunately, probing directly the nature of these objects via optical identification and spectroscopic follow-up turned out to be very difficult, in spite of the substantial efforts dedicated. The SCUBA diffraction-limited HPBW at 850 μm is large, ~ 15 arcsec FWHM, and the difficulty of the identification is further exacerbated by the usual extreme faintness of the optical counterparts, as demonstrated in the (few) cases in which the identification has been possible (see e.g. Figure 13).

The reliability of the identification has been evaluated by computing the probability that the nearest member of a population of candidate identifications with surface density n falls by chance within a distance d from the SCUBA source: $P = e^{-\pi n d^2}$. For a sample of size N of SCUBA detections, the product NP gives the number of spurious identifications (Lilly et al. 1999). This analysis has shown that the situation is not quite comfortable for the SCUBA surveys, essentially because of the faintness of the optical counterparts: roughly 50% of all identifications may be spurious.

Two approaches have been followed to improve the identification and try to characterize the population. One was to systematically survey spectroscopically all optical sources falling in the SCUBA beam, the other was to exploit cross-identifications with ultra-deep radio catalogues. Particularly well studied are the fields in the Cluster Lens Survey (Smail et al. 1997), exploiting the flux-amplification by massive foreground galaxy clusters. The current situation about redshift measurements in this survey is: the 16 SCUBA sources have 24 possible counterparts with spectroscopic redshifts, 6 reliable z estimates (a $z = 2.8$ combined AGN/starburst, a $z = 2.6$ galaxy pair, 2 galaxies with AGN signatures at $z = 1.16$ and $z = 1.06$, and finally 2 foreground cD cluster members [Barger et al. 1999]). Note that the identification with the galaxy pair has been later confirmed by CO mm observations (Frayer et al. 1999).

An interesting case is illustrated in Fig. 13, showing the brightest object HDF-850.1 in the Hughes et al. (1998) survey, confirmed by IRAM interferometry as a probable ultra-luminous lensed starburst with $L_{bol} \sim 2\ 10^{12} L_\odot$ at $z_{photom} \simeq 1.7$

The difficulty of the identification process is illustrated by the recent finding (Smail et al. 1999) of the presence of two Extremely Red Objects (ERO's) as probable counterparts of two SCUBA sources. Given the faintness of optical counterparts and the extreme difficulty to get the redshift from optical spectroscopy, some millimeter estimators of the redshift have been devised to override optical measurements. Hughes et al. (1998) use the S_{450}/S_{850} flux ratio as a measure of z. However, given the rather wide temperature-distribution of cosmic dust (see e.g. the three quite different spectral templates, for Arp220, M82, and M51 in Fig. 9), this test proved to be very uncertain. Much more reliable the technique proposed by Carilli & Yun (1999) to exploit the $S_{850\mu m}/S_{20cm}$ flux ratio, which has the advantage to rely on very robust mm spectral shapes at 850 μm ($S_\nu \propto \nu^{3.5}$, see §3) and in the radio (typical power-law synchrotron spectra), with opposing spectral slopes. Assuming an Arp 220 spectral template they got:

$$S_{850\mu m}/S_{20cm} = 1.1 \times (1+z)^{3.8}$$

whose small scatter mostly reflects the tight FIR to radio correlation.

Population constraints on the z-distributions have been derived in this way, and the basic result (still tentative and requiring confirmation) is that *faint SCUBA sources are mostly ultra-luminous galaxies at typical $z \sim 1$ to ~ 3* (e.g. Barger et al. 1999). Clearly, the details of the z-distribution cannot yet be quantified with precision, this will likely require new instrumentation (mm interferometers – e.g. ALMA – are particualrly needed, in addition to space FIR observatories).

As suggested by many authors, the similarity in properties between this high-z population and local ultra-luminous IR galaxies argues in favour of the idea that these represent the long-sought "primeval galaxies", those in particular originating the local massive elliptical and S0 galaxies. This is also supported by estimates of the volume density of these objects in the field $\sim 2 - 4 \times 10^{-4}\ Mpc^{-3}$, high enough to allow most of the field E/S0 to be formed in this way (Lilly et al. 1999). As for the E/S0 galaxies in clusters, a very interesting result was the recent discovery by SCUBA of a significant excess of very luminous ($L \sim 10^{13} L_\odot$) sources at 850 μm close to the $z = 3.8$ radiogalaxy 4C41.17 (Ivison et al. 2000), which parallels the evidence of a similar excess of EROs and Lyman-break galaxies in this area. It is tentalizing to interpret these data as indicative of the presence of a forming cluster surrounding the radiogalaxy, where the SCUBA sources would represent the very luminous ongoing starbursts.

By continuity, the less extreme starbursts ($L \sim 10^{11}-10^{12}\ L_\odot$) discovered by ISOCAM at lower redshifts can possibly originate the spheroidal components in later morphological type galaxies (see more in §13.2.4 below).

12.4. *AGN contribution to the energetics of the faint IR sources*

Within this interpretative scheme, a margin of uncertainty still exists about the possible contribution by gravitational accretion from a nuclear quasar to the energy budget in these high-z IR-mm sources. While stellar energy production provides a modest overall efficiency for baryon transformations of quite less than a percent at most, the theory of gravitational accretion predicts values in the range $\epsilon \sim 5-40\%$. A natural question then arises as of how much of the bolometric flux in these sources is contributed by an AGN. Unfortunately, the optical–UV–soft-X ray primary source spectrum in the high-redshift IR-mm sources is almost completely re-processed by dust into an IR spectrum largely insensitive to the properties of the primary incident one.

As for SCUBA sources, there have been indications for AGN activity for at least a fraction ($20-30\%$) of them. Indeed, since SCUBA selects the top luminosity end of the IR population, and considering the local evidence of a larger incidence of AGNs among ULIRGs, an important AGN contribution to the SCUBA sources would be expected (potentially biasing our conclusions about their contribution to the SFR history). Risaliti et al. (2000) and Bassani et al. (2001) claim evidence for a significant AGN contribution in the large majority ($>60\%$) of the local ULIRGs based on hard X-ray data, something confirmed also by high spatial resolution IR imaging by Soifer et al. (2000).

Since its launch the last year, the CHANDRA X-ray observatory (the ultimate imager in hard X-rays) has allowed to probe very deeply into the nature of the high-z SCUBA sources, using the hard X-ray flux as diagnostic tool (SB are weaker X-ray emitters than any kind of AGNs). Among several tens of hard X-ray and 850 μm sources detected in various independent survey areas, (Fabian et al. 2000, Hornschemeier et al. 2000, Barger et al. 2001), only very few are in common, the two samples being essentially orthogonal. Unless all these are Compton-thick and any hard X-ray scattered photons are also photoelectrically absorbed, the conclusion is that the bulk of the emission by high-luminosity SCUBA sources is due to star formation (in agreement with a dominant stellar emission in local ULIRGs found by Genzel et al. 1998).

While the detailed interplay between starburst and AGN is still an open issue even for local sources, the estimated fraction of the CIRB at 850 μm due to AGNs is not larger than 10% (Barger et al. 2001). Preliminary results of spectroscopic studies of the $H\alpha$ line properties in faint ISO mid-IR sources (D. Rigopoulou, private communication) seem also to indicate a modest incidence of AGN, which would imply that the overall AGN contribution to the bolometric CIRB is likely around 10% or so.

12.5. *Discussion*

ISO and SCUBA surveys have proven nicely complementary capabilities to explore, within the limitations of the current instrumentation, long-wavelength emission by galaxies over most of the Hubble time, up to z of several. Unfortunately, this has been possible only at the short- and long-wavelength tails of the CIRB background spectrum: a bad coincidence makes the wavelength interval including peak emission by distant dusty galaxies ($\lambda \sim 30$ to 300 μm) hardly accessible at present.

All mentioned exploratory surveys of the distant universe have indicated that the overall volume emissivity of galaxies at long wavelengths drastically increases as a function of redshift, to explain the very steep observed multi-wavelength counts and the redshift distributions showing substantial high-z tails. This evolution, however, should level off by $z \sim 1$ (see Fig. 14 below) to allow consistency with the observed z-distributions (Franceschini et al. 2001) and the CIRB spectral shape.

A spectacular finding by the deep SCUBA surveys was the discovery of ultra-luminous galaxies at high-redshifts, mostly emitting in the far-IR and possibly at the origin of present-day galaxy spheroids. However, the most precise quantification of the cosmic history of the IR population comes at the moment from the ISO deep and ultra-deep surveys, which provide very detailed constraints on the counts (Fig. 5) and also allow to unambiguously identify in the optical the faint IR sources (Fig. 4). The outcome of our spectroscopic observations is that the faint population making up the CIRB in the mid-IR is dominated by actively star-forming galaxies with substantial $H\alpha$ emission (§12.1.1). Preliminary inspection of $H\alpha$ line profiles and constraints set by the 15 to 7 micron flux ratio indicate that the majority of sources are powered by a SB rather than an AGN.

Mid-IR ISO counts and the redshift distributions of the sources require extremely high rates of evolution of the 15 μm luminosity function up to $z \sim 1$. Taking into account all effects due to the detector spectral response function to the complex mid-IR spectral features, the observable statistics may be explained in terms of a strong evolution for a population of IR starbursts contributing little to the local LF. Consequently, a plausible evolution pattern should involve both the source luminosities and spatial densities.

A natural way to account for this very high dependence on redshift of the IR starburst population is to assume that it consists of otherwise normal galaxies, but observed during a dust-extinguished luminous starburst phase, and that its extreme evolution is due to *an increased probability with z to observe a galaxy during a starburst event.*

The common wisdom that SBs are triggered by interactions and merging suggest that the inferred strong number density evolution may be interpreted as an increased probability of interaction with z. Assuming that the phenomenon is dominated by interactions in the field and a velocity field constant with z, then this probability would scale roughly as $\propto n(z)^2 \propto (1+z)^6$, n being the number density in the proper (physical) volume. A more complex situation is likely to occur, as the velocity field evolves with z in realistic cosmological scenarios and if we consider that the most favourable environment for interactions are galaxy groups, which indeed are observed to include the majority of ISOCAM distant sources (Cohen et al. 2000). The increased luminosity with z of the typical starburst is due, qualitatively, to the larger amount of gas available in the past to make stars.

To note is that closed or zero-Λ world models require evolution rates quite in excess of those inferred from deep optical imaging (Le Fèvre et al. 2000), whereas our best-fit solution for $\Omega_\Lambda = 0.8$ and $\Omega_m = 0.2$ (eq. 11.28) is closer to the optical results.

How this picture of a 2-phase evolution of faint IR sources compares with results of optical and near-IR deep galaxy surveys is matter of debate. Since, because of dust, most of the bolometric emission during a starburst comes out in the far-IR, we would not expect the optical surveys to see much of this violent IR starbursting phase. Indeed, B-band counts of galaxies and spectroscopic surveys are interpreted in terms of number-density evolution, consequence of merging, and essentially no evolution in luminosity. The Faint Blue Object population found in optical surveys may be interpreted as the "post-starburst" population, objects either observed mostly after the major event of SF, or more likely ones in which the moderately extinguished intermediate age ($\sim 10^7\ yrs$) stars in a prolonged starburst (several $10^7\ yrs$) dominate the optical spectrum. In this sense optical and far-IR selections trace different phases of the evolution of galaxies, and provide independent sampling of the cosmic star formation.

A lively debate is currently taking place about the capabilities of UV-optical observations to map accurately by themselves the past and present star-formation, based on suitable corrections for dust extinction in distant galaxies. Adelberger et al. (2000) suggest that the observed 850 μm galaxy counts and the background could possibly be explained with the optical Lyman drop-out high-z population by applying a proportionality correction to the optical flux and by taking into account the locally observed distribution of mm-to-optical flux ratios.

On the other hand, a variety of facts indicate that optically-selected and IR/mm-selected faint high-redshift sources form almost completely disjoint samples. Chapman et al. (2000) observed with SCUBA a subset of z=3 Lyman-break galaxies having the highest estimated rates of SF as inferred from the optical spectrum, but detected only one object out of ten. For this single detected source the predicted SFR based on the extinction-corrected optical spectrum was 5 times lower than found by SCUBA. A similar behaviour is also shared by local luminous IR galaxies, whose bolometric flux is unrelated to the optical spectrum (Sanders & Mirabel 1996).

Finally, our previously mentioned observational results by Rigopoulou et al. (2000) and the theoretical ones by Poggianti & Wu (2000) and Poggianti et al. (2000) report *independent evidence from both local and high-z luminous starbursts that typically 70% to 80% of the bolometric flux from young stars leaves no traces in the UV-optical spectrum, because it is completely obscured by dust.* As there seems to be no "a priory" way to correct for this missing energy, we conclude that only long-wavelength observations, with the appropriate instrumentation, can eventually *measure* SF in galaxies at any redshifts.

13. Global Properties: The SFR Density and Contributions to the CIRB

13.1. *Evolution of the comoving luminosity density and SFR*

As illustrated in Fig. 3, the CIRB intensity and spectral distribution are in clear support of models for evolving starbursts discussed above.

Unfortunately, we are not yet in the position to derive an independent assessment of the evolutionary SFR density based on the available complete samples of faint IR sources: although a substantial effort to follow them up in the optical has started (particularly good chances are offered by ongoing spectroscopic follow-up of the statistically rich faint ISOCAM samples like the GITES and HDFs), the process is far from complete. As a consequence, no detailed conclusions can yet be drawn about the contribution of IR sources to the global comoving luminosity and SFR densities (Madau et al. 1996).

Only rather model-dependent estimates are possible at the moment, based for example on the evolution scheme described in §11 and whose predictions are summarized in Figure 14. There is a clear indication here that the contribution of IR-selected sources to the luminosity density at high-z should significantly exceed those based on optically selected sources, and that the excess may be progressive with redshift up to $z \sim 1$.

This evolution should however level off at higher z, to allow consistency with the observed z-distributions for faint ISOCAM sources (Franceschini et al. 2000) and with the estimates of the average time-dependent emissivity $j_{eff}(z)$ (eq. 8.21) based on deconvolution of the CIRB spectrum (Gispert et al. 2000).

Altogether *these results indicate that the history of galaxy long-wavelength emission does probably follow a path similar to that revealed by optical-UV observations, by*

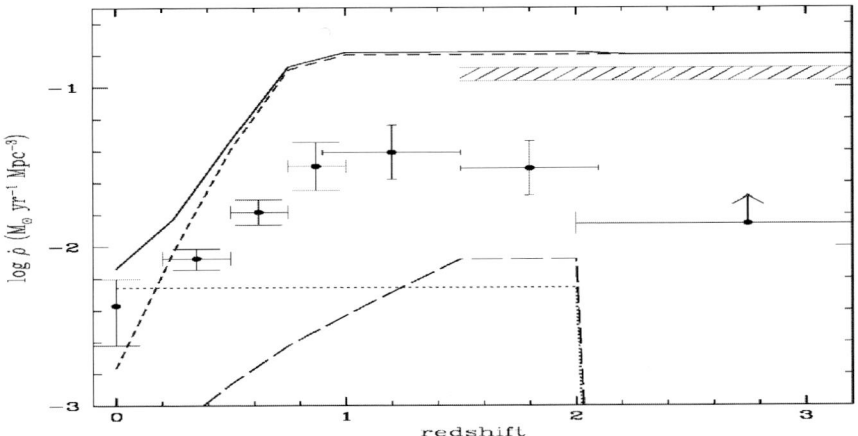

FIGURE 14. Evolution of the comoving luminosity density for the IR-selected population based on the model of IR evolution discussed in §5. The luminosity density is expressed here in terms of the star formation rate density (computed from the far-IR luminosity assuming a Salpeter IMF, according to the recipes reported in Rowan-Robinson et al. (1997, their eq. [7]). The IR evolution is compared with data coming from optical observations by Lilly et al. (1996), Connolly et al. (1997) and Madau et al. (1996), transformed to our adopted $\Omega_m = 0.3$, $\Omega_\Lambda = 0.7$ cosmology. Dotted line: quiescent population. Short-dash line: evolving starbursts. Long dashes: type-I AGNs (with arbitrary normalization). The continuous line is the total. The shaded horizontal region is an evaluation of the average SFR in spheroidal galaxies by Mushotzky & Loewenstein (1997).

showing a similar peak activity around $z \sim 1$, rather than being confined to the very high-z, as sometimes was suggested. This confirms that the bulk of the galaxy activity, and particularly the bulk of the energy released in the CIRB background, is to be placed around $z = 1$, which is obvious from Fig. 14 if the dependence of the cosmological timescale on redshift is considered (Harwit 1999; Haarsma & Partridge 1998).

These results can only be preliminary untill we will have more substantial identifications of existing IR-selected source samples, or, better, after the fleet of IR/mm facilities planned for this and the next decade will have eventually provided data of enough quality to allow a full long-wavelength complement to the optical-UV high-z observations.

13.2. Energy constraints from background observations

In the present situation, the most robust constraints on the high-redshift far-IR/sub-mm population come from observations of the global energetics residing in the CIRB and optical background radiations. The latter imply a very substantial demand on contributing sources, as detailed below in schematic terms.

Let us assume that a fraction f_* of the universal mass density in baryons

$$\rho_b = \frac{3H_0^2(1+z)^3}{8\pi G}\Omega_b \simeq 7\ 10^{10}(1+z)^3 \left(\frac{H_0}{50\ km/s/Mpc}\right)^2 \Omega_b\ [M_\odot/Mpc^3] \qquad (13.29)$$

undergoes a transformation (either processed in stars or by gravitational fields) with radiative efficiency ϵ, then the locally observed energy density of the remnant photons is

$$\rho_\gamma = \rho_b \frac{c^2 \epsilon f_*}{(1+z)^4} \simeq 5\ 10^{-30} \left(\frac{H_0}{50\ km/s/Mpc}\right)^2 \frac{\Omega_b}{0.05}\frac{f_*}{0.1}\frac{2.5}{(1+z_*)}\frac{\epsilon}{0.001}[gr/cm^3]. \quad (13.30)$$

For stellar processes, ϵ is essentially determined, within the moderate uncertainties of stellar models, by the IMF: $\epsilon = 0.001$ for a Salpeter IMF and a low-mass cutoff $M_{min} = 0.1\ M_\odot$ (see eq. 4.13), $\epsilon = 0.002$ and $\epsilon = 0.003$ for $M_{min} = 2$ and $M_{min} = 3$, while ϵ gets the usually quoted value of $\epsilon = 0.007$ only for $M_{min} > 10\ M_\odot$ [A. Bressan, private communication].

Note how the contribution to the photon background energy by very high redshifts is penalized in eq. (13.30) by the $(1+z)^{-1}$ factor: measurements of the photon background preferentially constrain source emission at moderate z, whereas estimates of the local average metal abundance (obviously much more difficult and indirect!) would in principle provide a less biased integral over the total stellar yield in the past.

13.2.1. Constraints from the integrated optical background

As already noticed (§8.3.1), the converging galaxy counts at faint magnitudes observed in the optical and near-IR allow to estimate with fair accuracy the total diffuse flux at these wavelengths (Fig. 3, see Madau & Pozzetti 2000). The bolometric emission from 0.1 to 7 μm by distant galaxies turns out to be

$$\nu I(\nu)|_{opt} \simeq (17 \pm 3)\ 10^{-9}\ Watt/m^2/sr, \qquad (13.31)$$

which in fact is a lower limit if we give credit to claims of a $(2-3)$ larger optical/NIR background, see §8.1 (but see also a counter-argument in §8.2).

We discussed evidence that for the most luminous starbursts the optical spectra are only moderately contributed by starburst emission, which is mostly hidden in the far-IR. Accordingly, let us assume that the optical/NIR BKG mostly originates by quiescent SF in spiral disks and by intermediate and low-mass stars. As observed in the Solar Neighborhood, a good approximation to the IMF in such relatively quiescent environments is the Salpeter law with standard low-mass cutoff, corresponding to a mass–energy conversion efficiency $\epsilon \sim 0.001$. With these parameter values, we can reproduce the whole optical BKG intensity of eq.(13.31) by transforming a fraction $f_* \simeq 10\%$ of all nucleosynthetic baryons into low-mass stars, assumed the bulk of this process happened at $z_* \sim 1.5$ and 5% of the closure value in baryons (for our adopted $H_0 = 50\ km/s/Mpc$, or $\Omega_b h^2 = 0.012$, consistent with the theory of primordial nucleosynthesis):

$$\nu I(\nu)|_{opt} \simeq 20\ 10^{-9} \left(\frac{H_0}{50\ km/s/Mpc}\right)^2 \frac{\Omega_b}{0.05} \frac{f_*}{0.1} \left(\frac{2.5}{1+z_*}\right) \frac{\epsilon}{0.001}\ Watt/m^2/sr.$$

It is generated in this way a local density in low-mass stars consistent with the observations (based on photometric surveys, Ellis et al. 1996, and assuming standard mass to light ratios):

$$\rho_b(stars) \simeq 7\ 10^{10} f_* \Omega_b \simeq 3.4\ 10^8\ M_\odot/Mpc^3, \qquad (13.32)$$

which, assuming typical solar metallicities, corresponds to a local density in metals of

$$\rho_Z(stars) \simeq 1.6\ 10^9 f_* \frac{Z}{Z_\odot} \Omega_b\ M_\odot/Mpc^3 \simeq 7.7\ 10^6 M_\odot/Mpc^3. \qquad (13.33)$$

Note that a factor 2-3 larger optical/NIR background than in eq. (13.31) could still be consistent with the present scheme if a similar scaling factor would also apply to eqs. (13.32) and (13.33): that is, if both the excess background and low-mass stars and stellar metals would be due to extended low-brightness halos, unaccounted for by deep HST imaging as well as by local photometric surveys.

13.2.2. Explaining the CIRB background

The total energy density between 7 and 1000 μm contained in the CIRB, including modellistic extrapolations as in Fig. 3 consistent with the constraints set by the cosmic opacity observations, amounts to

$$\nu I(\nu)|_{FIR} \simeq 40 \ 10^{-9} \ Watt/m^2/sr. \quad (13.34)$$

Following our previous assumption that luminous starbursting galaxies emit negligible energy in the optical-UV and most of it in the far-IR, we coherently assume that the energy resident in the CIRB background originates from star-forming galaxies at median $z_* \simeq 1.5$. The amount of baryons processed in this phase and the conversion efficiency ϵ have to account for the combined constraint set by eqs.(13.32) and (13.34), that is to provide a huge amount of energy with essentially no much stellar remnant in the local populations. The only plausible solution is then to change the assumptions about the stellar IMF characterizing the starburst phase, for example to a Salpeter distribution cutoff below $M_{min} = 2 \ M_\odot$, with a correspondingly higher efficiency $\epsilon = 0.002$ (see discussion in §13.2). This may explain the energy density in the CIRB:

$$\nu I(\nu)|_{FIR} \simeq 40 \ 10^{-9} \left(\frac{H_0}{50 \ km/s/Mpc} \right)^2 \frac{\Omega_b}{0.05} \frac{f_*}{0.1} \left(\frac{2.5}{1+z_*} \right) \frac{\epsilon}{0.002} \ Watt/m^2/sr,$$

assumed that a similar amount of baryons, $f_* \simeq 10\%$, as processed with low efficiency during the "inactive" secular evolution, are processed with higher efficiency during the starbursting phases, producing a two times larger amount of metals: $\rho(metals) \sim 1.4 \ 10^7 \ M_\odot/Mpc^3$. Note that by decreasing M_{min} during the SB phase would decrease the efficiency ϵ and increase the amount of processed baryons f_*, hence would bring to exceed the locally observed mass in stellar remnants (eq. 13.32).

The above scheme is made intentionally extreme, to illustrate the point. The reality is obviously more complex than this, e.g. by including a flattening at low mass values in the Salpeter law (see Zoccali et al. 2000) for the solar-neighborhood SF and, likewise, a more gentle convergence of the starburst IMF than a simple low-mass cutoff.

13.2.3. Galactic winds and metal pollution of the inter-cluster medium

A direct prediction of our scheme above is that most of the metals produced during the starburst phase have to be removed by the galaxies to avoid largely exceeding the locally observed metals in galaxies. As discussed in §6.4, there is clear evidence in local starbursts, based on optical and X-ray observations, for large-scale super-winds out-gassing high-temperature enriched plasmas from the galaxy. Our expectation would be that a substantial amount of metals, those originating from the same SF processes producing the CIRB background, are hidden in the hot inter-cluster medium.

But where all these metals are?

While densities and temperatures of the polluted plasmas in the diffuse (mostly primordial and un-processed) inter-cluster medium are such to hide easily these products of the ancient SB phase, an interesting support to the above scheme comes from consideration of the metal-enriched intra-cluster plasma (ICP) in clusters of galaxies. Rich clusters are considered to constitute a representative sample of the universe, while at the same time – given their deep gravitational potential – they are to be considered from a chemical point of view as closed boxes (all metals produced by cluster galaxies are kept inside the cluster itself).

The mass of metals in the ICP plasma is easily evaluated from the total amount of ICP baryons (measured to be ~ 5 times larger than the mass in galactic stars) and from their average metallicity, $\sim 40\%$ solar. The mass of ICP metals is $M_{metals,ICP} \simeq 5 \times 0.4 \ (Z/Z_\odot) \ M_{stars}$, which is two times larger than the mass of the metals present in galactic stars and consistent with the mass in metals produced during the SB phase.

Then the same starbursts producing the ICP metals are also likely responsible for the origin of the CIRB. As mentioned, the starburst enrichment process could have been pictured in a deep SCUBA image of the candidate proto-cluster surrounding the $z = 3.8$ radio-galaxy 4C41.17. In a similar fashion, Mushotzky & Loewenstein (1997) used their metallicity measurements in clusters to estimate the contribution of spheroidal galaxies to the SFR density (see Fig. 14).

13.2.4. *A two-phase star-formation: origin of galactic disks and spheroids*

The above scheme, best-fitting the available IR data as discussed in §11, implies that *star formation in galaxies has proceeded in two phases: a quiescent one taking place during most of the Hubble time, slowly building stars with standard IMF from the regular flow of gas in rotational supported disks; and a transient actively starbursting phase, recurrently triggered by galaxy mergers and interactions.* During the merger, *violent relaxation* redistributes old stars, producing de Vaucouleur profiles typical of galaxy spheroids, while young stars are generated following a top-heavy IMF.

Because of the geometric (thin disk) configuration of the diffuse ISM and the modest incidence of dusty molecular clouds, the quiescent phase is only moderately affected by dust extinction, and naturally originates most of the optical/NIR background (included early-type galaxies completely deprived of an ISM).

The merger-triggered active starburst phase is instead characterized by a large-scale redistribution of the dusty ISM, with bar-modes and shocks, compressing a large fraction of the gas into the inner galactic regions and triggering formation of molecular clouds. As a consequence, this phase is expected to be heavily extinguished and the bulk of the emission to happen at long wavelengths, naturally originating the cosmic CIRB background. Based on dynamical considerations, we expect that during this violent SB phase the elliptical and S0 galaxies are formed in the most luminous IR SBs at higher-z (corresponding to the SCUBA source population), while galactic bulges in later-type galaxies likely originate in lower IR luminosity, lower-z SBs (the ISO mid-IR population).

The presently available IR data cannot assess if the different luminosity ranks of SCUBA and ISO selected sources are characterized also by different formation timescales (SF activities being confined to the higher-z for the former and to lower-z for the latter), since the present samples are far dominated by K-correction and selection effects. Assumed however this is indeed the case, this could still be reconciled with the expectations of hierarchical clustering models if we consider that SCUBA sources likely trace the very high-density (galaxy clusters) environment with an accelerated merging rate at high-z, while ISO sources are likely related with lower-density environments (galaxy groups or the field) entering the non-linear collapse phase at later cosmic epochs (e.g. Franceschini et al. 1999).

Finally, *if indeed the IMF characteristic of the SB phase is deprived of low-mass stars, as suggested in the previous paragraphs, a consequence would be that the excess blue stars formed during the SB would quickly disappear, leaving the colors of the emerging remnant as typically observed for early-type galaxies and keeping consistent with the evidence that the stellar mass content in galaxies does not change much for $z < 1$.*

13.3. *Contribution by gravitational accretion to the global energetics*

The remarkable similarities between the cosmic evolution of galaxy and AGN emissivities have been taken as evidence that the same processes triggering SF also make a fraction of the gas to accrete and fuel the AGN (Hasinger 1998, Franceschini et al. 1999). Furthermore, detailed studies of local high-luminosity IR galaxies are showing that SF and AGN activities happen very often concomitantly in the same object (Genzel et al. 1998; Risaliti et al. 2000; Bassani et al. 2001). After all, this is a natural outcome of the scheme discussed in previous Sections, the violent radial inflow of gas following the merger/interaction should likely fuel not only nuclear star-clusters, but the BH itself at some stage.

Waiting for forthcoming and future powerful instrumentation (X-ray observatories CHANDRA and XMM, Constellation-X and XEUS in the future, and large space IR observatories like NGST and *Herschel*) to have a detailed quantification of the relative merits of the two fundamental baryon drivers, some order-of-magnitude estimates may be useful as a guideline. From a combined analysis of the AGN and starburst average bolometric emissivities as a function of redshift, Franceschini et al. (1999) infer a relationship between the mass M_{BH} of the local remnant super-massive BH after the AGN phase to the mass M_* in galactic stars from the SB phase:

$$M_{BH} \simeq 0.001 \left(\frac{\epsilon}{0.001}\right) \left(\frac{0.1}{\eta}\right) \left(\frac{n[type\ II]/n[type\ I]}{5}\right) M_*, \qquad (13.35)$$

where η is the radiative efficiency by BH accretion and $n[type\ II]/n[type\ I]$ is the ratio of the absorbed to unabsorbed AGNs (which should be close to 3-5 to explain the local AGN statistics and the observed intensity of the XRB). On the other hand, observations of supermassive BH's in local spheroidal galaxies (Magorrian et al. 1998, Faber et al. 1997) indicate a quite higher mass in the BH accreted material with respect to that in stars: $M_{BH} \simeq (0.002 - 0.006)\ M_*$. Assumed that η should not be lower than 0.1, this may require a stellar mass-energy conversion efficiency $\epsilon >> 0.001$, which is further independent support to the idea of a top-heavy IMF during the SB phase.

14. Conclusions

During the last few years a variety of observational campaigns, in particular by ISO from space in the far-IR and by large mm telescopes from ground, have started to provide a complementary view of the distant universe at long wavelengths with respect to that offered by standard optical-UV-NIR deep explorations. Also of crucial importance in this context was the discovery of an intense diffuse background radiation in the far-IR/sub-mm of extragalactic origin, the CIRB. These results are challenging those obtained from optical-UV observations only, by revealing luminous to very luminous phases in galaxy evolution at substantial redshifts, likely corresponding to violent events of star-formation in massive systems. In the most extreme of these sources, however, a quasar contribution cannot be excluded, and sometimes has indeed been proven.

Whereas the process of optical identification and spectroscopic characterization of the long-wavelength selected high-redshift sources is only at the beginning (and will keep being a challenging task for the next several years because of the faintness of the optical counterparts), some interesting constraints on the cosmic evolution can already been inferred from observations of the CIRB spectral intensity and the multi-wavelength source counts. The most robust conclusions at the moment appear to be those of a very rapid increase of galaxy long-wavelength emissivity with redshift, paralleled by an increased incidence in high-redshift sources of dust extinction and thermal dust reprocessing with respect to locally observed sources.

A way to interpret these results is to consider as a crucial cosmogonic ingredient the role of galaxy interactions and merging. The strong increases with redshift of the *probability* of interactions (as partly due to a plain geometrical effect in the expanding universe) and of the *effects* of interactions (due to the more abundant fuel avaliable in the past), likely explain the observed rapid evolution.

Altogether, the large energy content of the CIRB is not easily explained, unless the powerful infrared starburst phase is characterized by a stellar IMF somewhat deprived in low-mass stars.

Although the subject is presently subject to some controversies, we think we have provided enough evidence, based on pioneering efforts of deep sky surveys in the IR and mm, that only such long wavelengths contain the clue to an exhaustive description of the star formation phenomenon, now and in the past. It seems clear that there are no alternatives, neither in X-rays, optical nor radio, to the IR/mm flux measurement for a reliable determinantion of the rate of SF in galaxies, simply because it is there that a dominant fraction of photons from young very luminous stars emerges, and no ways are available to determine "a priory" what precisely this fraction is. Fundamental aspects of galaxy formation and evolution (e.g. the origin of galaxy spheroids, and the onset of quasar activity) can effectively be observed at long wavelengths. In this sense the variety of ground-based and space projects in this field planned for the present decade promises extremely rewarding benefits for observational cosmology.

This paper has benefited by a large collaboration, in particular concerning items discussed in the last chapters, including some yet unpublished results. I want to mention the people who have particularly contributed: H. Aussel, S. Bressan, C. Cesarsky, D. Clements, FX. Desert, D. Elbaz, D. Fadda, R. Genzel, G.L. Granato, M. Harwit, S. Oliver, B. Poggianti, J.L. Puget, D. Rigopoulou, M. Rowan-Robinson, L. Silva. I am also glad to thank L. Danese, G. De Zotti for a long-standing collaboration in this field, A. Cavaliere and C. Chiosi for many fruitful discussions. Finally, I want to warmly thank the organizers of the Canary Islands Winter School on "High-Redshift Galaxies" for their kind invitation.

REFERENCES

ADELBERGER, K. L. & STEIDEL, C. C. 2000, ApJ, 544, 218.
ALTIERI, B., METCALFE, L., KNEIB, J.P. et al. 1999, A&A, 343.
ANDREANI, P. & FRANCESCHINI, A. 1996, MNRAS, 283, 85.
AURIEMMA, C., PEROLA, G. C., EKERS, R. D., FANTI, R., LARI, C., JAFFE, W. J. & ULRICH, M. H. 1977, A&A, 57, 41.
AUSSEL, H. 1998, PhD Thesis, CEA Saclay, Paris.
AUSSEL, H. et al. 2001, in preparation.
AUSSEL H., CESARSKY C., ELBAZ D. & STARCK, J.L. 1999, A&A, 342, 313.
BARGER, A., COWIE, L. L., MUSHOTZKY, R. & RICHARDS, E. 2001, AJ 121, 662.
BARGER, A. J., COWIE, L. L., SANDERS, D. B., FULTON, E., TANIGUCHI, Y., SATO, Y., KAWARA, K. & OKUDA, H. 1998, Nature, 394, 248.
BARGER, A. J., COWIE, L. L., SMAIL, I., IVISON, R. J., BLAIN, A. W. & KNEIB, J.-P. 1999, AJ, 117, 2656.
BARNES, J. E. & HERNQUIST, L. E. 1992, ARA&A, 30, 705.
BARNES, J. E. & HERNQUIST, L. E. 1996, ApJ, 471, 115.
BASSANI, L., FRANCESCHINI, A., MALAGUTI, G., CAPPI, M. & DELLA CECA, R. 2001, A&A, submitted.
BERNSTEIN 1998, PhD Thesis, California Institute of Technology.
BERTIN, E., DENNEFELD, M. & MOSHIR, M. 1997, A&A, 323, 685.
BERTOLDI, F. et al. 2000, A&A, 360, 92.
BOHREN, C.F. & HUFFMAN, D.R. 1983, "Absorption and scattering of light by small particles", John Wiley, New York.

BOULANGER, F., BOISSSEL, P., CESARSKY, D. & RYTER, C. 1998, A&A, 339, 194.
BRUZUAL, A.G. 1983, ApJ, 273, 105.
CALZETTI, D. 1997, AJ, 113, 192.
CAPPI, M. et al. 1999, A&A, 350, 777.
CARILLI, C.L. & YUN, M.S. 1999, ApJ, 513, L13.
CAVALIERE, A. & VITTORINI, V. 2000, ApJ, 543, 599.
CHAPMAN, S.C. 2000, MNRAS, 319,318.
CHINI, R., KRUGEL, E., LEMKE, D. & WARD-THOMPSON, D. 1995, A&A, 295, 317.
CLEMENTS, D. & BAKER, A. 1997, in "Extragalactic Astronomy in the Infrared", Edited by G. A. Mamon, Trinh Xuan Thuan, and J. Tran Thanh Van. Paris, Editions Frontieres, p.347.
COHEN, J. et al. 2000, ApJ, 538, 29.
CONDON, J. J. 1974, ApJ, 188, 279.
CONDON, J. J. 1992, ARA&A, 30, 575.
CONNOLLY, A., SZLAY, A., DICKINSON, M., SUBBARAO, M. & BRUNNER, R. 1997, ApJ, 486, L11.
COPPI, P. & AHARONIAN, F. 1999, Astropart. Phys., 11, 35-39
DE JONG, T., KLEIN, U., WIELEBINSKI, R. & WUNDERLICH, E. 1985, A&A, 147, L6.
DE ZOTTI, G., FRANCESCHINI, A., TOFFOLATTI, L., MAZZEI, P. & DANESE, L. 1996, Astro. Lett. and Communications, 35, 289.
DESERT F.X. et al. 1999, A&A, 342, 363.
DEVRIENDT, J. E. G., GUIDERDONI, B. & SADAT, R. 1999, A&A, 350, 381.
DISNEY, M., DAVIES, J. & PHILLIPPS, S. 1989, MNRAS, 239, 939.
DOLE, H., GISPERT, R., LAGACHE, G. et al. 2000, in *ISO Beyond Point Sources, Studies of Extended Infrared Emission*, R. Laureijs, K. Leech and M. Kessler Eds., ESA-SP 455, 167.
DOWNES, D. et al. 1999, A&A, 347, 809.
DRAINE, B. T. & LEE, H. M. 1984, ApJ, 285, 89.
DUNNE, L., EALES, S. EDMUNDS, M., IVISON, R., ALEXANDER, P. & CLEMENTS, D. 2000, MNRAS, 315, 115.
DWEK, E. 1998, ApJ, 501, 634.
DWEK, E. & ARENDT, R. G. 1998, ApJ, 508, L9.
EALES, S. et al. 1999, ApJ, 515, 518.
ELBAZ, D. 1999, Proceedings of Rencontres de Moriond, "Building Galaxies, from the Primordial Universe to the Present", F. Hammer, et al. Eds, (Ed. Frontieres, astro-ph/9911050).
ELBAZ, D. 2001, Deep Fields, Proceedings of the ESO/ECF/STScI Workshop held at Garching, Germany, 9-12 October 2000. Stefano Cristiani, Alvio Renzini, Robert E. Williams (eds.). Springer, 2001, p. 369.
ELBAZ, D. et al. 1999, A&A, 351, L37.
ELLIS, R. S. 1997, ARA&A, 35, 389.
ELLIS, R. S., COLLESS, M., BROADHURST, T., HEYL, J. & GLAZEBROOK, K. 1996, MNRAS, 280, 235.
FABER, S.M. et al. 1997, AJ, 114, 1771.
FABIAN, A. et al. 2000, MNRAS, 315 8.
FANG F., SHUPE, D.L., XU, C. & HACKING, P.B. 1998, ApJ, 500, 693.
FINKBEINER, D.P., DAVIES, M. & SCHLEGEL, D.J. 2000, ApJ, 544, 81.
FIXSEN, D.J. et al. 1998, ApJ, 508, 123.
FLORES H., HAMMER F., THUAN T. et al. 1999, ApJ, 517, 148.
FRANCESCHINI, A. et al. 2001, A&A, 378, 1.
FRANCESCHINI, A., ANDREANI, P., DANESE, L. 1998, MNRAS, 296, 709.

Franceschini, A., Aussel, H., Elbaz, D. et al. 2000, Mem Sait, in press.
Franceschini, A., Danese, L., de Zotti, G., Xu, C. 1988, MNRAS, 233, 175.
Franceschini, A., Hasinger, G., Miyaji, T., Malquori, D. 1999, MNRAS, 310, L5.
Franceschini, A., La Franca, F., Cristiani, S., Martin-Mirones J. 1994a, MNRAS, 269, 683.
Franceschini, A., Mazzei, P., De Zotti, G., Danese, L. 1994b, ApJ, 427, 140.
Frayer, D.T. et al. 1999, ApJ, 514, 13L.
Genzel, R. et al. 1998, ApJ, 498, 579.
Genzel, R. & Cesarsky, C.J. 2000, ARA&A, 38, 761.
Gibson, B. K., Loewenstein, M. & Mushotzky, R. F. 1997, MNRAS, 290, 623.
Gispert, R., Lagache, G. & Puget, J. L. 2000, A&A, 360, 1.
Gordon, K.D., Calzetti, D. & Witt, A.N. 1997, ApJ, 486, 625.
Gorjian, V., Wright, E. L. & Chary, R. R. 2000, ApJ, 536, 550.
Granato, G.L. et al. 2000, ApJ, 542, 710.
Granato, G.L.& Danese, L. 1994, MNRAS, 268, 235.
Granato, G.L., Danese, L. & Franceschini, A. 1997, ApJ, 486, 147.
Gregorich, D. T., Neugebauer, G., Soifer, B. T., Gunn, J. E. & Herter, T. L. 1995, AJ, 110, 259.
Guhathakurta, P. & Draine, B.T. 1989, ApJ, 345, 230.
Guiderdoni, B. et al. 1997, Nature, 390, 257.
Haarsma, D. B. & Partridge, R. B. 1998, ApJ, 503, L5.
Hacking, P. & Houck, J. R. 1987, ApJS, 63, 311.
Hacking, P., Houck, J. R. & Condon, J. J. 1987, ApJ, 316, L15.
Harwit, M. 1999, ApJ, 510, L83.
Harwit, M. & Pacini, F. 1975, ApJ, 200, L127.
Harwit, M. & Protheroe, R. J., Biermann, P. L. 1999, ApJ, 524, 91.
Hasinger, G. 1998, Astronomische Nachrichten, 319, 37.
Hauser, M.G., Arendt, R.G., Kelsall, T. et al. 1998, ApJ, 508, 25.
Heckman, T.M., Armus, L. & Miley, G. 1990, ApJS, 74, 833.
Helou, G., Soifer, B. T. & Rowan-Robinson, M. 1985, ApJ, 298, L7.
Hornschemeier, A.E., Brandt, W. N., Garmire, G.P. et al. 2000, ApJ, 541, 49.
Hoyle, F. & Wickramasinghe, N. C. 1991, *The theory of cosmic grains*, Kluwer Academic Publishers.
Hughes, D. et al. 1998, Nature, 394, 241.
Ivison, R. J., Dunlop, J. S., Smail, I., Dey, A., Liu, M. & Graham, J. 2000, ApJ, 542, 27.
Jimenez, R., Padoan, P., Dunlop, J., Bowen, D., Juvela, M. & Matteucci, F. 2000, ApJ, 532, 152.
Juvela, M., Mattila, K., & Lemke, D. 2000, A&A, 360, 813.
Kennicut, R. C. 1998, ARA&A, 36, 189.
Kormendy, J. & Sanders, D. 1992, ApJ, 390, 53.
Krawczynski, H., Coppi, P. S., Maccarone, T. & Aharonian, F. A. 2000, A&A, 353, 97.
Lagache G., Abergel, A., Boulanger, F., Desert, F. X. & Puget J. L. 1999, A&A, 344, 322L.
Lagache, G. & Puget, J. L. 2000, A&A, 355, 17L.
Lagage, P. O. et al. 1996, A&A, 315, L273.
Lari, C. et al. 2001, MNRAS, 325, 1173.

LE FÈVRE, O. et al. 2000, MNRAS, 311, 565.
LEHNERT, M. & HECKMAN, T. M. 1996, ApJ, 462, 651.
LEITHERER, C. et al. 1999, ApJS, 123, 3.
LEITHERER, C. & HECKMAN, T. M. 1995, ApJS, 96, 9.
LILLY, S.J. et al. 1999, ApJ, 518, 641.
LILLY, S.J., LE FÈVRE, O., HAMMER, F. & CRAMPTON, D. 1996, ApJ, 460, L1.
LONSDALE, C.J., HACKING, P.B., CONROW, T. P. & ROWAN-ROBINSON, M. 1990, ApJ, 358, 60.
LUTZ, D., SPOON, H. W. W., RIGOPOULOU, D., MOORWOOD, A. F. M. & GENZEL, R. 1998, ApJ, 505, L103.
MADAU, P. et al. 1996, MNRAS, 283, 1388.
MADAU, P. & POZZETTI, L. 2000, MNRAS, 312, 9.
MAGORRIAN, J. et al. 1998, AJ, 115, 2285.
MATSUHARA, H. et al. 2000, A&A, 361, 407.
MAZZEI, P., AUSSEL, H., XU, C., SALVO, M., DE ZOTTI, G. & FRANCESCHINI, A. 2001, NewA, 6, 265.
MAZZEI, P., DE ZOTTI, G. & XU, C. 1994, ApJ, 422, 81.
MEURER, G. R., HECKMAN, T. M., LEHNERT, M. D., LEITHERER, C. & LOWENTHAL, J. 1997, AJ, 114, 54.
MIE, G. 1908, Ann. Physik, 25, 377.
MIYAJI, T., HASINGER, G. & SCHMIDT., M. 2000, A&A, 353, 25.
MOORWOOD, A. 1996, Space Science Reviews, 77, 303.
MOORWOOD, A., ORIGLIA, L., KOTILAINEN, J. & OLIVA, E. 1996, in Proceedings of the ESO Symposium on Spiral Galaxies, Springer-Verlag, Berlin, 299.
MUSHOTZKY, R. F. & LOEWENSTEIN, M. 1997, ApJ, 481, 63.
NORMAN, C. A. & SCOVILLE, N.Z. 1988, ApJ, 332, 124.
OLIVER, S. et al. 1996, MNRAS, 280, 673.
OLIVER, S. et al. 2000a, MNRAS, 316, 749.
OLIVER, S. et al. 2000b, MNRAS, submitted.
PADOAN, P., JUVELA, M., BALLY, J. & NORDLUND, A. 1998, ApJ, 504, 300.
PEARSON, C. & ROWAN-ROBINSON, M. 1996, MNRAS, 283, 174.
PIER, E.A. & KROLIK, J.H. 1992, ApJ, 401, 99.
POGGIANTI, B. M., BRESSAN, A. & FRANCESCHINI, A. 2000, ApJ, in press.
POGGIANTI, B. M. & WU, H. 2000, ApJ, 529, 157.
PUGET J. L. et al. 1996, A&A, 308, L5.
PUGET, G. L., LAGACHE, G., CLEMENTS, D., REACH W., AUSSEL, H., BOUCHET, F., CESARSKY, C., DESERT, F., DOLE, H., ELBAZ, D., FRANCESCHINI, A., GUIDERDONI, B. & MOORWOOD, A. 1999, A&A, 345, 29.
PUGET, J. L. & LÉGER, A. 1989, ARAA, 27, 161.
PUGET, J. L., LÉGER, A. & BOULANGER, F. 1985, A&A, 142, L19.
RIEKE, G. H., LEBOFSKY, M. J., THOMPSON, R. I., LOW, F. J. & TOKUNAGA, A. 1980, ApJ, 238, 24.
RIEKE, G. H., LEBOFSKY, M. J. & WALKER, C. E. 1988, ApJ, 325, 679.
RIGOPOULOU, D., FRANCESCHINI, A., AUSSEL, H. et al. 2000, ApJ, 537, L85.
RISALITI, G., GILLI, R., MAIOLINO, R. & SALVATI, M. 2000, A&A, 357, 13
ROWAN-ROBINSON, M. 1992, MNRAS, 258, 787.
ROWAN-ROBINSON, M. et al. 1997, MNRAS, 289, 482.
ROWAN-ROBINSON, M., SAUNDERS, W., LAWRENCE, A. & LEECH, K. 1991, MNRAS, 253, 485.

Rush, B., Malkan, M. A. & Spinoglio, L. 1993, ApJS, 89, 1.
Rybicki, G. B. & Lightman, A. P. 1979, "Radiative Processes in Astrophysics", J. Wiley & Sons.
Sanders, D. et al. 1988, ApJ, 325, 74.
Sanders, D. & Mirabel, I. F. 1996, ARA&A, 34, 749.
Saunders, W. et al. 1997, in *Extragalactic Astronomy in the Infrared*, G. A. Mamon, Trinh Xuan Thuan & J. Tran Thanh Van. Eds, Paris, Editions Frontieres, 431.
Saunders, W., Rowan-Robinson, M., Lawrence, A., Efstathiou, G., Kaiser, N., Ellis, R. S. & Frenk, C. S. 1990, MNRAS, 242, 318.
Scheuer, P. A. G. 1957, Proc. Cambridge Phil. Soc., 53, 764.
Scott, D. et al. 2000, A&A, 357, L5.
Scoville, N. Z., Yun, M. S. & Bryant, P. M. 1996, in *Cold gas at high redshifts*, M. Bremer et al. Eds., Kluwer Academic Publishers, 25.
Siebenmorgen, R. & Krugel, E. 1992, A&A, 259, 614.
Siebenmorgen, R., Rowan-Robinson, M. & Efstathiou, A. 2000, MNRAS, 313, 734.
Silva, L., Granato, G. L., Bressan, A. & Danese, L. 1998, ApJ, 509, 103.
Smail, I. et al. 1999, MNRAS, 308, 1061.
Smail, I., Ivison, R. J. & Blain, A. W. 1997, ApJ, 490, L5.
Soifer, B. T., Neugebauer, G., Matthews, K. et al. 2000, AJ, 119, 509.
Spinoglio, L. & Malkan, M. A. 1992, ApJ, 399, 504.
Stanev, T. & Franceschini, A. 1998, ApJ, 494, L159.
Starck, J. L., Aussel, H., Elbaz, D., Fadda, D. & Cesarsky, C. 1999, A&AS, 138, 365.
Stecker, F., De Jager, O. & Salamon, M. 1992, ApJ, 390, L49.
Stickel, M., Bogun, S., Lemke, D. et al. 1998, A&A, 336, 116.
Taniguchi, Y., Cowie, L. L., Sato, Y., Sanders, D. & Kawara, K. 1997, A&A, 328, L9
Telesco, C. 1988, ARA&A, 26, 343.
Tinsley, B.M. 1977, ApJ, 211, 621.
Toffolatti, Franceschini A., De Zotti G. & Danese L. 1987, A&A, 184, 7.
Toomre, A. 1977, in *The evolution of galaxies and their stellar populations*, Yale University Observatory, 401.
van der Werf, P. 1996, in *Cold Gas at High Redshift*, M. Bremer et al. Eds, Kluwer Academic Publishers, 37.
Vigroux, L. et al. 1998, in *The Universe as seen by ISO*, ESA-SP 427, 805.
Xu, C. 1997, in *Extragalactic Astronomy in the Infrared*, G. A. Mamon, Trinh Xuan Thuan, and J. Tran Thanh Van Eds., Editions Frontieres, Paris, 193.
Xu, C. 2000, ApJ, 541, 134.
Xu, C., Hacking, P., Fang, F. et al. 1998, ApJ, 508, 576.
Zoccali, M. et al. 2000, ApJ, 530, 418.

Quasar Absorption Lines

By JILL BECHTOLD

Steward Observatory and the Department of Astronomy, the University of Arizona, Tucson, AZ 85721, USA

"Analyzing a spectrum is exactly like doing a crossword puzzle, but when you get through with it, you call the answer research."
– *Henry Norris Russell.*

The absorption lines observed in quasar spectra have given us a detailed picture of the intergalactic medium and the metal abundance and kinematics of high redshift galaxies. In this review, we present an introduction to the field, starting with the techniques used for interpreting absorption line spectra. We then survey the observational and theoretical development of our understanding of the Lyman α forest, the metal absorbers, and the damped Lyα absorbers. We conclude with a discussion of some of the remaining outstanding issues, and prospects for the future.

1. Introduction

Absorption lines were seen in the earliest photographic spectra of quasars in the mid-1960's (Sandage 1965, Gunn & Peterson 1965, Kinman 1966, Burbidge, Lynds & Burbidge 1966, Burbidge 1967). By 1969, the concensus was that quasar redshifts are cosmological, and that many of the absorption lines in quasar spectra originate in intervening galaxies (see Bahcall & Salpeter 1966, Bahcall & Spitzer 1969). Subsequently, W. L. W. Sargent, Peter J. Young and collaborators wrote a series of papers presenting the first comprehensive studies of QSO absorption lines, using high quality spectra obtained with Boksenberg's photon counter, the IPCS, at the Double Spectrograph on the Palomar Hale telescope (Young et al. 1979; Sargent et al. 1979; Sargent, Young, Boksenberg & Tytler 1980; Young, Sargent & Boksenberg 1982ab; Sargent, Young & Schneider 1982).

A classification scheme was developed, summarized in an article in the Annual Review of Astronomy and Astrophysics, by Weymann, Carswell & Smith (1981). First, the so-called "metal-line absorbers" with redshifts, z_{abs}, much less than the quasar emission line redshift, z_{em}, were attributed to interstellar gas in intervening galaxies. Systems which were detected in Lyα and occasionally the Lyman series of hydrogen, were dubbed the Lyα forest. These showed no clustering like galaxies, nor detectable metals. They were thought to be primordial, pre-galactic, inter-galactic gas, pressure-confined by a hypothesized inter-galactic medium, and ionized by the integrated UV radiation of quasars themselves (Sargent, Young, Boksenberg & Tytler 1980). Finally, some quasars showed broad, P-Cygni like troughs of absorption at the emission line redshift; these were interpreted as radiatively driven winds associated with the quasar central engine (Turnshek 1984 and references therein).

Quasar absorbers are useful probes of the high redshift universe for many reasons. Atoms have a rich ultraviolet absorption spectrum; typically a single spectrum will contain lines from many elements in a range of ionization states. Lines like Lyα are extremely sensitive probes of small amounts of gas. Absorption spectra are relatively easy to interpret, since the lines arise from atoms in the ground state, and probe a pencil beam through the volume of gas. By contrast, emission lines are a complicated integral of density and other physical conditions over the emitting volume. Until recently, quasar

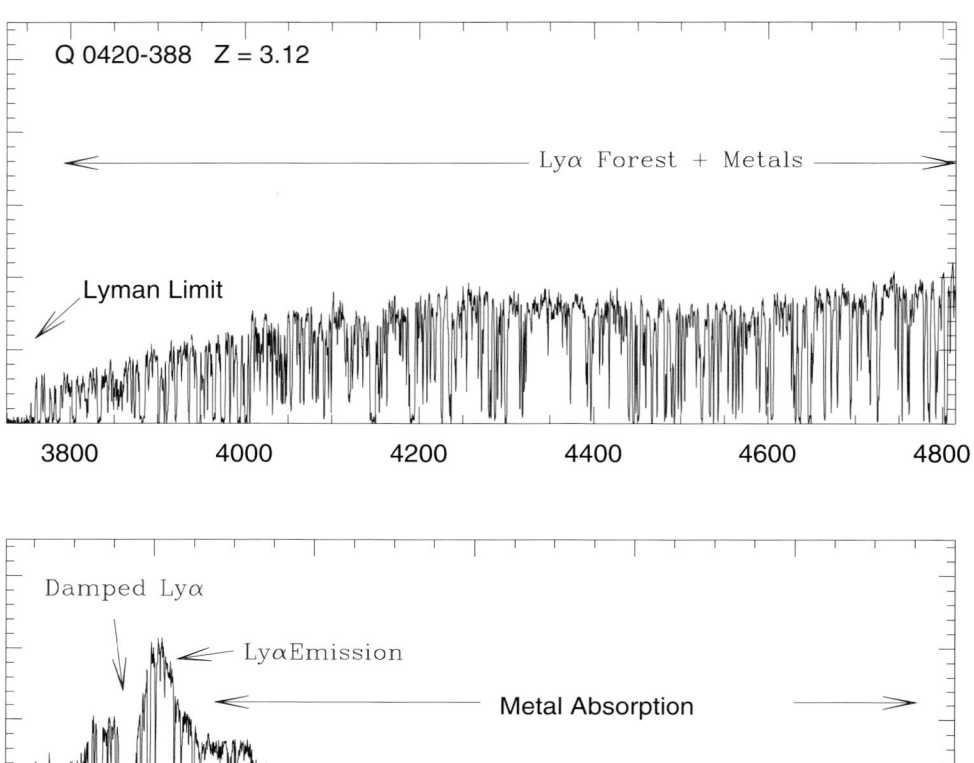

FIGURE 1. High resolution spectrum of a $z = 3.12$ quasar obtained with the Las Campanas echelle spectrograph, by the author and S. A. Shectman. Prominent features are indicated.

absorbers contained the only information about "normal", that is, non-active galaxies at high redshift. They still are the best way to study the detailed kinematics and metal abundances of high redshift galaxies.

Other reviews of the subject can be found in the conference proceedings edited by Meylan (1995) and Petitjean & Charlot (1997). The Lyα forest was reviewed by Rauch (1998) and damped Lyα absorber abundunces by Lauroesch et al (1996).

2. Analysis of Absorption Line Spectra

The spectrum of a z~3 quasar is shown in Figure 1. The broad emission line at $\lambda_{obs} = 5000$ Å is the Lyα emission associated with the quasar. The numerous absorption lines blueward the emission line are mostly Lyα lines at different redshifts $z_{abs} < z_{em}$, where

$$\lambda_{obs} = (1 + z_{abs})\lambda_{rest} \qquad (2.1)$$

and λ_{rest}= 1215.67 Å for Lyα. The broad feature just short of 5000 Å is a damped Lyα line arising from an intervening absorber, which also causes the observed continuum of the quasar to go to zero at the Lyman limit. The lines redward of the Lyα emission line are metal transitions associated with the damped Lyα absorber as well as other redshift systems along the line of sight.

Comprehensive lists of lines commonly seen in quasar spectra can be found in Morton (1991, 2000) and Verner, Barthel & Tytler (1994). Morton, York & Jenkins (1988) has a short list of the strongest lines suitable for many applications.

Spectra for wavelengths $\lambda_{obs} > 3200$ Å are observable with ground-based telescopes, whereas spectra at shorter wavelengths must be obtained with telescopes above the atmosphere. On the red end of the optical, CCDs become transparent to photons at ~ 9000 Å and although a few cases of absorption features have been reported in the near-IR, (e.g. Elston et al. 1996) the sky background in the IR makes observing faint quasars difficult from the ground. In the rocket UV, the *Hubble Space Telescope (HST)* and the *International Ultraviolet Explorer (IUE)* have been used for spectroscopy in the wavelength range 1150 − 3200 Å. The short wavelength limit arises from the cutoff in the MgFl coatings for the optics. Coatings for shorter wavelengths have relatively low reflectivity, so specialized missions have been launched for spectroscopy in the 911 − 1150 Å region – most recently, the *Hopkins Ultraviolet Telescope (HUT)* and the *Far Ultraviolet Spectroscopic Explorer (FUSE)*. The Milky Way is opaque between 911 Å and 62 Å or 0.2 keV in the X-rays. The *Chandra X-ray Observatory* and *XMM-Newton* are being used to study quasar absorption lines in the X-rays (Bechtold et al. 2001), but results were not available in time for inclusion in this review.

2.1. *Voigt Profile Fitting, the Curve of Growth and the Doublet Ratio*

How do we derive physical quantities of interest from the observed absorption line spectra? We summarize the basics in this section. The results are derived in full in a number of introductory texts, for example Gray (1992), Spitzer (1978), and Swihart(1976).

2.1.1. *Voigt profile fits*

If the spectral resolution of the observation is good enough that the intrinsic line widths are resolved, Voigt profiles can be fit to individual line profiles to derive properties of the absorbing gas. The observed line profile is the line profile for an individual atom convolved with (1) a function describing the broadening due to the distribution of atomic velocities (generally assumed to be Maxwellian with temperature T), and (2) the instrumental profile (generally modeled as a Gaussian). In addition, one expects that for a sight line passing through a galaxy, the absorption feature will be produced by several "clouds" or "velocity components", each with its own temperature T, and some overall velocity width characteristic of the kinematics of the galaxy, sometimes called the turbulent velocity. When the velocity dispersion of the clouds is comparable or smaller than the thermal widths of individual clouds, the individual velocity components will be blended.

The absorption line profile for an individual atom, called the natural line profile, is characterized by a cross-section which is a Lorentzian function of frequency ν

$$\sigma_\nu = \frac{\pi e^2}{m_e c} f \frac{\Gamma/4\pi^2}{(\nu - \nu_o)^2 + (\Gamma/4\pi)^2} \; cm^2 \qquad (2.2)$$

where f is the oscillator strength, Γ is the damping parameter, and ν_0 is the central frequency of the transition. The term "damping" arises from the semi-classical derivation of Equation 2.2 from consideration of a harmonic oscillator with a force that damps the particle oscillations arising from the inevitable emission of radiation by accelerating charges (e.g. Rybicki & Lightman 1979). The oscillator strength, or f-value, measures the quantum mechanical departure from the classical harmonic oscillator.

For the transitions of interest, the typical width of the Lorenzian is tiny, $\Delta\lambda \sim 10^{-4}$ Å full-width-half-maximum (FWHM) at $\nu_0 = 10^{14}$ Hz, or 0.006 km s^{-1} at 5000 Å. Thus the natural line width is much smaller than the thermal width of the absorbing Maxwellian velocity distribution of the atoms in the absorbing cloud, the instrumental line width, or the turbulent cloud velocity dispersion.

The convolution of the Lorentzian profile for an individual atom with a Maxwellian results in a cross-section for absorption described by the Voigt integral

$$\sigma_\nu = a_{\nu_o} \, H(a,x) \qquad (2.3)$$

where $H(a,x)$ is the Hjerting function

$$H(a,x) = \frac{a}{\pi} \int_{-\infty}^{\infty} \frac{\exp(-y^2)}{(x-y)^2 + a^2} dy \qquad (2.4)$$

and

$$a_{\nu_o} = \frac{\sqrt{\pi} e^2}{m_e c} \frac{f}{\Delta \nu_D} \qquad (2.5)$$

where we have defined the Doppler frequency, $\Delta\nu_D$,

$$\Delta\nu_D = \frac{1}{\lambda_o} \sqrt{\frac{2kT}{m_{atom}}} \qquad (2.6)$$

which is the central frequency of the Lorentzian of an atom with the RMS velocity of the Maxwellian. The dimensionless frequency x is

$$x = \frac{\nu - \nu_o}{\Delta\nu_D} = \frac{\lambda - \lambda_o}{\Delta\lambda_D} \qquad (2.7)$$

the difference between the frequency or wavelength and the line center in units of the Doppler frequency or Doppler wavelength,

$$\Delta\lambda_D = \frac{\lambda_o^2}{c} \Delta\nu_D = \frac{\lambda_o}{c} \sqrt{\frac{2kT}{m_{atom}}} \qquad (2.8)$$

The Voigt integral cannot be evaluated analytically and so is generated numerically. Humlicek (1979) gives a fortran subroutine for $H(a,x)$, and numerical tables are given by Finn & Mugglestone (1965). In the quasar absorption line literature, the program *VPFIT* is in widespread use for fitting Voigt profiles. *VPFIT* was developed by Bob Carswell and several generations of students, who have generously made it available to the community (Carswell et al. 1995). Theoretical Voigt profiles are shown in Figure 2.

The observed line profile depends on the dimensionless optical depth at line center, τ_{ν_0}, which is

$$\tau_{\nu_o} = \int n \, \sigma_{\nu_o} dl \qquad (2.9)$$

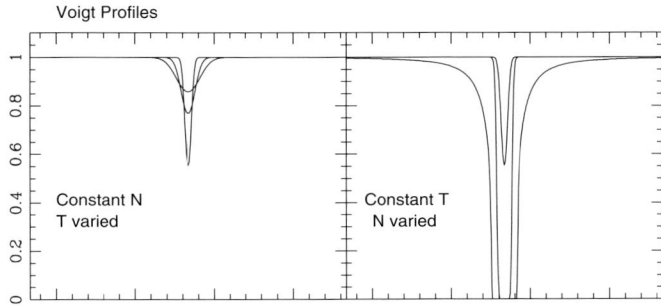

FIGURE 2. Theoretical Voigt profiles. Left side shows lines on the linear part of the curve of growth, $\tau_{\nu_o} < 1$, for constant column density N and $T = 10,000$; 50,000 and 150,000 K. Right hand side shows profiles for constant T and increasing N: τ_{ν_o} =0.5, 23 (flat or saturated) and 23,000 (damped).

where the right hand side is the integral through the absorbing cloud of the volume density n (units are cm^{-3}) times the cross-section σ_{ν_0} (units are cm^2) at line center for absorption. One can show that

$$\sigma_{\nu_0} = \frac{\sqrt{\pi}e^2}{m_e c} \frac{f}{\Delta \nu_D} \qquad (2.10)$$

so if the volume density n is uniform, we can write

$$\tau_{\nu_o} = N \frac{\sqrt{\pi}e^2}{m_e c} \frac{f}{\Delta \nu_D} \qquad (2.11)$$

where N is the *column density* with units cm^{-2}. Note that the optical depth at line center, τ_{ν_o}, increases with increasing column density, and is higher for higher f-values. As the temperature of the atoms increases, τ_{ν_o} decreases. If $\tau > 1$ then the line is said to be saturated. Increasing the column density of a saturated line has little effect on the line except to make it slightly wider, so it is difficult to measure the column density when this is the case.

For most metal-line transitions, and typical interstellar conditions, $T \sim 10,000$ K, so that most metal lines have very narrow line widths, and are almost always unresolved, even at high spectral resolution. Because the line width is inversely proportional to the square root of the mass of the absorbing atom, hydrogen lines are much wider for the same temperature.

2.1.2. *The curve of growth*

If the spectral resolution of the data is not good enough to resolve the line profiles, we can still learn a lot by measuring the "equivalent width" of the absorption line, which is defined as the width of a rectangular line with area equal to the absorbed area of the actual line (and typically is expressed with units of wavelength). The equivalent width reduces the information in the line to one number, proportional to the area of the line, or its strength; the line profile information is lost. However, if you measure lines from the

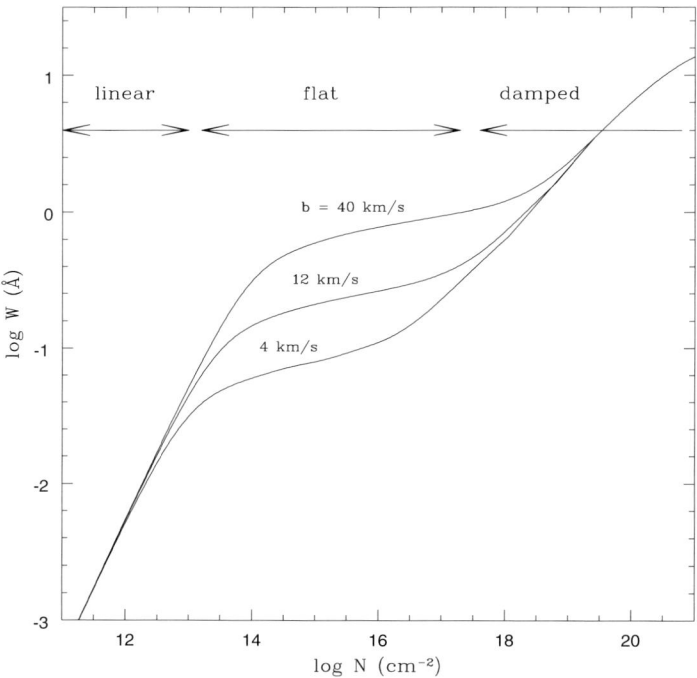

FIGURE 3. Theoretical curve of growth for hydrogen Lyα. Equivalent width W in Å versus column density N. Linear, flat and damping parts of the curve of growth are indicated.

same ion with different f-values, then you can construct a "curve-of-growth" and deduce the column density, N, and temperature, T, of the ions in the absorbing cloud.

A theoretical curve-of-growth is show in Figure 3. On the linear part of the curve of growth, the equivalent width increases linearly with increasing column density. On the flat part of the curve of growth, the lines are saturated at line center. The equivalent width is then insensitive to column density but has some sensitivity to temperature, or equivalently, the b-value of the line, where

$$b = \sqrt{\frac{2kT}{m}} \qquad (2.12)$$

and has units of km s^{-1}.

A convenient rule of thumb is that for a line on the linear part of the curve of growth

$$\log(\frac{W_\lambda}{\lambda}) = \log(\frac{N}{N_H}) + \log \lambda f + \log N_H - 20.053 \qquad (2.13)$$

where W_λ is the equivalent width, $\frac{N}{N_H}$ is the abundance of the atom relative to hydrogen, N_H is the total hydrogen column density and λ is the wavelength of the transition in Angstroms.

From equation 2.13 we see that ultraviolet absorption lines are very sensitive to small amounts of gas. For example, C III is the dominant ion for many interstellar clouds, and has a strong line at λ=977 Å, with $f = 0.768$, so that $\log \lambda f = 2.876$. The abundance, $\log$ C/H = -3.35 for solar abundance gas. One can easily detect a line with $W_\lambda = 0.1$Å, which corresponds to a log $N_H = 16.5$ cm^{-2}. This is a very small column compared to that easily probed by 21-cm or other emission line techniques. For hydrogen Lyα, column densities of log $N_H = 12.5$ cm^{-2} are easily measured.

2.1.3. Doublet Ratios

If a full curve-of-growth cannot be constructed, it is often sufficient to measure the *doublet ratio* of so-called H and K lines of a particular ion. These are pairs of lines which are near each other in wavelength so are easy to observe simultaneously, but far enough apart that they are easily resolved, even at moderate spectral resolution. Because they unambiguously identify redshift systems with two lines, doublets have been exploited to survey metal absorbers – particularly the very strong Mg II $\lambda\lambda$ 2803, 2796 Å and C IV $\lambda\lambda$ 1548, 1550 Å transitions. Other doublets include Si IV $\lambda\lambda$ 1393, 1402 Å N V $\lambda\lambda$ 1238, 1242Å O VI $\lambda\lambda$1031, 1037 Å. In the optical, Ca II H&K $\lambda\lambda$3933, 3968 Å and Na I D $\lambda\lambda$5889,5895 Å are doublets seen in absorption.

The shorter wavelength member (H) of the doublet has an f-value which is twice the f-value of the longer wavelength member (K) of the doublet. Thus, if both lines are on the linear part of the curve of growth, the ratio of the equivalent widths of the two lines will be exactly the ratio of the f-values, or 2. In this case, one can derive the column density from the equivalent width of either line, but not the b-value. At higher columns, the K line can be on the flat part of the curve of growth, with the H line is on the linear part. Then the ratio of the equivalent widths will be between 1 and 2. The column density may be derived from the equivalent width of the H line, and the b-value from the equivalent width of the K line. At still higher columns, both H and K are saturated. The equivalent widths of both lines are equal and the column density is not possible to derive (although one can measure a lower limit assuming a b-value). The doublet ratio and curve-of-growth are useful checks even if the spectra have sufficient resolution to fit Voigt profiles.

The wavelengths of doublets in redshifted quasar spectra can be used to investigate whether the fine structure constant, α, changes with cosmic epoch (Savedoff 1956; Bahcall, Sargent & Schmidt 1967; Wolfe, Brown & Roberts 1976; Ivanchik, Potekhin & Varshalovich 1999; Levshakov 1994; Cowie & Songaila 1995; Webb et al. 2000; Dzuba, Flambaum, Murphy & Webb 2001, Murphy et al. 2000abc and references therein). Murphy et al. (2000a) at al find a change significant at the 4σ level, with α being smaller in the past. The implications for physics are described in Murphy et al. (2000b). Various sources of systematic error may be important for assessing the security of this result (Murphy et al. 2000b). Radio observations of 21cm hyperfine splitting in redshifted absorption also put limits on the evolution of α (Drinkwater et al. 1998; Carilli et al. 2000 and references therein).

2.1.4. Limitations

Although ultraviolet absorption lines are remarkably sensitive probes of high redshift gas, there are several limitations to absorption line studies worth keeping in mind. Heavily reddened lines of sight are impossible to observe in the rest frame UV, and most absorption line studies have used quasars selected to be bright and blue. Note that half the mass of the interstellar medium in the Milky Way at the solar circle is in molecular clouds; these are completely inaccessible to ultraviolet absorption studies. Thus we expect quasar absorbers to select against reddened lines of sight (Malhotra 1997 and references therein). Likewise, very hot gas is also selected against, since the atoms are all ionized and so there are no accessible absorption lines.

Although many absorption features are well fit by a Voigt profile with some column density N and temperature T, the absorption is really a pencil beam average of the density and temperature along the line of sight. The volume density may be estimated by measuring fine-structure lines from excited states such as Si II * or C II* and comparing them to the columns of the ground state (Bahcall & Wolf 1968; Sarazin et al. 1979). If the excitation is caused by collisions, the ratio of the excited state columns to ground state columns will be sensitive to density. However, these lines are usually heavily saturated. Also, the excited states may be populated by direct excitation by infrared photons, or indirect excitation by ultraviolet photons into excited states which subsequently decay. Thus the interpretation of the observations requires some knowledge of the radiation field. In fact, usually the excited fine structure lines in quasar absorbers are used to estimate the radiation field, not the density (see below).

There is a small literature on the validity of curve of growth techniques and Voigt profile fitting generally, and their application to quasar absorbers in particular. At issue is the ability to properly account for multiple velocity components, particularly for saturated lines. The perils of applying the curve of growth to blended features are described by Nachman & Hobbs (1973), Crutcher (1975), Gómez-Gónzalez & Lequeux (1975), Parnell & Carswell 1988). A more optimistic view is described by Jenkins (1986). Levshakov and collaborators have written a series of papers emphasizing the effect of turbulent velocities on the interpretation of absorption line data (Levshakov & Kegel 1997; Levshakov et al. 1999 and references therein). As with all astronomical measurements, the limitations of the observations need to be understood and dealt with as best as possible.

Voigt profiles fit observed absorption line spectra quite well (Figure 4); however, they may be regarded as just one convenient parameterization of the data. The traditional approach has been to construct a list of lines, measure their redshifts, equivalent widths, column densities and b-values, and then base subsequent analysis on the line list. Alternatively, one can compare optical depths of the different transitions of the absorbing line profile on a pixel-by-pixel basis (Savage & Sembach, 1991). In the case of the intergalactic medium, it has been productive to mitigate the effects of blending by comparing the observed spectra on a pixel-to-pixel basis with the simulated "observations" of theoretical models for the intergalactic medium (Press & Rybicki 1993; Press, Rybicki & Schneider 1993; Dobrzycki & Bechtold 1996; Weinberg et al. 1999). Outram, Carswell & Theuns (2000) argue that Voigt profiles are a poor representation of Lyα forest lines and do not represent the proper physical picture. We discuss the Lyα forest further below.

2.2. Echelle Spectrographs

In order to obtain high resolution spectra (spectral resolution $R = \lambda/\Delta\lambda \sim 30,000$ or greater) astronomers use echelle spectrographs, where the dispersing element is a grating used in high spectral order and angle of diffraction. A second prism or grating serves as a cross-dispersing element which separates the closely spaced spectral orders perpendicular to the dispersion direction of the echelle grating. A good description of astronomical spectrographs is given in Schroeder's textbook *Astronomical Optics* (2000), and a review of the echelles planned or in use at large telescopes can be found in Pilachowski et al. (1995). Most of the results reviewed here were obtained with Vogt's HIRES spectrograph on the Keck 10m telescope (Vogt et al. 1994).

3. Absorption by Material Associated with the QSO

About 10% of QSOs show absorption attributable to gas associated with the QSO central engine or its immediate environs. In fact, the first quasar absorber reported turned out to be an "associated" C IV absorber (Sandage 1965). This category includes

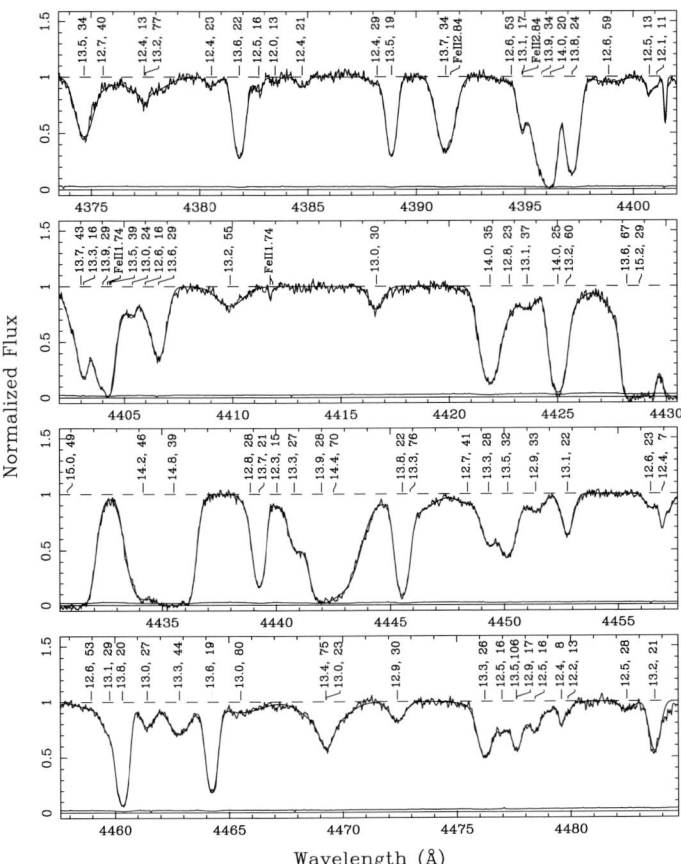

FIGURE 4. Voigt profile fits to a quasar spectrum. The spectrum is normalized to a continuum level of one. Wavelengths are vacuum heliocentric values. The log column density (in log cm^{-2}) and velocity dispersion (in km s^{-1}) of each line are displayed above a tick mark indicating the center of the line. Identified metal ions are indicated along with the redshift of the metal system. The calculated 1σ error spectrum (per pixel) is shown just above zero on the same scale. The pixel size is 2 km s^{-1}. From Kirkman & Tytler (1997).

the radiatively driven winds producing the broad absorption line (BAL) QSOs (Turnshek 1995 and references therein, see also Becker et al. 2000; Green et al. 2001), the associated C IV absorbers (Foltz et al. 1986, Anderson et al. 1987), the associated Mg II absorbers (Low, Cutri, Kleinman & Huchra 1989; Aldcroft, Bechtold & Elvis 1994), and the X-ray warm (that is, ionized) absorbers (Fiore et al. 1993, Mathur et al. 1994, George et al. 1998). An attempt to understand the various observations in terms of the unified model for the central engine is given by Elvis (2001). We have not reviewed the subject of associated absorption in radio galaxies here, but the interested reader can start with Arav, Shlosman & Weymann 1997.

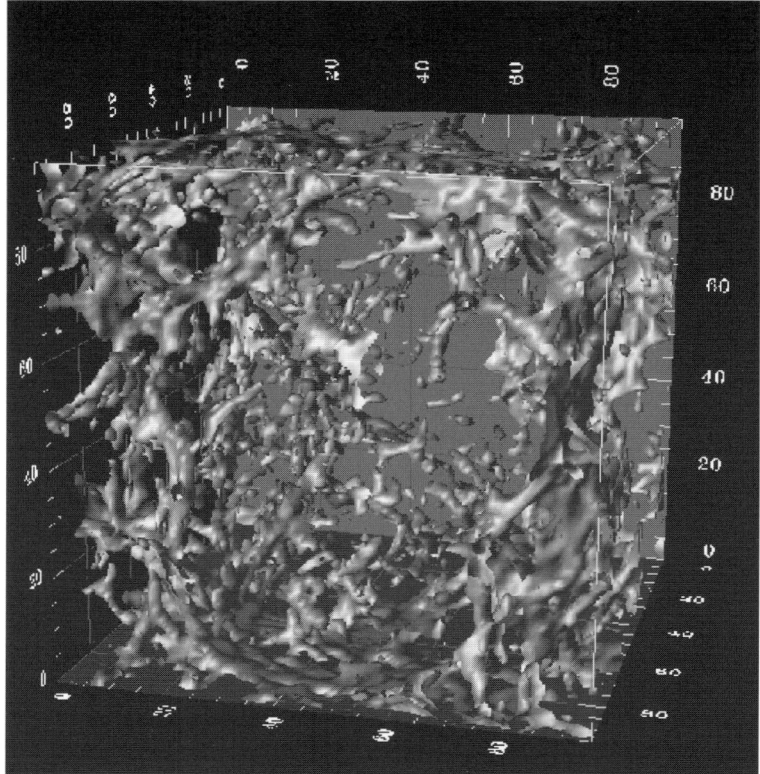

FIGURE 5. Simulated Lyα forest at $z = 3$ showing non-spherical geometry for the absorbing gas. Three-dimensional isodensity surfaces for baryon density three times the local mean. From Cen & Simcoe (1997).

4. The Lyman Alpha Forest

The techniques described above have been applied to large numbers of spectra of high redshift quasars in order to study the weak Lyα lines – those with log N(HI) $\sim$ 12.5-16.5 cm^{-2}. Prior to about 1995, the Lyα forest was defined as "Lyα lines with no metal absorption at the same redshift". However, it had long been suspected that the quasar absorbers discovered by their Lyα absorption should not be divided rigidly between "metal absorbers" and "forest" – that is, absorbers associated with galaxies and absorbers associated with intergalactic gas (e.g. Tytler 1987). The data obtained with 4m class telescopes would not have detected metal line absorption for the low column Lyα clouds even if they had solar metallicities – the strong lines of C III, O VI, C IV or Si IV were just too weak (Chaffee et al. 1986). When C IV and Si IV metal absorption was detected for the stronger Lyα forest lines (Cowie, Songaila, Kim & Hu 1995; Womble, Sargent & Lyongs 1996), the distinction between "Lyα forest" and "metal absorber" became less clear. However, as described below, the low column Lyα absorbers differ in several key observed properties from those at the opposite end of the column density distribution, the damped Lyα absorbers. The distinction between forest and metals may be a function of redshift as well.

Around the same time, numerical codes by several groups produced the first simulations of the evolution of the intergalactic medium and gas in galaxies at high redshift (Cen, Miralda-Escude, Ostriker & Rauch 1994; Zhang, Annions & Norman 1995; Hernquist et al. 1996; Wadsley & Bond 1996; Theuns et al. 1998). These allowed a detailed examination of the structure and dynamics of quasar absorbers, and produced a visual picture of what the structures are: typically sheets and filaments, aligned with the sites of future galaxies (see e.g. Figure 5). Here we focus on the observations, and the theoretical results which have shaped our understanding of the quasar absorbers.

4.1. The Column Density Distribution

One basic observed quantity is the distribution of column densities in the Lyα clouds. Voigt fits to high dispersion spectra have been used to derive the distribution from log $N(HI) = 12.3\text{-}14.5$ cm^{-2} (Hu et al. 1995; Lu, Sargent, Womble & Takada-Hidai 1996; Kim, Hu, Cowie & Songaila 1997; Kirkman & Tytler 1997). The distribution is a power-law, written

$$f(N_{HI}) = 4.9 \times 10^7 \, N_{HI}^{-1.46} \tag{4.14}$$

where $f(N_{HI})$ is defined as the number of absorbing systems per unit redshift path per unit column density as a function of neutral hydrogen column density, N_{HI}. The redshift path $X(z)$ is

$$X(z) = \frac{2}{3}[(1+z)^{3/2} - 1] \tag{4.15}$$

for $q_o = 0.5$. The data are shown in Figure 6.

Remarkably, the extension of this power law connects pretty well through the metal absorbers down to the damped Lyα absorbers, with log $N(HI) = 22.0$ cm^{-2} (Tytler 1987, Petitjean et al. 1993; Storrie-Lombardi & Wolfe 2000).

4.2. The b-value distribution

The distribution of Doppler b-values are roughly Gausians with a mean of about 30 km s^{-1} and width 8 km s^{-1} (Hu et al. 1995; Lu et al. 1996, Kim et al. 1997; Figure 7). The distributions must be corrected slightly for blending via analysis of simulated spectra. There is also a cut-off b-value, b_c, on the low side, indicating that below 15 km s^{-1} or so, no narrower Voigt components are needed to fit the observations.

There is some indication that the mean b and cut-off b decrease with increasing redshift (Williger et al. 1994; Lu et al. 1996; Kim et al. 1997; Schaye et al. 2000). On the face of it, this would suggest that the temperatures and/or the turbulent broadening of the absorbing gas is increasing with decreasing redshift.

However, there is more to the story (Schaye et al. 1999, 2000; Ricotti, Gnedin & Shull, 2000). On small spatial scales, the temperature of the gas is determined by a balance between photoionization heating by the UV background, and adiabatic cooling from the expansion of the Universe. As the Universe expands, the overdensity corresponding to a fixed observed column density decreases with redshift (Schaye et al. 1999, 2000; Ricotti, Gnedin & Shull 2000). The temperatures and densities obey what these authors call an "equation of state",

$$T = T_o(\frac{\rho}{\bar{\rho}})^{\gamma-1}. \tag{4.16}$$

If the IGM is reionized by sources of UV radiation quickly with respect to the timescales for cosmic expansion, the gas is isothermal and $\gamma \sim 1$. Subsequently, the mean temperature T_0 decreases as the Universe expands, and the slope γ increases because higher density regions expand less rapidly than average regions and photoionization heating is

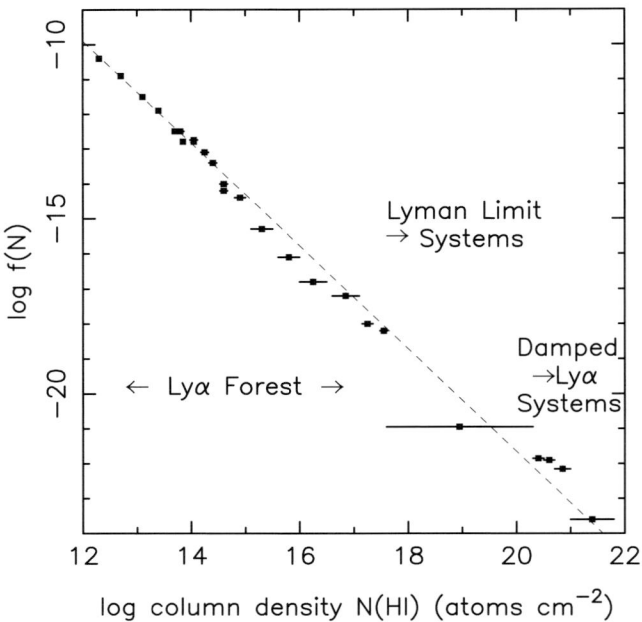

FIGURE 6. The column density distribution function of neutral hydrogen for $12 \leq \log N(HI) \leq 22$, for the Ly$\alpha$ forest, Lyman limit systems and the damped Lyα absorbers. The dashed line is a power law, $f(N) \propto N^{-1.46}$, which fits the data reasonably well, over 10 orders of magnitude in column density. From Storrie-Lombardi & Wolfe (2000).

more effective. The data show that the gas was nearly isothermal at $z \sim 3$, corresponding to the epoch of He II reionization (seen in the observations, section 4.8 below).

4.3. Evolution with Redshift

The number of lines per redshift observed, dN/dz, can be written

$$\frac{dN}{dz} = \frac{cn_o(z)\sigma(z)}{H_o}\frac{(1+z)}{(1+q_o z)^{1/2}} \tag{4.17}$$

where
$n_o(z)$ = the comoving number density of absorbers, and
$\sigma(z)$ = the geometric cross-section for absorbtion.
Note that

$$n_o(z)\sigma(z) = \frac{1}{l} \tag{4.18}$$

where l = the mean free path for absorption.

For objects with no intrinsic evolution of $n_o(z)\sigma(z)$, and $\Lambda=0$, we have

$$\frac{dN}{dz} \propto 1+z \tag{4.19}$$

for $q_0 = 0$ and

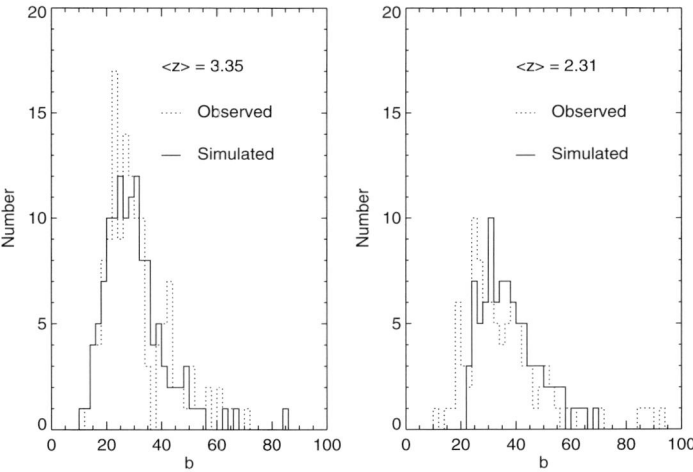

FIGURE 7. The b-value distribution of the Lyα forest. Observed (dotted) and simulated (solid) histograms of the b-values derived from Voigt profile fitting of Lyα forest clouds. From Kim et al. (1997).

$$\frac{dN}{dz} \propto (1+z)^{1/2} \qquad (4.20)$$

for $q_0 = 0.5$. The observations can be fit with a function

$$\frac{dN}{dz} = N_o(1+z)^\gamma \qquad (4.21)$$

The results are shown in Figure 8. The extension of dN/dz from $z = 2$, observable from the ground, to $z = 0$ was one of the long-anticipated results of the HST quasar absorption line Key Project (Bahcall et al. 1993; Schneider et al. 1993; Savage et al. 1993; Bergeron et al. 1994; Stengler-Larrea et al. 1995; Bahcall et al. 1996; Jannuzi et al. 1998; Weymann et al. 1998; Savage et al. 2000). Owing to the small numbers of lines in early ground-based samples, it has become customary to fit the observed dN/dz without binning into redshift bins. The best fit γ for lines stronger than some threshold equivalent width is solved for using maximum likelihood techniques (Murdoch, Hunstead, Pettini & Blades 1986; Weymann et al. 1998).

Figure 8 shows that for $z > 2$, $\gamma \sim 2-3$, that is the evolution is very steep, in the sense that there was more absorption in the past. There is some indication that γ steepens for $z > 4$ (Williger et al. 1994). At $z = 1 - 2$ the distribution flattens, and at $z < 1.5$, dN/dz is consistent with "no evolution".

Qualitatively, the evolution can be understood in terms of three main factors. One expects that the collapse of structures and the mergers of protogalactic fragments decreases both n_o and σ, so that dN/dz decreases with z. Second, the UV radiation field is expected to decrease from $z \sim 2$ to $z = 0$, whether it is comprised of photons from quasars or star-forming galaxies. Thus as the UV field decreases the neutral fraction increases and the number of detected lines increases. This tendency is balanced by the third effect, which is that as cosmic expansion proceeds, the density of the gas decreases

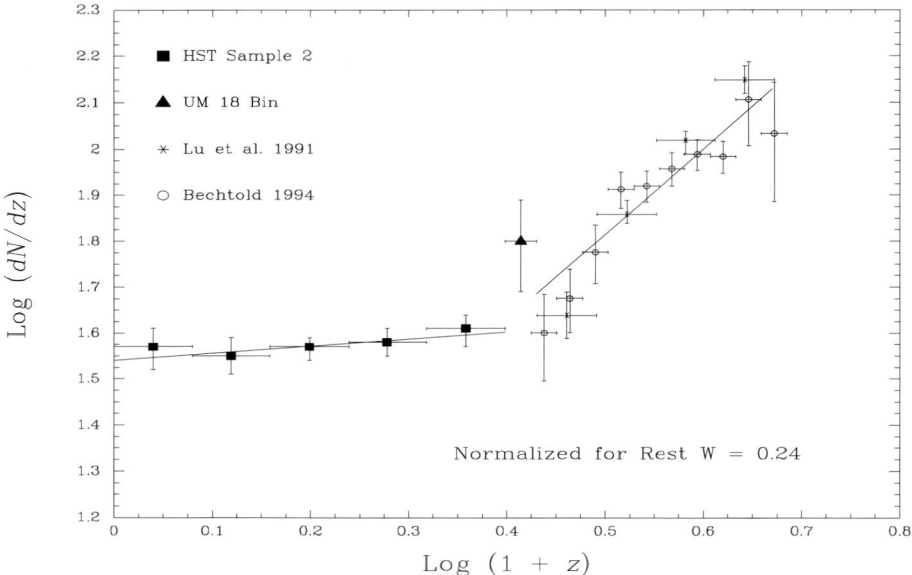

FIGURE 8. Evolution of the Lyα forest. At $z < 1.6$ the data come from the HST Key Project catalogue of Lyα absorbers, Bahcall et al. 1996, Jannuzi et al. 1998. At higher redshifts two ground-based samples are shown (Lu, Wolfe & Turnshek 1991; Bechtold 1994). Results from high dispersion data (Kim et al. 1997) are similar. The separate fits to the HST and ground-based data are shown as the two solid lines, with slopes of $\gamma = 0.5$ and 1.85 respectively. From Weymann et al. (1998).

and even if the photoionizing UV field were constant, the neutral fraction would decrease and the number of Lyman alpha lines detected will increase. Undoubtedly the evolution at low z is even more complicated than this.

4.4. The UV Radiation Field and the Proximity Effect

A key quantity for understanding the Lyα forest is the ionizing ultraviolet radiation field. Arons & McCray (1969) and Rees and Setti (1970) suggested that quasars are likely to be a significant source of hard UV photons at high redshifts. Star forming galaxies may also contribute (Bechtold et al. 1987; Haardt & Madau 1996; Steidel, Pettini & Adelberger 2001). The number counts of Lyα forest clouds themselves provide an independent way to estimate the UV background by analysis of the so-called proximity effect.

Weymann, Carswell & Smith (1981) first pointed out that the number density of Lyα forest lines is systematically low near the quasar being used to probe them. They suggested that photoionization by the quasar's UV light might be responsible. Carswell et al. (1987) pointed out that this effect could be used to estimate the ambient UV background. Qualitatively, the extent of the proximity effect indicates roughly how far the ionizing radiation from the quasar exceeds that of the background. By measuring the brightness of the quasar (often the Lyman limit flux is directly observable), one can therefore estimate the value of the specific intensity of the background, J_ν.

Initially, the estimates of J_ν derived from proximity effect analysis of various high redshift samples indicated high values. For example, Bajtlik, Duncan & Ostriker (1988) found that log $J_\nu = -21.0 \pm 0.5$ erg s^{-1} cm^{-2} Hz^{-1} sr^{-1} for the redshift range of their sample, $1.7 < z < 3.8$. This was significantly larger than estimates of the quasar background, particularly at $z > 3$ (e.g. Bechtold et al. 1987). Bajtlik, Duncan & Ostriker

(1988) suggested that there was another source of ionizing radiation, young hot stars in galaxies (Bechtold et al. 1987, Miralda-Escude & Ostriker 1990) or a population of quasars which we don't see today because they are obscured by dust in intervening galaxies (Ostriker & Heisler 1984; Wright 1986; Heisler & Ostriker 1988ab; Boisse & Bergeron 1988; Najita, Silk & Wachter 1990; Wright 1990; Ostriker, Vogeley & York 1990; Fall, Pei & McMahon 1989; Pei, Fall & Bechtold 1991; Pei & Fall 1995; Pei, Fall & Hauser 1999).

Subsequently, many quasars with z>3 were found, and the luminosity function for high redshift quasars became better defined. The value for the quasar contribution to the background increased (Miralde-Escude & Ostriker 1990; Haardt & Madau 1992), and is now much closer to the proximity effect estimates. Also, the sources of various systematic effects in proximity effect estimates of J_ν have been dealt with (Bechtold 1995; Scott, Bechtold & Dobrzycki 2000a). The proximity effect estimates of J_ν are now in broad agreement with the predictions of the contribution of known quasars to the ionizing background (Bechtold 1994, Lu et al. 1991; Scott et al. 2000b).

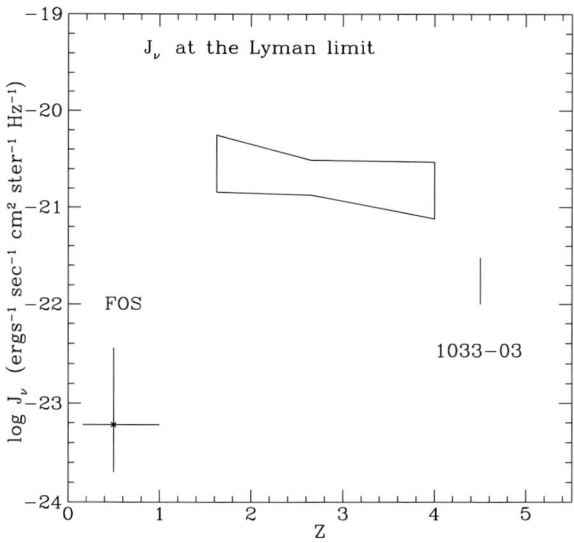

FIGURE 9. Evolution of the ionizing UV radiation field. J_ν, as derived from the proximity effect, as a function of redshift, z. From Bechtold (1995).

At $z < 2$, one expects J_ν to decline because of the decline in the population of quasars and star-bursting galaxies. Kulkarni & Fall (1993) derived a value based on early HST Key Project samples, and a larger sample has been analyzed recently by Scott et al. (2001). A summary of J_ν at the Lyman limit as derived from the proximity effect is shown in Figure 9. While the uncertainties are still large, broadly speaking, the results are consistent with the predicted contribution from quasars. The general trend that the ionizing UV radiation field is smaller at $z \sim 0$ and $z \sim 4.5$ than at $z \sim 3$ is probably secure.

If quasars happen to lie along the line-of-sight they could cause observable voids in the Lyα forest far from the quasar (Bajtlik, Duncan & Ostriker 1988). Although luminous quasars are rare objects, the path length sampled by a typical quasar spectrum is long, and such chance coincidences are common enough to be interesting. Kovner & Rees (1989) discussed possible "clearings" in the Lyα forest from the "two quasar proximity effect" and concluded that the unsuccessful searches for voids in the Lyα forest up to that time were consistent with expectations (Carswell & Rees 1987; Ostriker, Bajtlik & Duncan 1988; Duncan, Ostriker & Bajtlik 1989). In principle, one could use the "two quasar proximity effect" to estimate quasar lifetimes since a population of numerous short-lived ionizing sources would cause fewer large voids than rare, long-lived ones. Also, if the UV light from quasars is strongly beamed, one would expect to see examples of bright quasars along the line of sight with no associated proximity-induced void, or voids in the Lyα forest with no observable quasar to cause them. Of course voids could also be present from real underdensities in the total hydrogen gas, and these would be difficult to distinguish from a proximity-induced void. Dobrzycki & Bechtold (1991ab) found one significant void in the Lyα forest. There is a known quasar nearly but not *quite* at the right redshift, close enough to the line-of-sight, with the right brightness, to cause it. Subsequent searches for a more suitable ionizing quasar have been fruitless. Fernandez-Soto et al. (1995) looked for proximity-induced voids in the spectra of 3 quasars with known quasars along the line-of-sight, but with only 3 objects, they were not able to tell one way or the other whether significant voids are present.

Zuo (1992ab) generalized these ideas and looked for underdense regions in the forest, rather than complete voids. He pointed out that in principle, the nature of the sources of the ionizing background could be deduced from the statistics of underdense regions. A fixed *integrated* background could be made up of rare, bright sources (e.g. luminous quasars) or numerous, fainter ones (e.g. star-forming galaxies): these would cause different signatures in the distribution of Lyα clouds through the proximity effect. Unfortunately, the data available did not allow strong statements to be made.

4.5. *Characteristic Size and Geometry*

The characteristic size and geometry of the absorbing structures can be estimated by searching for common absorption in the spectra of pairs or groups of quasars which are serendipitously near each other on the sky. Individual images of gravitationally lensed quasars can also be used, provided the lensing geometry is known, since each image traverses a different path through space.

Lenses probe very small separations, and generally show that all the lines seen in one image are seen in the other, with very strongly correlated equivalent widths (Young et al. 1981; Weymann & Foltz 1983; Foltz, Weymann, Roser & Chaffee 1984; Smette et al. 1992, 1995; Turnshek & Bohlin 1993; Bechtold & Yee 1995; Zuo et al. 1997; Michalitsianos et al. 1997; Petry, Impey & Foltz 1998; Lopez et al. 1999; Rauch, Sargent & Barlow 1999). Thus the characteristic sizes must be larger than the typical separations. For very wide pairs of quasars, where the projected separation is 20" to arcminutes, few if any lines are seen in common (Sargent et al. 1980; Crotts & Fang 1998). Thus the sizes are smaller than the separations probed by large pairs. Pairs and groups of quasars with separations of 10" to 1 arcminute contain lines in common, and lines which are not in common, and so are probing the characteristic sizes of the absorbing structures (Bechtold et al. 1994; Dinshaw et al. 1994, 1995; Fang et al. 1996; Monier, Turnshek & Lupie 1998; Monier, Turnshek & Hazard 1999).

Unfortunately, the number of favorable groupings of quasars bright enough to observe is extremely small, since luminous quasars are themselves quite rare objects. One of the useful products of the Sloan digital Sky Survey will be the discovery of groups and pairs of quasars to observe for common absorption. Nevertheless, it is clear from existing observations that the Lyα clouds have sizes 200-500 h_{100}^{-1} kpc, which are very large.

Note that common absorption measures essentially *transverse* sizes. Rauch & Haehnelt (1995) pointed out that if the clouds are spherical and ionized as expected by the UV radiation field of quasars (so that the measured N(HI) is a small fraction of the total H), then the mass density of the forest would exceed the baryon density derived from primordial nucleosynthesis. Therefore, the geometry is more likely sheetlike, as suggested already by numerical simulations.

4.6. *Clustering of the Lyα forest clouds*

One of the early results which distinguished the Lyα forest clouds from higher column density clouds was the apparent weak clustering in their redshift distribution (Sargent et al. 1980; Rauch et al. 1992; Hu et al. 1995; Cristiani et al. 1995; Lu et al. 1996; Fernandez-Soto et al. 1996; Kirkman & Tytler 1997).

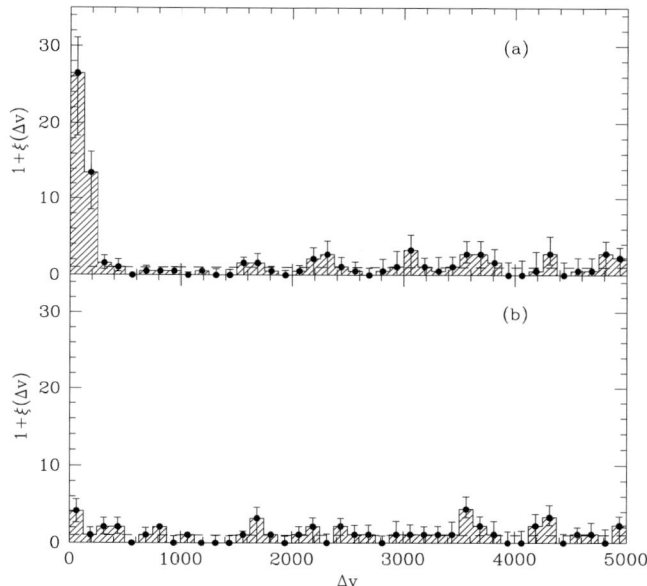

FIGURE 10. Clustering of absorbers at $z \sim 2.5$. Two-point correlation function as a function of velocity splitting, Δv, for C IV absorbers (a), and Lyα absorbers (b). From Fernandez-Soto et al. (1996).

However, the b-values of the lines are ~ 25 km s^{-1}, compared to the expected velocity dispersion of collapsed protogalaxies of perhaps a few hundred km s^{-1}. Thus, any signal in the two-point correlation function may be weakened by blending. For the metals with b-values of only a few km s^{-1} it is easier to resolve separate components with small velocity splittings, so signal in the two point correlation function is easier to detect. Results for C IV absorbers and Lyα absorbers are shown in Figure 10. In either case, the clustering amplitude is small compared to that of local galaxies.

At low redshift, the *HST FOS* data set has been analyzed by Ulmer (1996) who claims that there is strong clustering of the forest lines, but this claim has been questioned by Dobrzycki et al. (2001) who found weaker clustering in a larger sample. The Key Project made the interesting claim that Lyα forest lines were correlated with metal absorbers, but this was based on a very small number of objects, and statistical analysis of a larger sample again failed to confirm this result.

4.7. *The Low Redshift Lyα Forest*

One of the most anticipated observations to be made with HST was the search for Lyα forest clouds at low redshift, so one could carry out a detailed study of the relationship between the clouds and galaxies. A number of authors carried out deep galaxy redshift surveys in the directions of quasars with high quality HST FOS and GHRS spectra, both optically (Morris et al. 1993; Stocke et al. 1995; Lanzetta et al. 1995; Shull, Stocke & Penton 1996; Chen et al. 1998; Tripp, Lu & Savage 1998; Ortiz-Gil et al. 1999; Penton, Stocke & Shull 2000; Penton, Shull & Stocke 2000; Figure 11) and with 21-cm techniques (Van Gorkom et al. 1996).

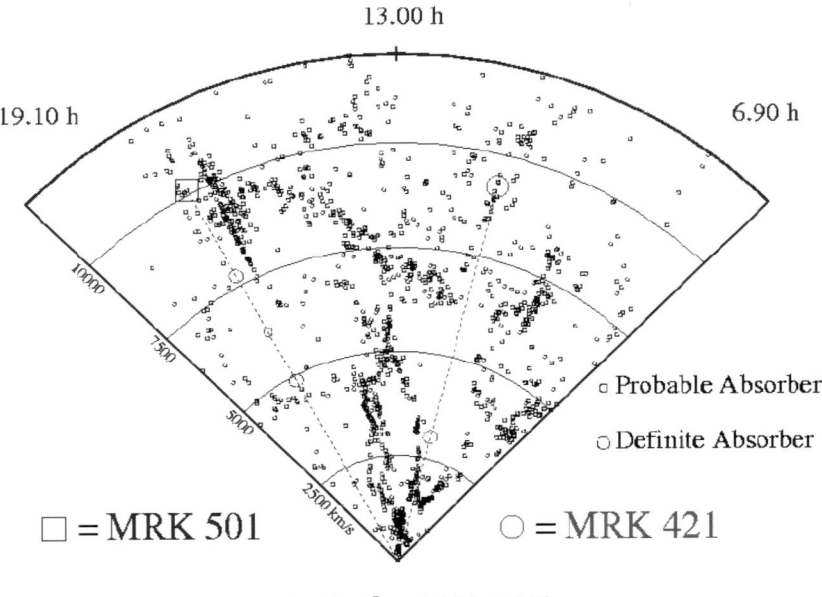

FIGURE 11. Pie-diagram distributions of recession velocity and right ascension of bright (CfA survey) galaxies and four Lyα absorbers toward Mrk 501 and Mrk 421. Two of these systems lie in voids; the nearest bright galaxies lie $> 4h_{75}^{-1}$ Mpc from the absorber. From Shull, Penton & Stocke (1999).

The results show that the forest clouds at low redshift arise in a variety of places. The clouds show some correlation with galaxies but are not as strongly correlated as galaxies are to each other (Stocke et al. 1995). There are some clouds at redshifts which place them in voids of the galaxy distribution. Others may be associated with tidal debris of galaxies or groups of galaxies. Although Lanzetta et al. (1995) have argued that the Lyα clouds arise in huge halos of individual galaxies, a consensus is growing that they are associated more often with structures in the galaxy distribution, such as superclusters and sheets.

4.8. The Gunn-Peterson Effect; Metals and the Ionization of the IGM

Gunn & Peterson (1965) pointed out that any generally distributed neutral hydrogen would produce a broad depression in the spectrum of high redshift quasars at wavelengths shortward of Lyα $\lambda 1215$. Since such a depression is not seen, they concluded that the intergalactic medium must be ionized – most of the hydrogen is H II. In the current view, the intergalactic medium is not uniform – it has clumped into sheets and filaments that we call the Lyα forest. Several authors have used high resolution data to put limits on any diffuse intercloud medium (Steidel & Sargent 1987; Webb, Barcons, Carswell & Parnell 1992; Giallongo et al. 1994, Songaila et al. 1999).

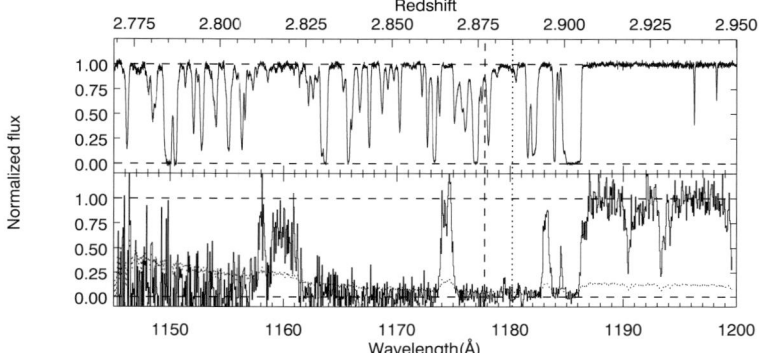

FIGURE 12. Spectra of HE 2347–4342 showing variation of HI/He II at $z \sim 2.8$. Top: Keck-HIRES spectrum of H I Lyα. Bottom: *HST*/STIS FUV MAMA G140M spectrum of He II. In both panels, the vertical dashed line at ~ 1178 Å corresponds to $z = 2.877$, redwards of which the spectra are likely to be affected by the known $z_{abs} \simeq z_{em}$ systems. From Smette et al. (2001).

Direct information on the ionization of the intergalactic gas may be derived from observing the neutral and singly ionized helium Lyα lines at $\lambda 584$ and $\lambda 304$, respectively. Tripp, Green & Bechtold (1990) used IUE to put an upper limit on He I in the IGM using the lack of a depression at $\lambda 584$ in a high redshift quasar. However, if the quasar radiation field is ionizing the IGM, most helium is expected to be He II or He III.

He II $\lambda 304$ is difficult to observe because it is so far in the far ultraviolet. The redshift of the quasar must be at least 3 to redshift this feature into the HST range, and at least 2 to redshift the feature into the HUT/FUSE range. Most $z = 2 - 3$ quasars have optically thick Lyman limits at other redshifts which wipe out the continuum so that $\lambda 304$ is not observable. To observe He II, one must survey many quasars to find the one or two which happen to be clear of Lyman limit absorption.

Jakobsen et al. (1994) succeeded in observing the He II $\lambda 304$ region in the $z = 3.2$ quasar Q 0302-003 with FOC on HST and found a sharp cut-off in the continuum flux – implying $\tau_{HeII} > 3$. Tytler et al. (1995) used FOS to observe a second $z = 3$ quasar, PKS 1935-692 and also saw that He II was optically thick. Remarkably, Davidson et al. (1996) used HUT to observe HS1700+6416, and found that He II was not completely absorbed at $z = 2.4$, $\tau_{HeII} = 1$. This would imply that the He II IGM opacity changed abruptly at $z \sim 2.7$. Subsequently, Reimers et al. (1997) used GHRS on HST to observe a fourth quasar, HE 2347-4342, and found that at $z \sim 2.8$ the He II ionization is patchy – that is, there is significant variation in H I / He II. Recent reobservation of these object with STIS (Heap et al. 2000; Smette et al. 2001) and FUSE (Kriss et al. 2001) confirm this – see Figure 12.

More information about the ionization state of the Lyα forest at high redshift was found when weak C IV, Si IV and O VI absorption lines were detected for Lyα forest lines (Songaila & Cowie 1996; Lu et al. 1996; Figure 13). Remarkably, about 75% of the Lyα lines with $N(HI) > 10^{14.5}$ cm^{-2} show C IV absorption. The inferred metallicity is low, a factor 100 – 1000 less than solar. Whether there is a change in metallicity with redshift is a difficult question to answer, owing to the large uncertainty in the ionization corrections. Also it is not clear whether the Lyα lines with lower columns have similar metallicities or not (Ellison et al. 2000; Schaye et al. 2000). Nonetheless, the discovery of metals associated with such low column density absorption means that some sort of chemical enrichment of the IGM has taken place by $z = 3$.

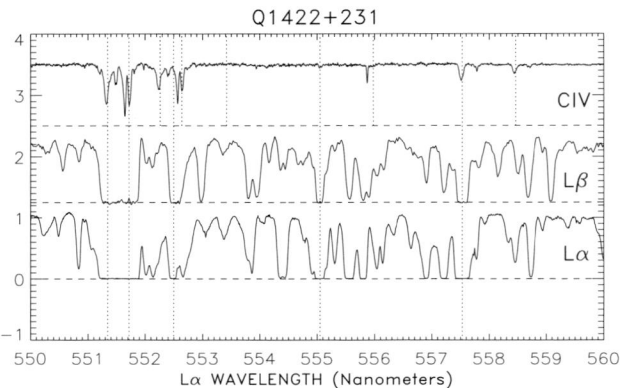

FIGURE 13. A 100Å portion of the spectrum of Q1422+231 (bottom) with the corresponding Lyβ (middle) and C IV λ1548Å (top) shifted in wavelength so they lie above Lyα. Dotted vertical lines indicate clouds that are saturated in both Lyβ and Lyα. Short dotted lines show the position of C IV λ1550Å corresponding to these clouds. From Songaila & Cowie (1996).

At low redshift, O VI absorption may be common as well (Bergeron et al. 1994; Burles & Tytler 1996; Verner, Tytler & Barthel 1994; Davé et al. 1998; Tripp, Savage & Jenkins 2000; Tripp & Savage 2000; Reimers et al. 2001; Figure 14) and the warm gas which produces O VI may be an important component of the baryons. The low redshift systems may not be simply photoionized by the ambient UV background, however, but may represent shock heated gas associated with infalling material in groups and clusters of galaxies.

After this review was completed, Becker et al. (2001) presented spectra of the first z~6 quasars, discovered in the Sloan Survey (see http://www.sdss.org). The very strong Lyα absorption in the highest redshift quasar now known, at $z = 6.28$, suggests that the IGM was reionized at $z \sim 6$ (Figure 15). This relatively low redshift for reionization has interesting implications for formation scenarios for galaxies, and strategies for deep surveys with NGST. Also, since early reionization would have erased more of the cosmic microwave background anisotropies (Ostriker & Vishniac 1986; Hu & White 1997; Liddle & Lyth 2000 and references therein), reionization at $z \sim 6$ means that more cosmological information may potentially be derived from satellite experiments such as MAP and Planck.

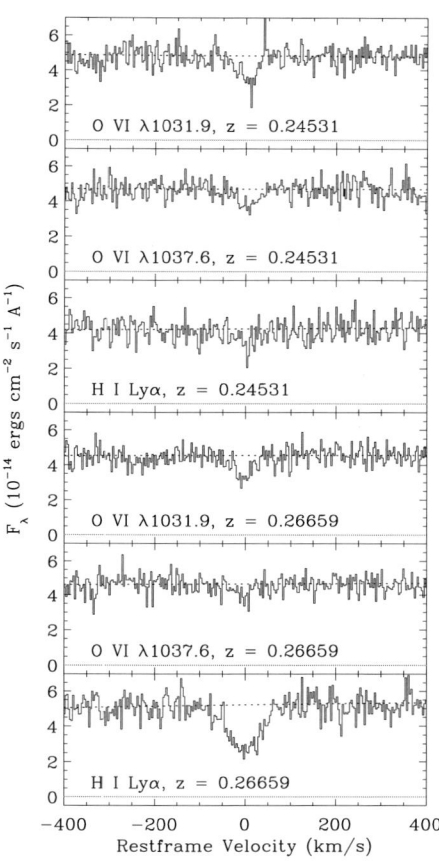

FIGURE 14. Portion of the STIS E140M spectrum of H1821+643 showing O VI absorption at $z(abs)=0.24531$ (upper three panels) and at $z(abs) = 0.26659$ (lower three panels), plotted versus restframe velocity. From Tripp, Savage & Jenkins (2000).

4.9. *Simulations of the IGM*

The first hydrodynamic cosmological simulations with sufficient resolution and dynamic range to model the intergalactic gas showed that the Lyα forest is a natural consequence of the growth of structure in the CDM hierarchical gravitational collapse scenario. All models were remarkably successful in reproducing the basic observations (Cen et al. 1994; Zhang, Anninos & Norman 1995; Hernquist et al. 1996; Miralda-Escude, Cen, Ostriker & Rauch 1996; Weinberg, Hernquist & Katz 1997; Wadsley & Bond 1996; Theuns et al. 1998; Haehnelt & Steinmetz 1998), including the column density distribution and b-value distribution. There were few free parameters – an essential one being the assumed UV radiation field which photoionizes the clouds.

At high redshifts, $z \sim 2-5$, the forest is produced by gas with low overdensities, $\rho/\bar{\rho} \leq 10$, photoionized by a diffuse ultraviolet background radiation field. Virtually all the baryons collapse into structures which can be identified with observable lines – there is very little diffusely distributed gas, or Gunn-Peterson effect. The high column density lines ($N_{HI} > 10^{17}$ cm^{-2}) are identified with high density regions at the intersections of

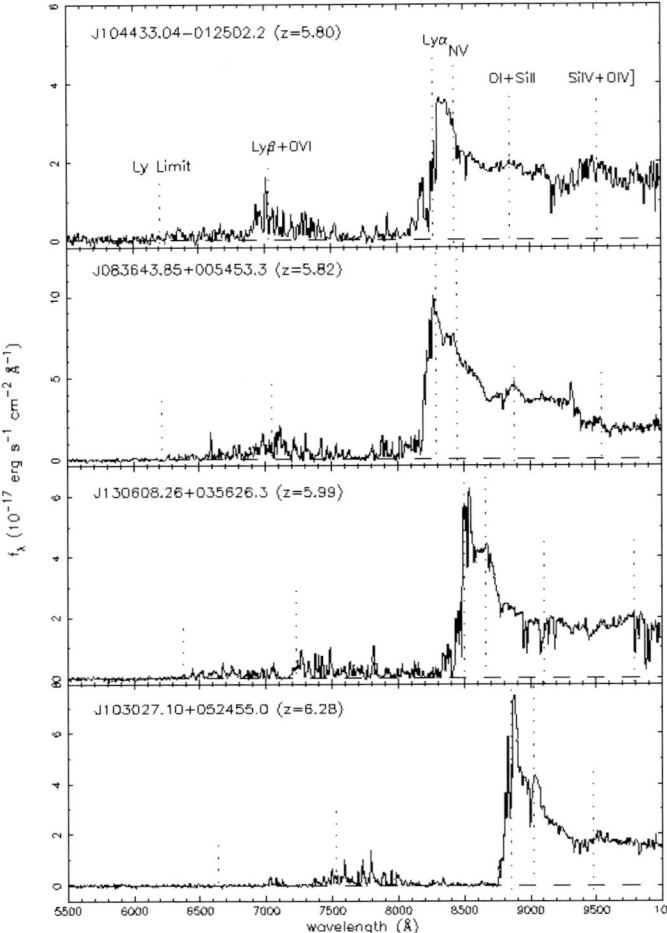

FIGURE 15. Optical spectra of $z > 5$, 8 quasars observed with Keck/ESI, in the observed frame. The strong absorption in the $z = 6.28$ quasar suggests that reionization of the intergalactic medium took place at $z \sim 6$. From Becker et al. (2001).

sheets and filaments, where the density is high and star-formation is presumed to have started. The low column density lines ($N_{HI} < 10^{12} - 10^{15}$ cm^{-2}) arise from a range of physical conditions and scales, at different stages of collapse – including warm gas in filaments and sheets, cool gas near shocks, and underdense regions megaparsecs across (voids).

Since the overdensities producing the Lyα forest lines are only close to linear, they can also be approximated with semi-analytical models (McGill 1990; Bi & Davidsen 1997; Hui, Gnedin & Zhang 1997). These are easier to calculate than the full 3-dimensional models.

The success of the models by different groups using different detailed cosmological assumptions meant that the observed properties of the Lyα forest clouds that were being used for comparison depended only on the assumptions that are common to all models, that is, the forest provides little leverage to distinguish among them (see Weinberg et al. 1999 and references therein, also Machacek et al. 2000). However, the forest observations do allow one to measure the mass power spectrum, $P(k)$, of fluctuations (Croft et al. 1999), the mass density parameter Ω_M (Weinberg et al. 1999) and the spectrum of the UV radiation field which turned out to be surprisingly soft (Zhang et al. 1997; Rauch et al. 1997; Zhang et al. 1998).

At $z < 2$ the models of the intergalactic medium must take into account that an increasing fraction of the baryon mass is participating in star-formation (that is, has collapsed into galaxies), or has fallen into the cores of clusters of galaxies and been heated to X-ray temperatures. Details of gas dynamics, radiative cooling, and "feedback" or galactic winds become important. One salient feature at low redshift, the change in dN/dz evolution at $z \leq 2$ is easily reproduced by a modest change in Jν, expected from the changing luminosity function of quasars and starburst galaxies. Implications derived from simulations and a discussion of the observations is given by Davé et al. (1999).

5. Damped Lyman Alpha Absorbers and other Metal-Line Absorbers

The look-back time for quasar absorbers is comparable to the age of the oldest stars in the Milky Way, so we expect that quasar absorbers will arise in gas associated with galaxies in their earliest stages of formation. The damped Lyα absorbers have proven particularly valuable for look-back studies, since the strong absorption of Lyα may be detected with moderate spectral resolution data, and the damping profile allows an accurate measurement of $N(HI)$, a first step in metallicity studies. Art Wolfe and collaborators have undertaken extensive surveys to find damped Lyα absorbers and study their properties (Wolfe, Turnshek, Smith & Cohen 1986; Turnshek et al. 1989; Lanzetta et al. 1991; Wolfe, Turnshek, Lanzetta & Lu 1993; Lu, Wolfe, Turnshek & Lanzetta 1993; Lu & Wolfe 1994; Lanzetta, Wolfe, & Turnshek 1995; Storrie-Lombardi & Wolfe 2000). Other metal absorption surveys include those for Mg II (Lanzetta, Wolfe & Turnshek 1987; Sargent, Steidel & Boksenberg 1988a; Barthel, Tytler & Thomson 1990; Steidel & Sargent 1992; Aldcroft, Bechtold & Elvis 1994; Churchill, Rigby, Charleton & Vogt 1999), C IV (Sargent, Steidel & Boksenberg 1988b; Foltz et al. 1986) and Lyman limits (Tytler 1982; Bechtold et al. 1984; Sargent et al. 1989; Stengler-Larrea et al. 1995). A heterogeneous catalog of absorbers from the literature, which is nonetheless useful for some purposes, has been assembled by York et al. (1991).

5.1. *The Mean Free Path for Absorption and its Evolution*

The evolution of the number of metal absorbers per redshift can be parameterized by γ, for the Lyα forest clouds. In contrast to the steep evolution seen for the low column forest clouds, the Lyman limit absorbers and damped Ly α absorbers are consistent with little if any evolution from $z = 0$ to $z = 5$. Figure 16 shows the evolution of the number of absorbers per redshift versus redshift.

Assuming that the density of absorbers is given by the galaxy luminosity function, and that the cross-section for absorption, σ_o, scales with luminosity, it is straight-forward to calculate how big the gaseous extent of a typical (say L*) galaxy must be in order to account for the number of absorbers seen in quasar spectra (e.g. Burbidge et al. 1977). The relatively large sizes led to the notion that the quasar absorbers arise in huge gaseous halos. While at low redshift this may be a valid interpretation, at high redshift, the fact that the number density of absorbers is changing must be included. In light of other results which suggest that large galaxies assembled out of numerous protogalactic fragments at $z < 3$ (reviewed by Dickinson at this school), the statistics probably indicate instead that mergers of PGFs have taken place.

The evolution of the damped Lyα absorbers is of interest, because at low redshift

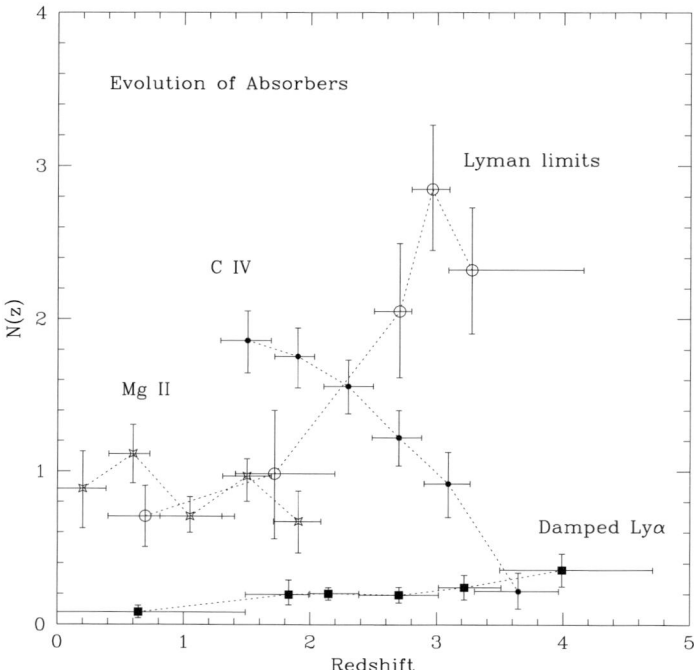

FIGURE 16. Number of absorbers per unit redshift, N(z), as a function of redshift, for damped Lyα systems (filled square), Mg II absorbers (cross), C IV absorbers (filled circles) and Lyman limits (open circles). Data from Storrie-Lombardi & Wolfe (2000), Stengler-Larrea et al. (1995), and the compilation of York et al. (1991).

they appear to arise in normal galaxies, and at high redshift, they appear to contain a large fraction of the available baryon density of the Universe (Lanzetta et al. 1991; Rao, Turnshek & Briggs 1995; Rao & Turnshek 2000; Storrie-Lombardi & Wolfe 2000 and references therein).

The column density distribution of the damped Lyα may be integrated to estimate the comoving mass density of the gas in units of the current critical density,

$$\Omega_g(z) = \frac{H_0}{c} \frac{\mu m_H}{\rho_{crit}} \frac{\sum_i N_i(HI)}{\Delta X(z)}. \tag{5.22}$$

Here μ is the mean particle mass per m_H where the latter is the mass of the H atom, ρ_{crit} is the current critical mass density, and the sum is over damped Lyα systems in the sample surveyed over distance X(z), given by Equation 4.15. $\Omega_g(z)$ is somewhat sensitive to the column density of the highest column density absorber in the sample, and there is a strong selection *against* the high column density absorbers in UV studies because of reddening. However, the results, shown in Figure 17, are interesting. The evolution of Ω_g as measured by quasar absorbers is compared to the mass density of stars in nearby galaxies, as estimated by Fukugita, Hogan & Peebles (1998). The point at $z = 0$ is from a 21cm emission survey of local galaxies (Zwaan et al. 1997). The suggestion is that the $z \sim 3$ damped Lyα gas has been converted into the stars we see in local galaxies today.

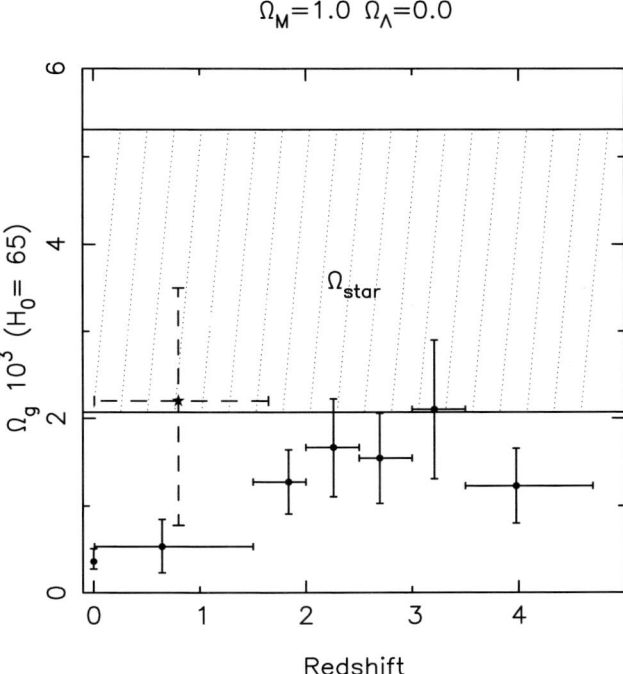

FIGURE 17. For $\Omega_{matter} = 1$, $\Omega_\Lambda = 0$, the comoving mass density in neutral gas contributed by damped Lyα absorbers, $\Omega_g(z)$, compared to Ω_{star}, the comoving mass density of stars. The dashed lines are a determination $\Omega_g(z)$ from the survey of Rao & Turnshek (2000) for $z < 1.65$. From Storrie-Lombardi & Wolfe (2000).

5.2. *The Damped Lyman Alpha Absorbers: Metallicity and Dust*

The aim of abundance studies of quasar absorbers were described by Trimble (1995) as an investigation of Z(Z(Z)) – that is, the abundance as a function of scale height as a function of redshift. The damped Lyα absorbers are the ideal absorbers for abundance studies since they have high total columns, so that weak transitions are detectable of rare elements on the linear part of the curve of growth.

A serious complication of all abundance studies based on UV absorption is that the elements of interest suffer depletion – they freeze out onto dust grains, so the gas phase column density measured by the absorption lines is less than the actual abundance of the gas (see review by Jenkins 1995). In the Milky Way, the refractory elements (Si, Fe, Ca, Ti, Mn, Al and Ni) show depletion factors of 10-1000, and the exact pattern of depletion depends on the shock history of the gas. For C, N and O, the depletions are factors of 2 to 3. S, Zn and Ar are relatively undepleted, and therefore can be used to measure metallicities.

Pettini and collaborators did a comprehensive survey of Zn II in damped Lyα absorbers, using the William Herschel Telescope on La Palma (Pettini, Smith, Hunstead & King 1994, 1997; Pettini, King, Smith & Hunstead 1997; Pettini & Bowen 1997; Pettini, Ellison, Steidel & Bowen 1999; Pettini et al. 2000; see also Meyer, Welty & York 1989; Meyer & Roth 1990; Meyer & York 1992; Sembach, Steidel, Macke & Meyer 1995; Meyer, Lanzetta & Wolfe 1995). The result was that the zinc to hydrogen abundance is about

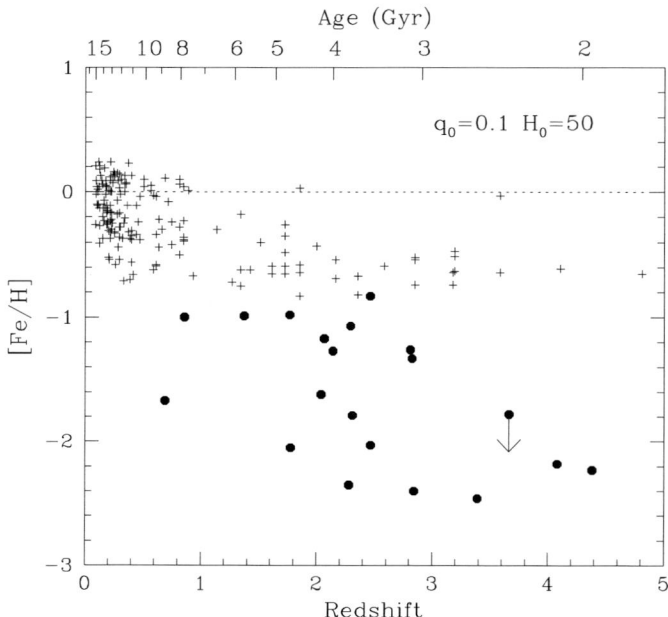

FIGURE 18. Distribution of [Fe/H] as a function of redshift for damped Lyα absorbers (solid circles). The age-metallicity relation for Galactic disk stars (Edvardsson et al. 1993) is shown with "+" symbols. The damped Lyα galaxies have Fe-metallicities in the range of 1/10 to 1/300 solar, or [Fe/H] between −1.0 and −2.5. From Lu et al. (1996).

1/10-1/50 solar, with a large scatter. They are generally more metal poor than Milky Way disk stars of the same age as the lookback time to the absorbers (Figure 18). The quasar absorbers are less metal-rich than thin or thick disk stars, but not as metal poor as halo stars or globular clusters in our Galaxy (Pettini et al. 1997; Figure 19). Zinc is an iron group element, so even though it is difficult to measure [Fe/H] directly, the evolution of the [Zn/H] ratio is probably indicative of evolution of [Fe/H].

Keck HIRES data allowed more detailed studies of the abundance patterns of the elements to be carried out (Prochaska & Wolfe 1996, 1997a, 1999; Lu et al. 1996). The overall result is that the extent of depletion onto dust grains is small, and the pattern of iron group to α rich nuclei suggests an origin similar to Population II Milky Way objects, which are enriched early by Type II supernovae. Ionization corrections may mimic halo α element enrichment, however, and should be accounted for (Howk & Sembach 1999).

The ratio of nitrogen to oxygen is also of interest, since whether or not N/O depends on O/H can distinguish between the "primary" or "secondary" production of nitrogen. Primary production of nitrogen results from the asymptotic branch phase of intermediate mass stars, which depends on CNO burning of seed carbon nuclei dredged up from the interior of the star – thus N/O should be independent of the initial metallicity of the star. If nitrogen is produced by the CNO cycle of main sequence stars ("secondary" production) then it depends on the initial metallicity. At low metallicity, one expects a large scatter in N/O because of the delay between the release of primary nitrogen from intermediate mass stars and oxygen from massive stars. The problem with quasar

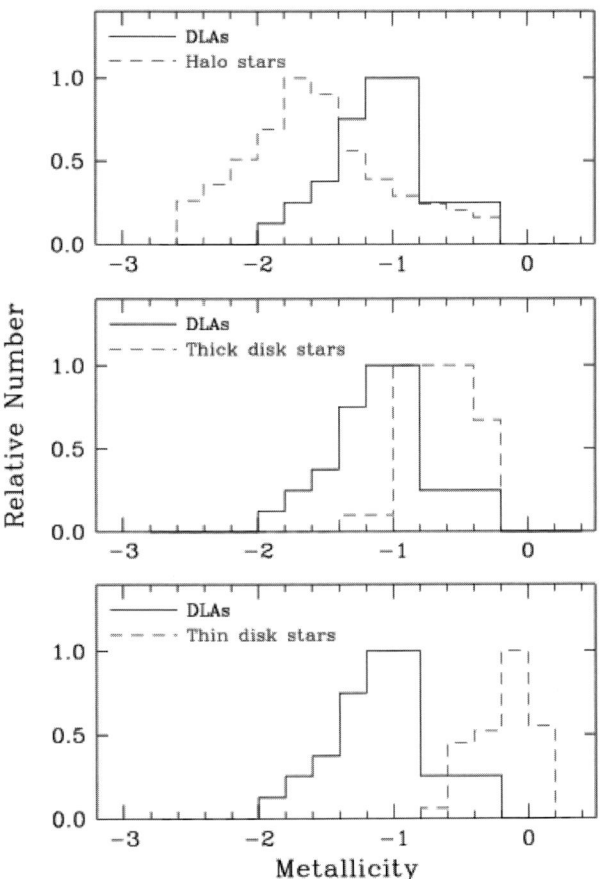

FIGURE 19. Metallicity distributions, normalised to unity, of damped Lyα systems and of stars belonging to the disk and halo populations in the Milky Way. From Pettini et al. (1997).

absorbers is that it is oxygen and nitrogen are both difficult to measure since the lines are saturated or in the Lyα forest. The preliminary results suggest that there is a large scatter at low metallicity, larger than that seen in low metallicity dwarf galaxies locally (Pettini, Lipman & Hunstead 1995; Lu, Sargent & Barlow 1998).

A summary of the heavy element abundance studies of quasar absorbers is shown in Figure 20.

5.3. *The Damped Lyman Alpha Absorbers: Kinematics*

The absorption profiles of metal lines which aren't too badly saturated give information on the kinematics of the gas in high redshift galaxies. If the galaxy has a rotating disk with a radial gradient of decreasing gas density with increasing radius, then the absorption profile of a typical line-of-sight has a distinctive signature (Weisheit 1978), with the strongest absorption velocity component on the extreme red or blue side of the profile.

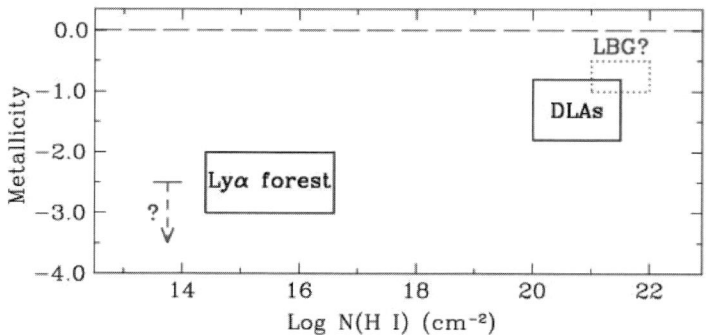

FIGURE 20. Summary of abundances relative to solar at high redshift. $N(HI)$ is the column density of neutral hydrogen, shown for the Lyα forest, damped Lyα absorbers (DLAs) and Lyman break galaxies. From Pettini (1999).

Prochaska & Wolfe (1997b, 1998) showed that indeed many damped Lyα systems have the expected absorption profile if the absorbing velocity clouds are in a rotating thick disk. A detailed analysis of one system at $z = 3.15$ with a detected emission line galaxy also was consistent with the expectations of a rotating disk (Djorgovski, Pahre, Bechtold & Elston 1996; Lu, Sargent & Barlow 1997). Prochaska & Wolfe further compared the profiles observed in a large sample of about 30 systems, with what you would expect to see if the quasar sight lines randomly sampled clouds with various geometries and kinematics, including isothermal halos, spherical distributions undergoing collapse, and thick and thin rotating disks. They argued that the distribution of line profiles favored a model where the damped Lyα absorbers originate in rapidly rotating disks, with maximum velocities, $\Delta v \sim 100\text{-}300$ km s^{-1} (Figure 21). Moreover, the average size and Δv of the disks was inferred to be larger than what was comfortably predicted by semi-analytical models for the evolution of galaxy disks in the cold-dark-matter hierarchical collapse picture, as described for example by Kauffmann (1996). However, Haehnelt, Steinmetz & Rauch (1998; Figure 22) showed that collapsing protogalactic fragments in their n-body/hydro simulations also rotate, and could reproduce the distribution of line profiles seen by Prochaska & Wolfe, without the need for large, rapidly rotating disks. The full-up simulations include non-linear effects which aren't included in the linear Press-Schechter theory. For merging protogalactic clumps, unlike the case for Lyα forest clouds, it is crucial to include non-linear effects in the models.

At lower redshift, Churchill, Steidel & Vogt (1996) and Churchill et al. (1999) have studied the kinematics of Mg II absorbers. The distribution of line profiles favors an origin in galaxies with large rotating disks and an infalling gaseous halo. These results are consistent with the direct imaging of the absorbing galaxies, discussed below, which suggest that the $z < 1$ absorbers are relatively mature galaxies, similar to ones seen locally.

5.4. Absorption by Molecules

Since high redshift quasar absorbers are pointers to gas-rich galaxies with active star-formation, they undoubtedly also contain molecular gas, mostly H_2. Warm H_2 can be

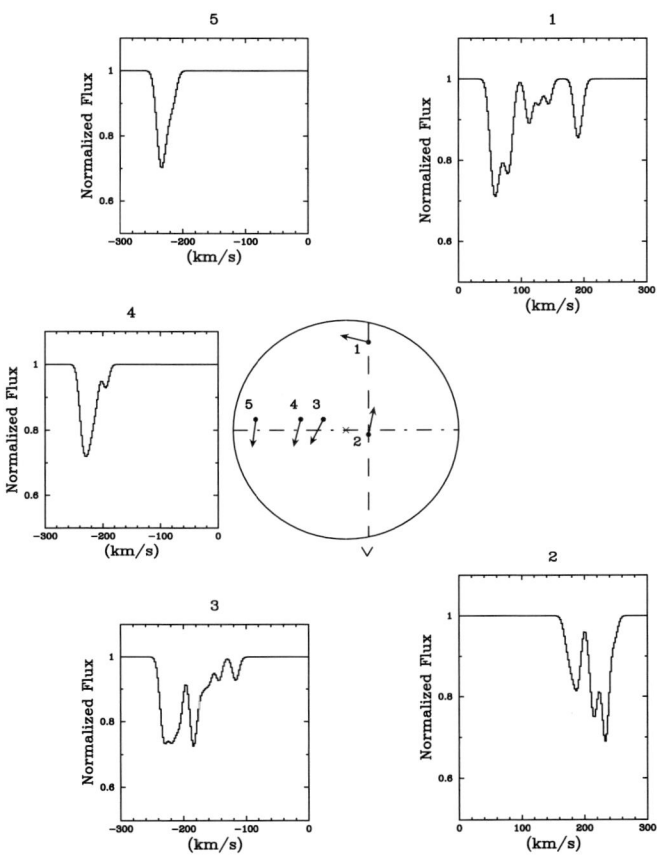

FIGURE 21. Expected absorption line profiles from a rotating disk of gas. The center circle represents an inclined disk (top view) rotating counter clockwise with rotation speed $v_{rot} = 250$ km s^{-1}. The solid dots represent the intersection points for 5 different sightlines with the midplane of the disk. The solid arrows indicate the direction of the rotation vector. The sightlines are inclined by 70 deg with respect to the normal of the disk and yield the profiles labeled 1–5. From Prochaska & Wolfe (1997b).

detected through IR emission, but most H$_2$ is probably contained in a cold phase, measurable by ultraviolet Lyman and Werner absorption (Black & Dalgarno 1976). Shortly after the detection of H$_2$ absorption in local interstellar clouds with Copernicus in the 1970's (Spitzer et al. 1973), searches for H$_2$ in quasar absorbers began (Aaronson, Black & McKee 1974; Carlson 1974). The UV H$_2$ lines have a distinctive pattern which is straightforward to distinguish from the Lyα forest (Figure 23).

However, evidence for molecular gas or dust in quasar absorbers has been scant (Black, Chaffee & Foltz 1987; Foltz, Chaffee & Black 1988; Lanzetta, Wolfe & Turnshek 1989; Levshakov, Chaffee, Foltz & Black 1992; Bechtold 1996, 1999; Ge & Bechtold 1997, 1999; Petitjean, Srianand & Ledoux 2000; Levshakov et al. 2000; Ge, Bechtold & Kulkarni 2001; Figure 24). The inclusion of quasars for absorption studies that are bright and

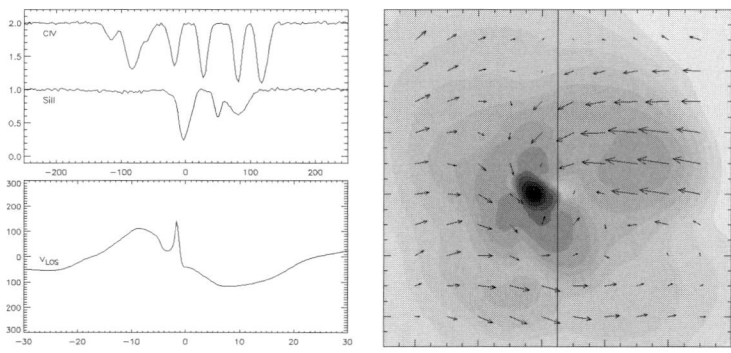

FIGURE 22. Velocity field of a collapsing protogalactic fragment which shows the same absorption line profile as a larger, rotating disk. From Steinmetz (2001).

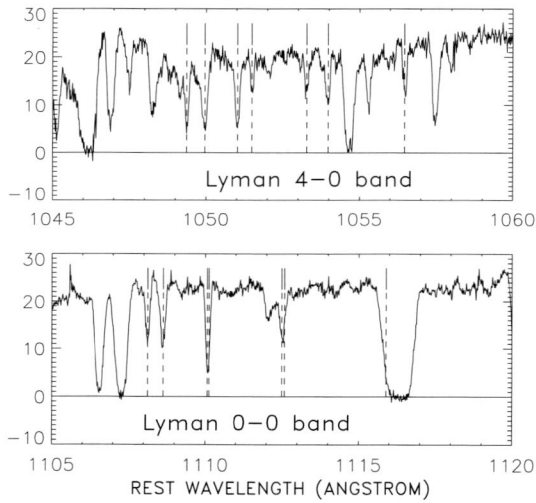

FIGURE 23. Spectra of the 0-0 and 4-0 vibration transitions of the Lyman series of molecular hydrogen at $z = 2.811$ in PKS 0528-250. The spectrum is shown in the rest frame, and the dashed lines indicate the expected wavelengths of the $J \leq 2$ transitions. From Cowie & Songaila (1995).

blue selects against lines-of-sight which contain dust, and hence detectable molecular absorption.

For the absorbers with detected H_2, the inferred molecular fraction

$$f_{H_2} = \frac{2N(H_2)}{2N(H_2) + N(HI)} \qquad (5.23)$$

is $f_{H_2} = 0.07$-0.1 whereas the limits for systems where H_2 is not detected is typically $f_{H_2} < 10^{-5}$.

The bimodality of f is understood as the result of self-shielding. Once H_2 begins to form, it shields itself from subsequent photodissociation by ultraviolet radiation, and f increases.

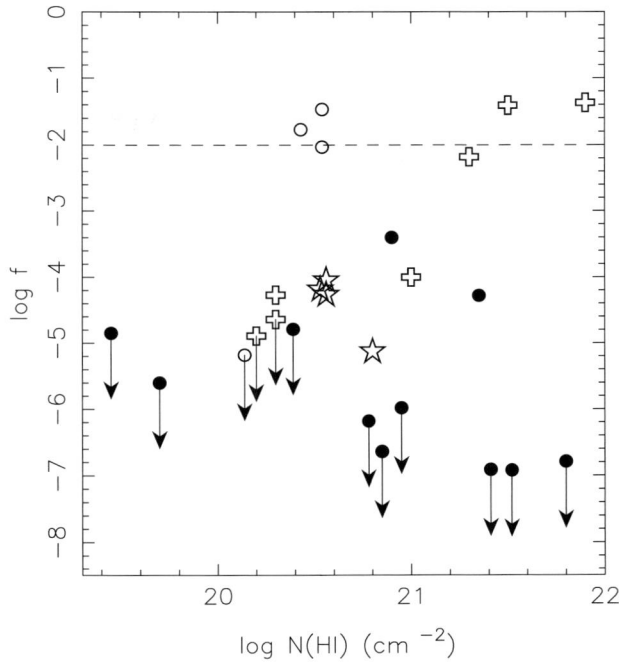

FIGURE 24. Molecular fraction, $f = 2N(H_2)/(2N(H_2)+N(H\ I))$, versus H I column density. Damped Lyman-α systems observed at high spectral resolution are indicated by filled circles (verticle arrows are for upper limits). Measurements in the Magellanic clouds are indicated by open crosses. Open circles and open stars are for, respectively, lines of sight perpendicular to the disk of our Galaxy and lines of sight toward the Magellanic clouds. The horizontal dashed line marks the transition from low to high molecular fraction in the local Galactic gas ($f = 0.01$). From Petitjean, Srianand & Ledoux (2000).

High resolution spectroscopy of the H_2 lines give estimates of excitation temperatures, volume densities (assuming some of the excitation is collisional), and the ultraviolet radiation field which can excite or photodissociate the molecules. The lack of molecules in some absorbers may be the result of an enhanced ultraviolet radiation field. For example, the well-studied absorber of PHL 957 (Q0100+13) has $N(HI) = 2.5 \times 10^{21}$ cm^{-2}, the column density is high enough that H_2 should have been detected (Black, Chaffee & Foltz 1987). They suggested that the lack of H_2 in that system resulted from photodissociation by the ultraviolet radiation field produced by the intense star-formation in the galaxy within which the absorption features arise.

A related observation is the measurement of the excited fine structure lines of ions such as Si II, C II and particularly C I, which can be used to place limits on the temperature of the cosmic microwave background as a function of redshift (Bahcall & Wolf 1968; Meyer et al. 1986; Songaila et al. 1994; Lu et al. 1996; Ge, Bechtold & Black 1997; Roth & Bauer 1999). For standard expanding Universes, the microwave background has temperature $T(z) = T_o(1 + z)$, where $T_o = 2.7$ K, the value measured today. C I * in particular has a small excitation energy and is well suited for this measurement. The problem is that C I can be ionized by photons longward of the Lyman limit and is rare in redshifted systems; those showing H_2 are generally shielded from UV radiation adequately to have detectable C I as well. Thus far, the few systems where C I has been

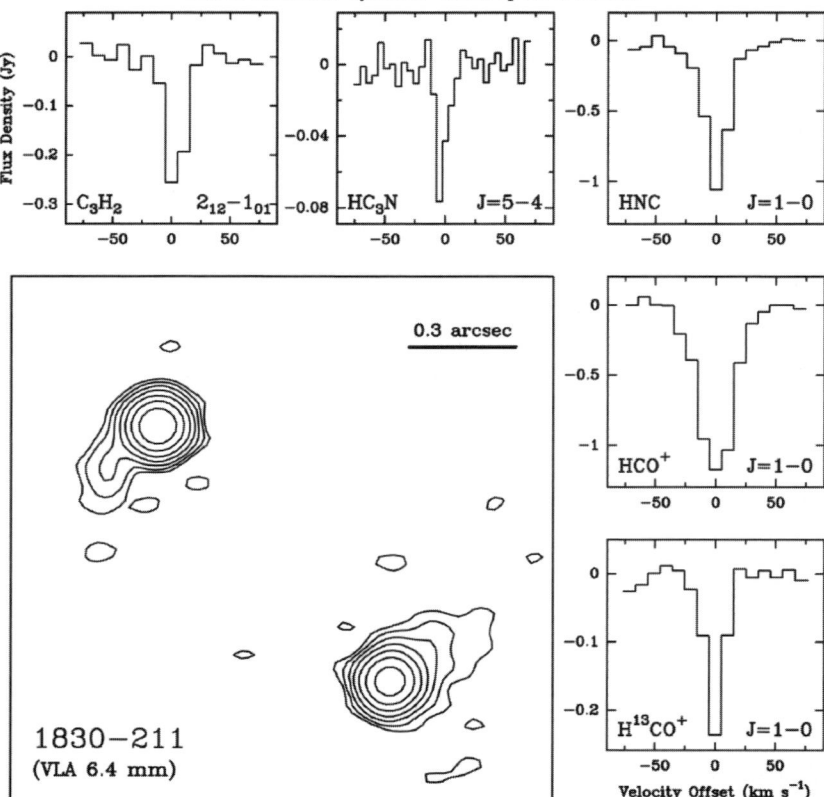

FIGURE 25. Molecular absorption associated with the Einstein ring, PKS 1830-211. Contour plot shows source continuum, spectra show molecular absorption by gas in the lensing galaxy toward the southwest component. Zero velocity corresponds to $z = 0.88582$. From Carilli & Menten (2000).

detected, and the excited states measured or limited, give $T(z)$ which (unsurprisingly) is consistent with the predictions of standard cosmology. As mentioned above, direct excitation by CMB photons is not the only source of excitation, and so these estimates of $T(z)$ are really upper limits to the cosmic microwave background temperature.

Molecular absorption in the millimeter (Figure 25) has been detected for many different molecules in redshift systems at $z = 0.2 - 0.9$ by Wiklind & Combes (1994ab, 1995, 1996ab, 1997, 1998), Combes & Wiklind (1997), and Carilli & Menten (2000). The millimeter transitions can be used to probe detailed molecular chemistry. Isotope ratios such as $^{12}C/^{13}C$ have been measured; this ratio is expected to decrease in time with increasing stellar processing since ^{13}C is produced in low and intermediate mass stars whereas ^{12}C is produced in massive stars as well (Wilson & Matteucci 1992). In one case, CO absorption identified the redshift of the intervening galaxy lensing a background radio AGN (Wiklind & Combes 1996a). These observations complement the UV studies, since so far detections of molecular absorption have been achieved for heavily reddened objects only, where the optical/UV continuum is too faint for absorption line spectroscopy.

Redshifted molecular transitions can also be used to measure the variation of fundamental physical constants with time, an observational constraint of interest to certain unified theories (Varshalovich & Potekhin 1995, 1996; Potekhin et al. 1998; Drinkwater et al. 1998). Specifically, the wavelengths of H_2 are sensitive to the ratio of electron to proton inertial mass (Thompson 1975) and observations of H_2 at $z = 2.8$ by Foltz, Chaffee & Black (1988) put an upper limit on the fractional variation of 2.0×10^{-4}. Cowie & Songaila (1995) improved on this by about a factor of 10.

Although all known damped Ly-α absorbers appear to be enriched with heavy elements (Pettini et al. 1999 and references therein), as are the quasars used to find them (Ferland et al. 1996, Hamann & Ferland 1999), someday one may be able to find UV bright objects at high enough redshift to probe truly primordial H_2. The chemistry of primordial molecule formation is important for models of the thermal history of the early intergalactic medium since H_2 cooling controls the collapse and fragmentation of the first PGFs. Observations of primordial H_2 would allow interesting measures of the ultraviolet radiation background, which may suppress the formation of low-mass objects (Couchman & Rees 1986; Efstathiou 1992; Haiman, Rees & Loeb 1996).

5.5. *Deuterium*

Deuterium was produced throughout the Universe about a minute after the Big Bang, and has been destroyed by astration ever since (Reeves, Audouze, Fowler & Schramm 1973; Schramm & Turner 1998). The primordial ratio of deuterium to hydrogen produced during Big Bang nucleosynthesis depends on the baryon-to-photon ratio, η; combined with the photon density of the cosmic microwave background radiation (measured precisely with $COBE$), a measurement of the primordial D/H therefore measures the cosmic baryon density, Ω_{baryon}. Since D/H is destroyed in stars, D/H decreases with time; any measurement is a lower limit to D/H and hence an upper limit to Ω_{baryon}. High redshift measurements of D/H should provide the best limits on Ω_{baryon}, being closer to the primordial, big-bang value. A decline of D/H with cosmic formation time would be a satisfying confirmation of the big bang nucleosynthesis theory.

Deuterium in quasar absorbers has been searched for by measuring the deuterium Lyα line, located 81 km s^{-1} blueward of hydrogen Lyα (Chaffee, Weymann, Strittmatter & Latham 1983; Webb, Carswell, Irwin & Penston 1991; Songaila, Cowie, Hogan & Rugers 1994; Carswell et al. 1994, 1996; Wampler et al. 1996; Tytler, Fan & Burles 1996; Rugers & Hogan 1996ab; Songaila, Wampler & Cowie 1997; Hogan 1998; Burles & Tytler 1998abc; Molaro, Boniracio, Centurion & Vladilo 1999; Burles, Kirkman & Tytler 1999; O'Meara et al. 2001; D'Odorico, Dessauges-Zavadsky & Molaro 2001; Kirkman et al. 2000; Tytler, O'Meara, Suzuki & Lubin 2000). There are two practical difficulties. First, to have detectable deuterium given the expected cosmic abundance, the hydrogen column density must be relatively high, so hydrogen Lyα and the Lyman series lines are very saturated, and N(H) is correspondingly uncertain. In damped Lyα systems, Lyα is too broad to allow deuterium to be measured at all. The systems with high enough column to detect deuterium, invariably contain metals, so some astration has already taken place. Second, metal absorbers typically have velocity substructure on scales of $\Delta v \sim 80$ km s^{-1} so one must search a large number of redshift systems to argue statistically that one is not seeing merely hydrogen Lyα interlopers (which should occur in equal numbers on the blue and red side of Lyα). On the other hand, for redshifted absorbers, the far ultraviolet lines of the Lyman series are accessible from the ground, making a larger number of systems measurable than with HST.

These difficulties have lead to some debate in the literature about the D/H ratio in quasar absorbers, and whether or not it is consistent with the abundances of ^{4}He and ^{7}Li, also produced by big bang nucleosynthesis. Most of the disagreements were the result of curve-of-growth uncertainties in the column of hydrogen, and other technical details, and have since been sorted out by the authors. Figure 26 summarizes the quasar absorber measurement of D/H with the predictions from the Standard Big Bang Nucleosynthesis calculations (Burles & Tytler 1998b), based on ground-based high resolution data for absorbers at $z \sim 3$. Absorbers at $z \sim 1$ have also been measured with HST (Webb et al. 1997; Shull et al. 1998).

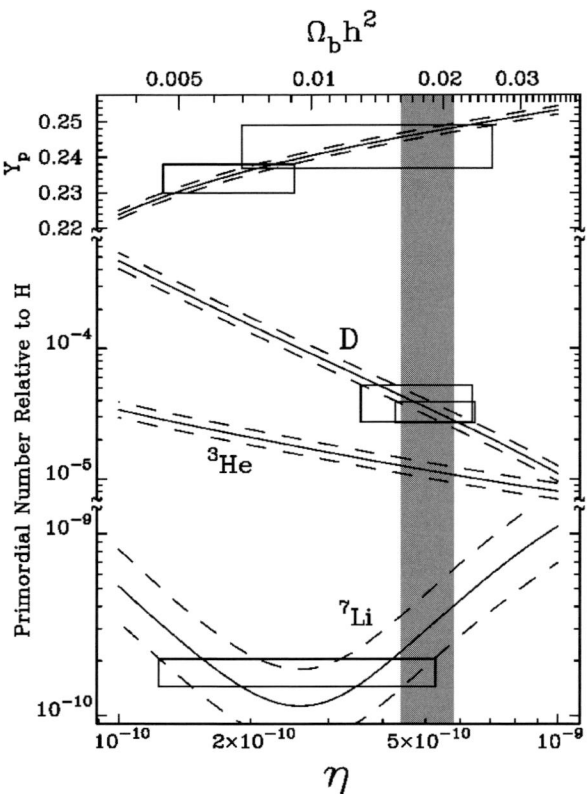

FIGURE 26. The predicted abundance ratios of the light elements from Standard Big Bang Nucleosynthesis (SBBN) as a function of η and $\Omega_b h^2$. ^{4}He is shown as primordial mass fraction, Y_p. Boxes represent 95% confidence levels of recent observational determinations. The width of the boxes include 95% confidence levels in the SBBN calculations. From Burles & Tytler (1998b).

The high redshift quasar measurements reflect D/H as it was about 10-20 Gyr ago, and can be compared to the pre-solar value 4.5 Gyr ago measured in terrestrial sea water, the solar wind, comets and the atmospheres of Jupiter and Saturn (by measuring HD or deuterated methane in the mid-IR). The value of D/H today can be measured by measuring the Lyman series in the Milky Way interstellar medium. Spectra from FUSE will be important to get the whole Lyman series, but the main limitation of the Milky Way measurements is the estimation of the stellar atmosphere Lyman series absorption from which the ISM absorption must be deblended. The various measurements of the D/H ratio have been reviewed by Lemoine et al. (1999). Indeed, within the major uncertainties of all the measurements, the D/H ratio appears to be declining with cosmic time, as predicted (Figure 27).

5.6. Redshifted 21-cm absorption

A small subset of UV discovered systems show absorption in the 21-cm transition of neutral hydrogen (Roberts et al. 1976; Wolfe & Davis 1979; Wolfe & Briggs 1981; Briggs

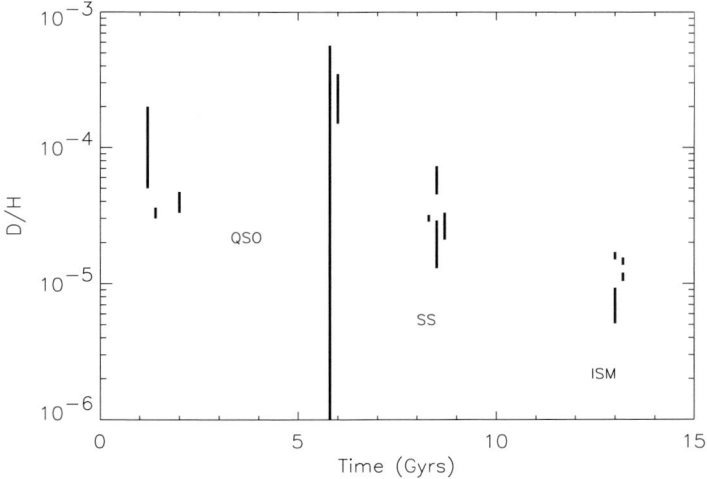

FIGURE 27. Deuterium to hydrogen ratio as a function of cosmic time (for $\Omega_0 = 1, q_o = 0.5, H_o = 50$ km/s/Mpc). Quasar absorbers are from Burles & Tytler 1998ab, Songaila (1998), Webb et al. (1997), Tytler et al. (1999). Presolar values are (SS) from the solar wind and Jupiter. Interstellar (ISM) values are from observations of stars in the Milky Way – for complete references, see Lemoine et al. (1999).

& Wolfe 1983; Briggs 1988; Taramopoulos et al. 1995; Carilli et al. 1996, 1997; Lane et al. 1998; Kanekar & Chengalur 2001). In an elegant experiment, Briggs et al. 1989 compared the 21-cm absorption along an extended radio source associated with a damped Lyα absorber at $z \sim 2$ and concluded that the absorber has a large, disk-like geometry. Comparison of the 21-cm absorption with the column density of $H\ I$ measured by the damped Lyα profile can measure the spin temperature of the absorbing gas (Lane et al. 1998 and references therein), which reflects some average temperature along the line of sight. The quasar absorbers seem to have spin temperatures which are larger than those seen in Milky Way clouds, suggesting warmer temperatures for the cold phase of the ISM at higher redshift (Wolfe et al. 1985; Lane et al. 1998).

6. Imaging of QSO Absorbers

They say a picture is worth a thousand words; for some, a picture is worth *ten* thousand echelle spectra. Despite the wealth of detailed information derived from quasar absorption line spectra, some feel that one won't understand what the absorbers "are" until you have a picture of them, and connect the absorbers to classes of known objects.

6.1. $z < 1$ Mg II Absorbers

Bergeron & Boisse (1991) and Steidel and collaborators (Steidel, Dickinson & Persson 1994; Steidel, Pettini, Dickinson & Persson 1994; Steidel 1995; Steidel, Bowen, Blades & Dickinson 1995; Steidel et al. 1997) undertook comprehensive imaging and spectroscopic surveys in order to identify the galaxies responsible for low redshift absorbers, primarily Mg II absorbers. The general result was that the absorbing galaxies are drawn from the same population of normal galaxies seen in galaxy redshift surveys, with relatively little evolution in color or luminosity. Deep imaging (Figure 28) showed that some absorbers appear to arise in the halos of disk galaxies, supporting the result from statistical studies

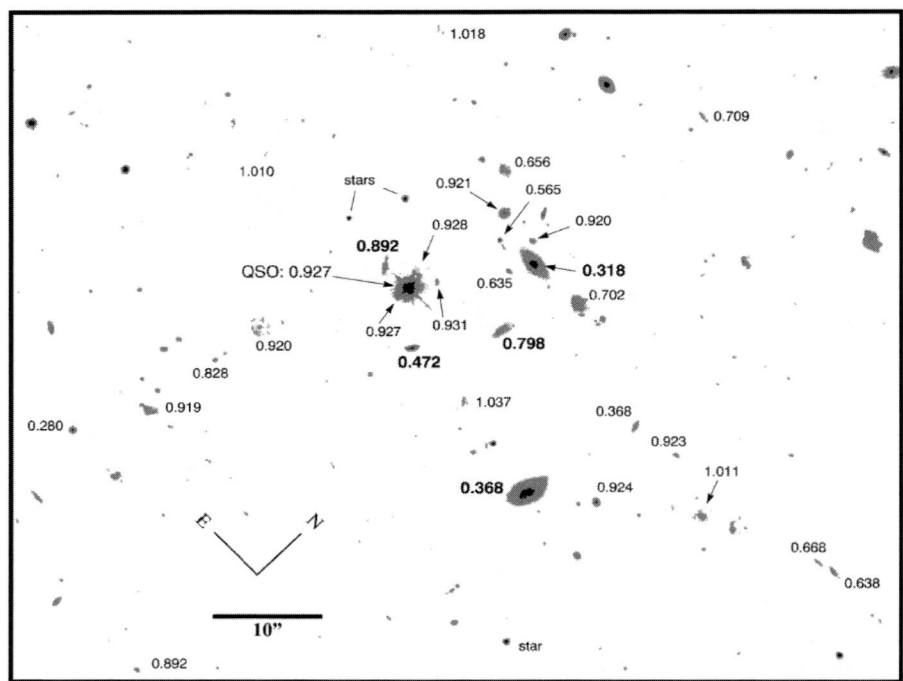

FIGURE 28. Deep WFPC2 image of the galaxies around quasar 3C 336. The bold numbers are redshifts of indicated galaxies which show absorption in the background quasar; the others are redshifts of galaxies which do not show absorption. From Steidel et al. (1997).

that the cross-section for absorption is large. Some galaxies do not produce absorption in background quasars at all.

With few exceptions, galaxies have been found at the redshift of intervening Mg II absorbers, supporting the interpretation that the absorbers are interstellar clouds of intervening galaxies. One suggestion has been made that large absorption cross-section is actually the result of the fact that absorbers often arise in dwarf irregular satellites of large galaxies (York, Dopita, Green & Bechtold 1986), analogues of the Large and Small Magellanic Clouds of the Milky Way. Small, faint galaxies with small separations from the quasar image would presumably be difficult to detect. This is not supported by an apparent anti-correlation of Mg II equivalent widths and increasing projected distance from the identified galaxy center (Steidel et al. 1994; Lanzetta et al. 1995).

6.2. Imaging of the Damped Lyman Alpha Absorbers

Deep galaxy redshift surveys typically find galaxies only to $z \sim 1.5$, so the galactic counterparts of high redshift damped Lyα absorbers are searched for by means of line emission from the H II regions associated with them, or Lyman break techniques. A great deal of effort went into searches for Lyα emission associated with damped Lyα absorbers at $z = 2-3$ with limited success (Hunstead, Fletcher & Pettini 1990; Lowenthal et al. 1990, 1991, 1995; Wolfe, Lanzetta, Turnshek & Oke 1992; Moller & Warren 1993; Giavalisco, Macchetto & Sparks 1994; Warren & Moller 1996; Djorgovski et al. 1996; Fynbo, Møller & Warren 1999). In many cases, Lyα may be multiply scattered by dust (Charlot & Fall 1991, 1993).

Infrared surveys have been more successful in identifying damped Lyα absorbers via

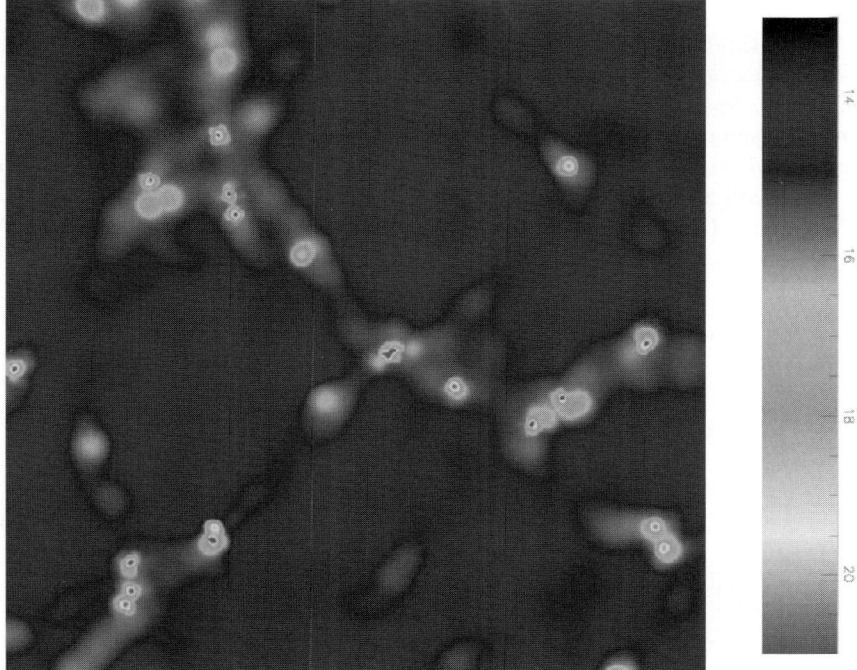

FIGURE 29. Distribution of neutral hydrogen from simulations. From Steinmetz (2001).

direct imaging and Hα emission (Elston et al. 1991; Aragon-Salamanca, Ellis & O'Brien 1996; Bechtold et al. 1998; Mannucci, Thompson, Beckwith & Williger 1998; Kulkarni et al. 2000, 2001; Kobulnicky & Koo 2000; Warren, Møller, Fall & Jakobsen 2001). Wide field infrared surveys show that the absorbers can be identified with the same population of small, star-forming galactic fragments found by other techniques, described in detail at this school by Mark Dickinson; see also Pettini et al. (2001). Figure 29 shows the simulation of what one expects. The bright clumps at the intersection of the filaments are the regions with high enough column to produce damped Lyα absorption, and these probably produce stars as well.

7. Future Prospects

As this review was being written, a number of new facilities were just beginning to produce results which will enhance our knowledge of quasar absorbers. *Chandra* and *XMM* are showing a surprising wealth of absorption line spectral features in Seyferts and low redshift quasars. *FUSE* is allowing the spectral region blueward of rest Lyα to be probed with great sensitivity, and large surveys of absorbers at echelle resolution are underway with STIS on HST. UVES on the VLT is being used to study quasars in the southern sky. The *Cosmic Origins Spectrograph* is planned to be installed on HST in a few years, and one of its main science drivers is the study of quasar absorbers, particularly He II λ304. The Arecibo upgrade and completion of the Greenbank Telescope will enable more sensitive radio observations of redshifted absorption. The Sloan Digital Sky Survey is beginning to produce new bright quasars for absorption line studies. Future CMB observations with MAP and Planck will probably identify the redshift of HI reionization, so that searches may target the ionizing objects effectively.

On the longer timescale, plans are being made to build a new generation of 30m-class ground-based telescopes, to be huge light-buckets for high dispersion spectroscopy. The *Next Generation Space Telescope* promises to be an important tool for studying the first generation of objects responsible for reionization of the intergalactic medium. A large, ultraviolet optimized telescope in space is probably the ultimate dream of workers in this field, which will allow high quality spectroscopy of fainter objects than are reachable with *HST*. Figure 30 shows a simulation of spectra expected with ST2010, or SUVO, a 6-8m class telescope, described by Morse, Shull & Kinney (1999).

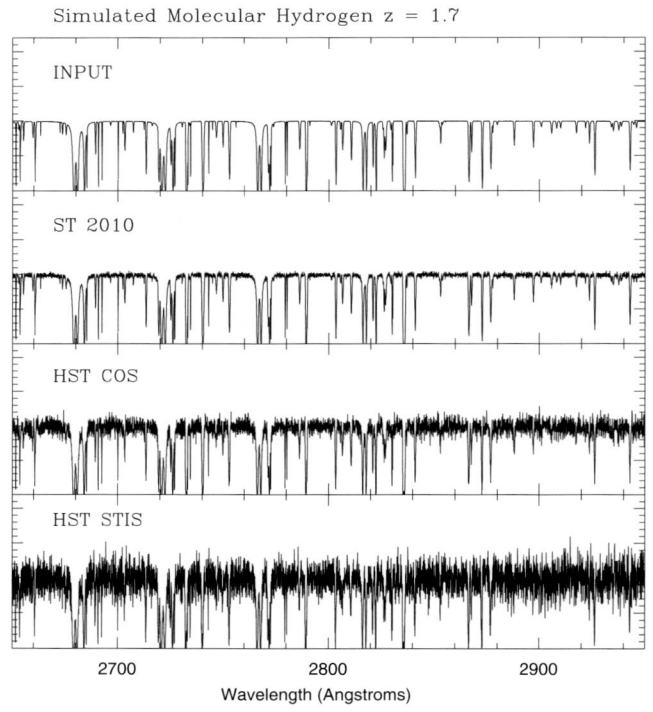

FIGURE 30. Simulated spectrum of molecular hydrogen and the Lyα forest at $z = 1.7$ as observed in 30 *HST* orbits with STIS, COS and a future mission, ST2010 or SUVO. From Bechtold (1999).

It is a pleasure to thank Ismael Pérez-Fournon, Marc Balcells, and the Director and staff of the IAC for their invitation to participate in the Winter School, and their gracious hospitality in Tenerife. I thank my collaborators whose work is quoted here in advance of publication, and longtime advice and support from W. Sargent, S. Shectman, A. Wolfe, M. Shull, and P. Jakobsen. Without Dr. Allison Stopeck I would not have participated in the Winter school. Support for the preparation of this review was provided in part by NSF grant AST-9617060.

REFERENCES

AARONSON, M., BLACK, J. H. & MCKEE, C. F. 1974, A Search for Molecular Hydrogen in Quasar Absorption Spectra, ApJ, 191, L53.

ALDCROFT, T. L., BECHTOLD, J. & ELVIS, M. 1994, MG II absorption in a sample of 56 steep-spectrum quasars, ApJS, 93, 1.

ANDERSON, S. F., WEYMANN, R. J., FOLTZ, C. B. & CHAFFEE, F. H. 1987, Associated C IV absorption in radio-loud QSOs - The '3C mini-survey', AJ, 94, 27.

ARAGON-SALAMANCA, A., ELLIS, R. S. & O'BRIEN, K. S. 1996, Faint galaxies close to QSOs with damped Lyα absorption systems, MNRAS, 281, 945.

ARAV, N., SHLOSMAN, I. & WEYMANN, R. J. 1997, ASP Conf. Ser. 128: *Mass Ejection from Active Galactic Nuclei*.

ARONS, J. & MCCRAY, R. 1969, Interaction of Cosmic Gamma Rays with Intergalactic Matter, ApJ, 158, L91.

BAHCALL, J. N. et al. 1993, The Hubble Space Telescope quasar absorption line key project. I - First observational results, including Lyα and Lyman-limit systems, ApJS, 87, 1.

BAHCALL, J. N. et al. 1996, The Hubble Space Telescope Quasar Absorption Line Key Project. VII. Absorption Systems at $z(abs) \leq 1.3$, ApJ, 457, 19.

BAHCALL, J. N. & SALPETER, E. E. 1966, Absorption Lines in the Spectra of Distant Sources, ApJ, 144, 847.

BAHCALL, J. N., SARGENT, W. L. W. & SCHMIDT, M. 1967, An Analysis of the Absorption Spectrum of 3C 191, ApJ, 149, L11.

BAHCALL, J. N., SPITZER, L. 1969, Absorption Lines Produced by Galactic Halos, ApJ, 156, L63.

BAHCALL, J. N. & WOLF, R. A. 1968, Fine-Structure Transitions, ApJ, 152, 701.

BAJTLIK, S., DUNCAN, R. C. & OSTRIKER, J. P. 1988, Quasar ionization of Lyα clouds - The proximity effect, a probe of the ultraviolet background at high redshift, ApJ, 327, 570.

BARTHEL, P. D., TYTLER, D. R. & THOMSON, B. 1990, Optical spectra of distant radio loud quasars. I - Data: Spectra of 67 quasars, A&AS, 82, 339.

BECHTOLD, J. 1994, The Lyα forest near 34 quasi-stellar objects with $z > 2.6$, ApJS, 91, 1.

BECHTOLD, J. 1995, The Proximity Effect, in QSO Absorption Lines, Proceedings of the ESO Workshop Held at Garching, Germany, 21 - 24 November 1994, edited by Georges Meylan. Springer-Verlag Berlin Heidelberg New York. Also ESO Astrophysics Symposia, 1995, p.299.

BECHTOLD, J. 1996, IAU Symp. 178: Molecules in Astrophysics: Probes & Processes, 178, 525.

BECHTOLD, J. 1999, H2 in Space, meeting held in Paris, France, September 28th - October 1st, 1999. Eds.: F. Combes, G. Pineau des Forêts. Cambridge University Press, Astrophysics Series, E51.

BECHTOLD, J., CROTTS, A. P. S., DUNCAN, R. C. & FANG, Y. 1994, Spectroscopy of the double quasars Q1343+266A, B: A new determination of the size of Lyα forest absorbers, ApJ, 437, L83.

BECHTOLD, J., ELSTON, R., YEE, H. K. C., ELLINGSON, E. & CUTRI, R. M. 1998, ASP Conf. Ser. 146: The Young Universe: Galaxy Formation and Evolution at Intermediate and High Redshift, 241.

BECHTOLD, J., GREEN, R. F., WEYMANN, R. J., SCHMIDT, M., ESTABROOK, F. B., SHERMAN, R. D., WAHLQUIST, H. D. & HECKMAN, T. M. 1984, IUE Observations of High-redshift Quasars, ApJ, 281, 76.

BECHTOLD, J., GREEN, R. F. & YORK, D. G. 1987, High-resolution spectroscopy of the $Z = 1.79$ absorption-line system toward B2 1225 + 317, ApJ, 312, 50.

BECHTOLD, J., SIEMIGINOWSKA, A., ALDCROFT, T. L., ELVIS, M., DOBRZYCKI, A. 2001, ApJ, in press, astro-ph/0107003.

BECHTOLD, J. & YEE, H. K. C. 1995, High Spatial Resolution Spectroscopy of the Lyα Absorption Towards the Gravitational Lens System B1422+2309, AJ, 110, 1984.

BECKER, R. H. et al. 2001, AJ in press, astro-ph/0108097.
BECKER, R. H., WHITE, R. L., GREGG, M. D., BROTHERTON, M. S., LAURENT-MUEHLEISEN, S. A. & ARAV, N. 2000, Properties of Radio-selected Broad Absorption Line Quasars from the First Bright Quasar Survey, ApJ, 538, 72.
BERGERON, J. et al. 1994, The Hubble Space Telescope quasar absorption line key project. VI: Properties of the metal-rich systems, ApJ, 436, 33.
BERGERON, J. & BOISSE, P. 1991, A sample of galaxies giving rise to MG II quasar absorption systems, A&A, 243, 344.
BI, H. & DAVIDSEN, A. F. 1997, Evolution of Structure in the Intergalactic Medium and the Nature of the Lyα Forest, ApJ, 479, 523.
BLACK, J. H., CHAFFEE, F. H. & FOLTZ, C. B. 1987, Molecules at early epochs. II - H2 and CO toward PHL 957, ApJ, 317, 442.
BLACK, J. H. & DALGARNO, A. 1976, Interstellar H2 - The population of excited rotational states and the infrared response to ultraviolet radiation, ApJ, 203, 132.
BRIGGS, F. H. 1988, Proceedings of the QSO Absorption Line Meeting, 275.
BRIGGS, F. H. & WOLFE, A. M. 1983, The incidence of 21 centimeter absorption in QSO redshift systems selected for MG II absorption - Evidence for a two-phase nature of the absorbing gas, ApJ, 268, 76.
BRIGGS, F. H., WOLFE, A. M., LISZT, H. S., DAVIS, M. M. & TURNER, K. L. 1989, The spatial extent of the Z = 2.04 absorber in the spectrum of PKS 0458-020, ApJ, 341, 650.
BURBIDGE, E. M. 1967, Quasi-Stellar Objects, ARA&A, 5, 399.
BURBIDGE, E. M., LYNDS, C. R. & BURBIDGE, G. R. 1966, On the Measurement and Interpretation of Absorption Features in the Spectrum of the Quasi-Stellar Object 3C 191, ApJ, 144, 447.
BURBIDGE, G. R., ODELL, S. L., ROBERTS, D. H. & SMITH, H. E. 1977, On the origin of the absorption spectra of quasistellar and BL Lacertae objects, ApJ, 218, 33.
BURLES, S., KIRKMAN, D. & TYTLER, D. 1999, Deuterium toward Quasar Q0014+813, ApJ, 519, 18.
BURLES, S. & TYTLER, D. 1996, The Cosmological Density and Ionization of Hot Gas: O VI Absorption in Quasar Spectra, ApJ, 460, 584.
BURLES, S. & TYTLER, D. 1998a, On the Measurements of D/H in QSO Absorption Systems Closing in on the primordial abundance of deuterium, Space Science Reviews, 84, 65.
BURLES, S. & TYTLER, D. 1998b, The Deuterium Abundance toward QSO 1009+2956, ApJ, 499, 699.
BURLES, S. & TYTLER, D. 1998c, The Deuterium Abundance toward Q1937-1009, ApJ, 507, 732.
CARILLI, C. L. et al. 2000 Astronomical Constraints on the Cosmic Evolution of the Fine Structure Constant and Possible Quantum Dimensions, Physical Review Letters, 85, 5511.
CARILLI, C. L., LANE, W., DE BRUYN, A. G., BRAUN, R. & MILEY, G. K. 1996, Redshifted H I 21 cm Line Observations of Damped Lyα Absorption Systems, AJ, 112, 1317.
CARILLI, C. L. & MENTEN, K. M. 2000, Molecular QSO Absorption Line Systems in *Cold Gas and Dust at High Redshift*, Highlights of Astronomy, Vol. 12. ed. D. Wilner.
CARILLI, C. L., MENTEN, K. M., REID, M. J. & RUPEN, M. P. 1997, Neutral Hydrogen 21 Centimeter Absorption at Redshift 0.673 toward 1504+377, ApJ, 474, L89.
CARLSON, R. W. 1974, Molecular-Hydrogen Absorption Features in the Spectrum of Quasi-Stellar Object 4C 05.34, ApJ, 190, L99.
CARSWELL, R. F. et al. 1996, The high-redshift deuterium abundance: the z=3.086 absorption complex towards Q 0420-388, MNRAS, 278, 506.
CARSWELL, R. F., RAUCH, M., WEYMANN, R. J., COOKE, A. J. & WEBB, J. K. 1994, Is There Deuterium in the Z=3.32 Complex in the Spectrum of 0014+813, MNRAS, 268, L1.
CARSWELL, R. F. & REES, M. J. 1987, Constraints on voids at high redshifts from Ly-alpha absorbers, MNRAS, 224, 13P.

CARSWELL, R. F., WEBB, J. K., BALDWIN, J. A., & ATWOOD, B. 1987, High-redshift QSO absorbing clouds and the background ionizing source, ApJ, 319, 709.

CARSWELL, R. F., WEBB, J. K., COOKE, A. J., IRWIN, M. J. 1995, http://www.ast.cam.ac.uk/~rfc/vpfit.html.

CEN, R., MIRALDA-ESCUDE, J., OSTRIKER, J. P. & RAUCH, M. 1994, Gravitational collapse of small-scale structure as the origin of the Lyα forest, ApJ, 437, L9.

CEN, R. & SIMCOE, R. A. 1997, Sizes, Shapes, and Correlations of Lyman Alpha Clouds and Their Evolution in the Lambda CDM Universe, ApJ, 483, 8.

CHAFFEE, F. H., FOLTZ, C. B., BECHTOLD, J. & WEYMANN, R. J. 1986, On the abundance of metals and the ionization state in absorbing clouds toward QSOs, ApJ, 301, 116.

CHAFFEE, F. H., WEYMANN, R. J., STRITTMATTER, P. A. & LATHAM, D. W. 1983 High-resolution spectroscopy of selected absorption lines toward quasi-stellar objects. I - Lyman-alpha toward PHL 957, ApJ, 267, 12.

CHARLOT, S. & FALL, S. M. 1991, Attenuation of Lyα emission by dust in damped Lyα systems, ApJ, 378, 471.

CHARLOT, S. & FALL, S. M. 1993, Lyα Emission from Galaxies, ApJ, 415, 580.

CHEN, H., LANZETTA, K. M., WEBB, J. K. & BARCONS, X. 1998, The Gaseous Extent of Galaxies and the Origin of Lyα Absorption Systems. III. Hubble Space Telescope Imaging of Lyα - absorbing Galaxies at z < 1, ApJ, 498, 77.

CHURCHILL, C. W., MELLON, R. R., CHARLTON, J. C., JANNUZI, B. T., KIRHAKOS, S., STEIDEL, C. C. & SCHNEIDER, D. P. 1999, The C IV Absorption-Mg II Kinematics Connection in $<z>\sim 0.7$ Galaxies, ApJ, 519, L43.

CHURCHILL, C. W., RIGBY, J. R., CHARLTON, J. C. & VOGT, S. S. 1999, The Population of Weak Mg II Absorbers. I. A Survey of 26 QSO HIRES/Keck Spectra, ApJS, 120, 51.

CHURCHILL, C. W., STEIDEL, C. C. & VOGT, S. S. 1996, On the Spatial and Kinematic Distributions of Mg II Absorbing Gas in $<z>$ approximately 0.7 Galaxies, ApJ, 471, 164.

COMBES, F. & WIKLIND, T. 1997, Detection of Water at z = 0.685 toward B0218+357, ApJ, 486, L79.

COUCHMAN, H. M. P. & REES, M. J. 1986, Pregalactic evolution in cosmologies with cold dark matter, MNRAS, 221, 53.

COWIE, L. L. & SONGAILA, A. 1995, Astrophysical Limits on the Evolution of Dimensionless Physical Constants over Cosmological Time ApJ, 453, 596.

COWIE, L. L., SONGAILA, A., KIM, T. & HU, E. M. 1995, The metallicity and internal structure of the Lyman-alpha forest clouds, AJ, 109, 1522.

CRISTIANI, S., D'ODORICO, S., FONTANA, A., GIALLONGO, E. & SAVAGLIO, S. 1995, The space distribution of the Lyα clouds in the line of sight to the z=3.66 QSO 0055-269, MNRAS, 273, 1016.

CROFT, R. A. C., WEINBERG, D. H., PETTINI, M., HERNQUIST, L. & KATZ, N. 1999, The Power Spectrum of Mass Fluctuations Measured from the Lyα Forest at Redshift z=2.5, ApJ, 520, 1.

CROTTS, A. P. S. & FANG, Y. 1998, Reobservation of Close QSO Groups: The Size Evolution and Shape of Lyα Forest Absorbers, ApJ, 502, 16.

CRUTCHER, R. M. 1975, Velocity component structure of the zeta Ophiuchi interstellar lines, ApJ, 200, 625.

DAVÉ, R., HELLSTEN, U., HERNQUIST, L., KATZ, N. & WEINBERG, D. H. 1998, Constraining the Metallicity of the Low-Density Lyα Forest Using O VI Absorption ApJ, 509, 661.

DAVÉ, R., HERNQUIST, L., KATZ, N. & WEINBERG, D. H. 1999, The Low-Redshift Lyα Forest in Cold Dark Matter Cosmologies, ApJ, 511, 521.

DINSHAW, N., FOLTZ, C. B., IMPEY, C. D., WEYMANN, R. J. & MORRIS, S. L. 1995, The Large Characteristic Size of Lyman Alpha Forest Clouds, Nature, 373, 223.

DINSHAW, N., IMPEY, C. D., FOLTZ, C. B., WEYMANN, R. J. & CHAFFEE, F. H. 1994, Common Lyman-alpha absorption toward the quasar pair Q1343+2640A, B: Evidence for large and quiescent clouds, ApJ, 437, L87.

DJORGOVSKI, S. G., PAHRE, M. A., BECHTOLD, J. & ELSTON, R. 1996, Identification of a galaxy responsible for a high-redshift Lyα absorption system, Nature, 382, 234.

DOBRZYCKI, A. & BECHTOLD, J. 1991a, The QSOs Influence on the Lymanα Forest BAAS, 23, 952.

DOBRZYCKI, A. & BECHTOLD, J. 1991b, A approximately 10 MPC void in the Ly-α forest at Z = 3.17 ApJ, 377, L69.

DOBRZYCKI, A. & BECHTOLD, J. 1996, Simulation Analysis of LY alpha Forest Spectra. I. Empirical Description at Z∼3, ApJ, 457, 102.

DOBRZYCKI, A., BECHTOLD, J., SCOTT, J., MORITA, M. 2001, A Uniform Analysis of the Lyα Forest at z = 0 - 5: IV. The clustering and evolution of clouds at z < 1.7 , ApJ, submitted.

D'ODORICO, S., DESSAUGES-ZAVADSKY, M. & MOLARO, P. 2001, A new deuterium abundance measurement from a damped Lyα system at z(abs) =3.025, A&A, 368, L21.

DRINKWATER, M. J., WEBB, J. K., BARROW, J. D. & FLAMBAUM, V. V. 1998, New limits on the possible variation of physical constants, MNRAS, 295, 457.

DUNCAN, R. C., OSTRIKER, J. P. & BAJTLIK, S. 1989, Voids in the Lyα forest, ApJ, 345, 39.

DZUBA, V. A., FLAMBAUM, V. V., MURPHY, M. T. & WEBB, J. K. 2001, Relativistic effects in Ni II and the search for variation of the fine-structure constant, Physics Review A, 63, 060d.

EDVARDSSON, B., ANDERSEN, J., GUSTAFSSON, B., LAMBERT, D.L., NISSEN, P.E. & TOMKIN, J. 1993, The Chemical Evolution of the Galactic Disk: Part One - Analysis and Results, A&A, 275, 101.

EFSTATHIOU, G. 1992, Suppressing the formation of dwarf galaxies via photoionization, MNRAS, 256, 43P.

ELLISON, S. L., SONGAILA, A., SCHAYE, J. & PETTINI, M. 2000, The Enrichment History of the Intergalactic Medium-Measuring the C IV/H I Ratio in the Lyα Forest, AJ, 120, 1175.

ELSTON, R., BECHTOLD, J., HILL, G. J. & GE, J. 1996, A Redshift 4.38 MG II Absorber toward BR 1202-0725, ApJ, 456, L13

ELSTON, R., BECHTOLD, J., LOWENTHAL, J. & RIEKE, M. 1991, Emission from the z = 2 damped Lyα absorber toward Q1215 + 333, ApJ, 373, L39.

ELVIS, M. 2001, A Structure for Quasars, ApJ, 545, 63.

FALL, S. M., PEI, Y. C. & MCMAHON, R. G. 1989, Detection of dust in damped Lyα systems, ApJ, 341, L5.

FANG, Y., DUNCAN, R. C., CROTTS, A. P. S. & BECHTOLD, J. 1996, The Size and Nature of Lyα Forest Clouds Probed by QSO Pairs and Groups, ApJ, 462, 77.

FERLAND, G. J., BALDWIN, J. A., KORISTA, K. T., HAMANN, F., CARSWELL, R. F., PHILLIPS, M., WILKES, B. & WILLIAMS, R. E. 1996, High Metal Enrichments in Luminous Quasars, ApJ, 461, 683.

FERNANDEZ-SOTO, A., BARCONS, X., CARBALLO, R. & WEBB, J. K. 1995, The proximity effect on the Lyman alpha forest due to a foreground QSO, MNRAS, 277, 235.

FERNANDEZ-SOTO, A., LANZETTA, K. M., BARCONS, X., CARSWELL, R. F., WEBB, J. K. & YAHIL, A. 1996, ApJ, 460, L85.

FINN, G. D. & MUGGLESTONE, D. 1965, Tables of the line broadening function H(a,v), MNRAS, 129, 221.

FIORE, F., ELVIS, M., MATHUR, S., WILKES, B. J. & MCDOWELL, J. C. 1993, The ROSAT spectrum of 3C 351 - A warm absorber in an X-ray-'quiet' quasar?, ApJ, 415, 129.

FOLTZ, C. B., WEYMANN, R. J., PETERSON, B. M., SUN, L., MALKAN, M. A. & CHAFFEE, F. H. 1986, C IV absorption systems in QSO spectra - Is the character of systems with Z(abs) ∼ Z(em) different from those with Z(abs) << Z(em)?, ApJ, 307, 504.

FOLTZ, C. B., CHAFFEE, F. H., & BLACK, J. H. 1988, Molecules at early epochs. IV - Confirmation of the detection of H2 toward PKS 0528 - 250 ApJ, 324, 267.

FOLTZ, C. B., WEYMANN, R. J., ROSER, H.-J. & CHAFFEE, F. H. 1984, Improved lower limits on Lyman-alpha forest cloud dimensions and additional evidence supporting the gravitational lens nature of 2345 + 007A,B, ApJ, 281, L1.

FUKUGITA, M., HOGAN, C. J. & PEEBLES, P. J. E. 1998, The Cosmic Baryon Budget, ApJ, 503, 518.

FYNBO, J. U., MØLLER, P. & WARREN, S. J. 1999, Extended Lyα emission from a damped Lyα absorber at z=1.93, and the relation between damped Lyα absorbers and Lyman-break galaxies, MNRAS, 305, 849.

GE, J. & BECHTOLD, J. 1997, Molecular Hydrogen Absorption in the $z = 1.97$ Damped Lyα Absorption System toward Quasi-stellar Object Q0013-004, ApJ, 477, L73.

GE, J. & BECHTOLD, J. 1999, ASP Conf. Ser. 156: *Highly Redshifted Radio Lines*, p. 121.

GE, J., BECHTOLD, J. & BLACK, J. H. 1997, A New Measurement of the Cosmic Microwave Background Radiation Temperature at Z = 1.97, ApJ, 474, 67.

GE, J., BECHTOLD, J. & KULKARNI, V. P. 2001, H2, C I, Metallicity, and Dust Depletion in the $z = 2.34$ Damped Lyα Absorption System toward QSO 1232+0815, ApJ, 547, L1.

GEORGE, I. M., TURNER, T. J., NETZER, H., NANDRA, K., MUSHOTZKY, R. F. & YAQOOB, T. 1998, ASCA Observations of Seyfert 1 Galaxies. III. The Evidence for Absorption and Emission Due to Photoionized Gas, ApJS, 114, 73.

GIALLONGO, E., D'ODORICO, S., FONTANA, A., MCMAHON, R. G., SAVAGLIO, S., CRISTIANI, S., MOLARO, P. & TREVESE, D. 1994, The Gunn-Peterson effect in the spectrum of the Z = 4.7 QSO 1202-0725: The intergalactic medium at very high redshifts, ApJ, 425, L1.

GIAVALISCO, M., MACCHETTO, F. D. & SPARKS, W. B. 1994, Narrow-band imaging of fields around optically-thick absorption systems: The line-of-sight towards Q 0000-2619, A&A, 288, 103.

GÓMEZ-GÓNZALEZ, J. & LEQUEUX, J. 1975, On the interpretation of Copernicus observations of interstellar absorption lines in front of XI Persei, A&A, 38, 29.

GRAY, D. 1992 The Observation and Analysis of Stellar Photospheres, New York: Cambridge University Press.

GREEN, P. J., ALDCROFT, T. L., MATHUR, S., WILKES, B. J., ELVIS, M. 2001, A Chandra Survey of Broad Absorption Line Quasars, ApJ, 558, 109.

GUNN, J. E., PETERSON, B. A. 1965 On the Density of Neutral Hydrogen in Intergalactic Space, ApJ, 142, 1633.

HAARDT, F. & MADAU, P. 1996, Radiative Transfer in a Clumpy Universe. II. The Ultraviolet Extragalactic Background, ApJ, 461, 20.

HAEHNELT, M. G. & STEINMETZ, M. 1998, Probing the thermal history of the intergalactic medium with Lyα absorption lines, MNRAS, 298, L21.

HAEHNELT, M. G., STEINMETZ, M. & RAUCH, M. 1998, Damped Lyα Absorber at High Redshift: Large Disks or Galactic Building Blocks?, ApJ, 495, 647.

HAIMAN, Z., REES, M. J. & LOEB, A. 1996, H 2 Cooling of Primordial Gas Triggered by UV Irradiation, ApJ, 467, 522.

HAMANN, F. & FERLAND, G. 1999, Elemental Abundances in Quasistellar Objects: Star Formation and Galactic Nuclear Evolution at High Redshifts, ARA&A, 37, 487.

HEAP, S. R., WILLIGER, G. M., SMETTE, A., HUBENY, I., SAHU, M. S., JENKINS, E. B., TRIPP, T. M. & WINKLER, J. N. 2000, STIS Observations of HE II Gunn-Peterson Absorption toward Q0302-003, ApJ, 534, 69.

HERNQUIST, L., KATZ, N., WEINBERG, D. H. & JORDI, M. 1996, The Lyman-Alpha Forest in the Cold Dark Matter Model, ApJ, 457, L51.

HEISLER, J. & OSTRIKER, J. P. 1988a, Models of the quasar population. I - A new luminosity function, ApJ, 325, 103.

HEISLER, J. & OSTRIKER, J. P. 1988b, Models of the quasar population. II - The effects of dust obscuration, ApJ, 332, 543.

HOGAN, C. J. 1998, Extragalactic Abundances of Hydrogen, Deuterium and Helium: New Steps, Missteps and Next Steps, Space Science Reviews, 84, 127.

HOWK, J. C. & SEMBACH, K. R. 1999, Ionized Gas in Damped Lyα Systems and Its Effects on Elemental Abundance Studies, ApJ, 523, L141.

HU, E. M., KIM, T., COWIE, L. L., SONGAILA, A. & RAUCH, M. 1995, The Distribution of Column Densities and B Values in the Lyα Forest, AJ, 110, 1526.

HU, W. & WHITE, M. 1997, The Damping Tail of Cosmic Microwave Background Anisotropies, ApJ, 479, 568.

HUI, L., GNEDIN, N. Y. & ZHANG, Y. 1997, The Statistics of Density Peaks and the Column Density Distribution of the Lyα Forest, ApJ, 486, 599.

HUMLICEK, R. 1979 J. Quant. Spec. & Radiative Transfer, 21, 30.

HUNSTEAD, R. W., FLETCHER, A. B. & PETTINI, M. 1990, Metal enrichment, dust, and star formation in galaxies at high redshifts. II - Lyα emission from the $z = 2.465$ absorber toward Q0836 + 113, ApJ, 356, 23.

IVANCHIK, A. V., POTEKHIN, A. Y. & VARSHALOVICH, D. A. 1999, The fine-structure constant: a new observational limit on its cosmological variation and some theoretical consequences A&A, 343, 439.

JAKOBSEN, P., BOKSENBERG, A., DEHARVENG, J. M., GREENFIELD, P., JEDRZEJEWSKI, R. & PARESCE, F. 1994, Detection of intergalactic ionized helium absorption in a high-redshift quasar, Nature, 370, 35.

JANNUZI, B. T. et al. 1998, The Hubble Space Telescope Quasar Absorption Line Key Project: XIII. A Census of Absorption-Line Systems at Low Redshift, ApJS, 118, 1.

JENKINS, E. B. 1986, The analysis of ensembles of moderately saturated interstellar lines, ApJ, 304, 739.

JENKINS, E. B. 1995, QSO Absorption Lines, Proceedings of the ESO Workshop Held at Garching, Germany, 21 - 24 November 1994, edited by Georges Meylan. Springer-Verlag Berlin Heidelberg New York. Also ESO Astrophysics Symposia, 1995, p.107, 107.

KAUFFMANN, G. 1996, Disc galaxies at z=0 and at high redshift: an explanation of the observed evolution of damped Lyα absorption systems, MNRAS, 281, 475.

KIM, T., HU, E. M., COWIE, L. L. & SONGAILA, A. 1997, The Redshift Evolution of the Lyα Forest, AJ, 114, 1.

KINMAN, T. D. 1966, Object PHL 938 in the Haro-Luyten Catalogue of Blue Stars, ApJ, 144, 1212.

KIRKMAN, D. & TYTLER, D. 1997, Intrinsic Properties of the $z \leq 2.7$ Lyα Forest from Keck Spectra of Quasar HS 1946+7658, ApJ, 484, 672.

KIRKMAN, D., TYTLER, D., BURLES, S., LUBIN, D. & O'MEARA, J. M. 2000, QSO 0130-4021: A Third QSO Showing a Low Deuterium-to-Hydrogen Abundance Ratio, ApJ, 529, 655.

KOBULNICKY, H. A. & KOO, D. C. 2000, Near-Infrared Spectroscopy of Two Galaxies at z=2.3 and z=2.9: New Probes of Chemical and Dynamical Evolution at High Redshift, ApJ, 545, 712.

KOVNER, I. & REES, M. J. 1989, Clearings in Lyman Alpha Forests, ApJ, 345, 52.

KRISS, G. A. et al. 2001 Resolving the Structure of Ionized Helium in the Intergalactic Medium with the Far Ultraviolet Spectroscopic Explorer, Science, 293, 5532.

KULKARNI, V. P. & FALL, S. M. 1993, The proximity effect and the mean intensity of ionizing radiation at low redshifts ApJ, 413, L63.

KULKARNI, V. P., HILL, J. M., SCHNEIDER, G., WEYMANN, R. J., STORRIE-LOMBARDI, L. J., RIEKE, M. J., THOMPSON, R. I. & JANNUZI, B. T. 2000, NICMOS Imaging of the Damped Lyα Absorber at z=1.89 toward LBQS 1210+1731: Constraints on Size and Star Formation Rate, ApJ, 536, 36.

KULKARNI, V. P., HILL, J. M., SCHNEIDER, G., WEYMANN, R. J., STORRIE-LOMBARDI, L. J., RIEKE, M. J., THOMPSON, R. I. & JANNUZI, B. T. 2001, A Search for the Damped Lyα Absorber at z=1.86 toward QSO 1244+3443 with NICMOS, ApJ, 551, 37.

LANE, W., SMETTE, A., BRIGGS, F., RAO, S., TURNSHEK, D. & MEYLAN, G. 1998 H I 21 Centimeter Absorption in Two Low-Redshift Damped Lyα Systems, AJ 116, 26.

LANZETTA, K. M., BOWEN, D. V., TYTLER, D. & WEBB, J. K. 1995, The gaseous extent of galaxies and the origin of Lyα absorption systems: A survey of galaxies in the fields of Hubble Space Telescope spectroscopic target QSOs, ApJ, 442, 538.

LANZETTA, K. M., MCMAHON, R. G., WOLFE, A. M., TURNSHEK, D. A., HAZARD, C. & LU, L. 1991, A new spectroscopic survey for damped Lyα absorption lines from high-redshift galaxies, ApJS, 77, 1.

LANZETTA, K. M., WOLFE, A. M. & TURNSHEK, D. A. 1987, An absorption-line survey of 32 QSOs at red wavelengths - Properties of the MG II absorbers, ApJ, 322, 739.

LANZETTA, K. M., WOLFE, A. M. & TURNSHEK, D. A. 1989, A search for molecular hydrogen and dust in the z = 2.796 damped Lyα absorber toward Q1337+113, ApJ, 344, 277.

LANZETTA, K. M., WOLFE, A. M. & TURNSHEK, D. A. 1995, The IUE Survey for Damped Lyα and Lyman-Limit Absorption Systems: Evolution of the Gaseous Content of the Universe, ApJ, 440, 435.

LAUROESCH, J. T., TRURAN, J. W., WELTY, D. E., YORK, D. G. 1996 QSO Absorption-Line Systems and Early Chemical Evolution, PASP, 108, 641.

LEMOINE, M. et al. 1999 Deuterium abundances, NewA, 4, 231.

LEVSHAKOV, S. A. 1994, Differential Measurements of the Fine-Structure Constant from Quasar Spectra: a Test of Kaluza-Klein Cosmologies MNRAS, 269, 339.

LEVSHAKOV, S. A., CHAFFEE, F. H., FOLTZ, C. B. & BLACK, J. H. 1992, Molecules at early epochs. VI - A search for the molecular hydrogen in the z = 3.391 damped Lyα system toward Q0000-263, A&A, 262, 385.

LEVSHAKOV, S. A. & KEGEL, W. H. 1997, New aspects of absorption line formation in intervening turbulent clouds - I. General principles, MNRAS, 288, 787.

LEVSHAKOV, S. A., MOLARO, P., CENTURIÓN, M., D'ODORICO, S., BONIFACIO, P. & VLADILO, G. 2000, UVES observations of QSO 0000-2620: molecular hydrogen abundance in the damped Lyalpha system at zabs = 3.3901, A&A, 361, 803.

LEVSHAKOV, S. A., TAKAHARA, F. & AGAFONOVA, I. I. 1999, Measurability of Kinetic Temperature from Metal Absorption-Line Spectra Formed in Chaotic Media, ApJ, 517, 609.

LIDDLE, A. R. & LYTH, D. H. 2000, *Cosmological inflation and large-scale structure*, New York: Cambridge University Press.

LOPEZ, S., REIMERS, D., RAUCH, M., SARGENT, W. L. W. & SMETTE, A. 1999, First Comparison of Ionization and Metallicity in Two Lines of Sight toward HE 1104-1805 AB at z=1.66, ApJ, 513, 598.

LOW, F. J., CUTRI, R. M., KLEINMANN, S. G., & HUCHRA, J. P. 1989, The properties of infrared color-selected quasars, ApJ, 340, L1.

LOWENTHAL, J. D., HOGAN, C. J., GREEN, R. F., CAULET, A., WOODGATE, B. E., BROWN, L. & FOLTZ, C. B. 1991, Discovery of a Lyα galaxy near a damped Lyα absorber at z = 2.3, ApJ, 377, L73.

LOWENTHAL, J. D., HOGAN, C. J., GREEN, R. F., WOODGATE, B., CAULET, A., BROWN, L. & BECHTOLD, J. 1995, Imaging and Spectroscopy of Damped Lα Quasi-stellar Object Absorption-Line Clouds, ApJ, 451, 484.

LOWENTHAL, J. D., HOGAN, C. J., LEACH, R. W., SCHMIDT, G. D. & FOLTZ, C. B. 1990, Spectroscopic Limits on High Redshift Lyα Emission, ApJ, 357, 3.

LU, L., SARGENT, W. L. W. & BARLOW, T. A. 1997, Evidence for Rotation in the Galaxy at z=3.15 Responsible for a Damped Lyα Absorption System in the Spectrum of Q2233+1310, ApJ, 484, 131.

LU, L., SARGENT, W. L. W. & BARLOW, T. A. 1998, The N/Si abundance ratio in 15 damped Lyα galaxies - Implications for the origin of nitrogen, AJ, 115, 55.

LU, L., SARGENT, W. L. W., BARLOW, T. A., CHURCHILL, C. W. & VOGT, S. S. 1996, Abundances at High Redshifts: The Chemical Enrichment History of Damped Lyα Galaxies, ApJS, 107, 475.

LU, L., SARGENT, W. L. W., WOMBLE, D. S. & TAKADA-HIDAI, M. 1996, The Lyα Forest at $z \sim 4$: Keck HIRES Observations of Q0000-26, ApJ, 472, 509.

LU, L. & WOLFE, A. M. 1994, More confirmed damped Lyα systems from the Las Campanas/Palomar survey, AJ, 108, 44.

LU, L., WOLFE, A. M. & TURNSHEK, D. A. 1991, The redshift distribution of Lyα clouds and the proximity effect, ApJ, 367, 19.

LU, L., WOLFE, A. M., TURNSHEK, D. A. & LANZETTA, K. M. 1993, A spectroscopic study of damped Lyα systems in the Las Campanas/Palomar survey, ApJS, 84, 1.

MACHACEK, M. E., BRYAN, G. L., MEIKSIN, A., ANNINOS, P., THAYER, D., NORMAN, M. & ZHANG, Y. 2000, Hydrodynamical Simulations of the Lyα Forest: Model Comparisons, ApJ, 532, 118.

MALHOTRA, S. 1997, Detection of the 2175 A Dust Feature in Mg II Absorption Systems, ApJ, 488, L101.

MANNUCCI, F., THOMPSON, D., BECKWITH, S. V. W. & WILLIGER, G. M. 1998, Infrared Emission-Line Galaxies Associated with Dampened Lyα and Strong Metal Absorber Redshifts, ApJ, 501, L11.

MATHUR, S., WILKES, B., ELVIS, M., & FIORE, F. 1994, The X-ray and ultraviolet absorbing outflow in 3C 351 ApJ, 434, 493.

MCGILL, C. 1990, The redshift projection. II - Caustics and the Lyα forest, MNRAS, 242, 544.

MEYER, D. M., LANZETTA, K. M. & WOLFE, A. M. 1995, The Iron Group Abundance Pattern of the Damped Lyα Absorber at z = 1.3726 toward the QSO 0935+417, ApJ, 451, L13.

MEYER, D. M. & ROTH, K. C. 1990, Observations of nickel, chromium, and zinc in QSO absorption-line systems, ApJ, 363, 57.

MEYER, D. M. & YORK, D. G. 1992, Observations of Zn II, CR II, Fe II, and CA II in the damped Lyman-alpha absorber at z = 0.692 toward 3C 286, ApJ, 399, L121.

MEYER, D. M., YORK, D. G., BLACK, J. H., CHAFFEE, F. H. & FOLTZ, C. B. 1986, An upper limit on the microwave background temperature at Z = 1.776, ApJ, 308, L37.

MEYER, D. M., WELTY, D. E. & YORK, D. G. 1989, Element abundances at high redshift, ApJ, 343, L37.

MEYLAN, G. 1995 Proceedings of the ESO Workshop Held at Garching, Germany, 21 - 24 November 1994 Springer-Verlag Berlin Heidelberg New York.

MICHALITSIANOS, A. G. et al. 1997, Lyα Absorption-Line Systems in the Gravitational Lens Q0957+561, ApJ, 474, 598.

MIRALDA-ESCUDE, J., CEN, R., OSTRIKER, J. P. & RAUCH, M. 1996, The Lyα Forest from Gravitational Collapse in the Cold Dark Matter + Lambda Model, ApJ, 471, 582.

MIRALDA-ESCUDE, J. & OSTRIKER, J. P. 1990, What produces the ionizing background at large redshift?, ApJ, 350, 1.

MOLARO, P., BONIFACIO, P., CENTURION, M. & VLADILO, G. 1999, Low deuterium abundance in the z_{abs}=3.514 absorber towards APM 08279+5255, A&A, 349, L13.

MOLLER, P. & WARREN, S. J. 1993, Emission from a damped Lyα absorber at z = 2.81, A&A, 270, 43.

MONIER, E. M., TURNSHEK, D. A. & HAZARD, C. 1999, Hubble Space Telescope Faint Object Spectrograph Observations of a Unique Grouping of Five QSOS: The Sizes and Shapes of Low-z Lyα Forest Absorbers, ApJ, 522, 627.

MONIER, E. M., TURNSHEK, D. A. & LUPIE, O. L. 1998, Hubble Space Telescope Observations of the Gravitationally Lensed Cloverleaf Broad Absorption Line QSO H1413+1143: Spectroscopy of the Lyα Forest and Metal-Line Systems, ApJ, 496, 177.

MORRIS, S. L., WEYMANN, R. J., DRESSLER, A., MCCARTHY, P. J., SMITH, B. A., TERRILE, R. J., GIOVANELLI, R. & IRWIN, M. 1993, The Environment of Lyα Absorbers in the Sight Line toward 3C 273, ApJ, 419, 524.

MORSE, J. A., SHULL, J. M. & KINNEY, A. L. 1999, *ASP Conf. Ser. 164: Ultraviolet-Optical Space Astronomy Beyond HST*.

MORTON, D. C. 1991, Atomic data for resonance absorption lines. I - Wavelengths longward of the Lyman limit, ApJS, 77, 119.

MORTON, D. C. 2000, Atomic Data for Resonance Absorption Lines. II. Wavelengths Longward of the Lyman Limit for Heavy Elements ApJS, 130, 403.

MORTON, D. C., YORK, D. G. & JENKINS, E. B. 1988, A search list of lines for quasi-stellar object absorption systems ApJS, 68, 449.

MURDOCH, H. S., HUNSTEAD, R. W., PETTINI, M. & BLADES, J. C. 1986, Absorption spectrum of the Z = 3.78 QSO 2000-330. II - The redshift and equivalent width distributions of primordial hydrogen clouds, ApJ, 309, 19.

MURPHY, M.T., WEBB, J. K., FLAMBAUM, V. V., CHURCHILL, C. W. & PROCHASKA, J. X. 2000b Possible Evidence for a variable fine structure constant from QSO absorption lines: systematic errors, astro-ph/0012420.

MURPHY, M.T., WEBB, J. K., FLAMBAUM, V. V., DZUBA, V. A., CHURCHILL, C. W., PROCHASKA, J. X., BARROW, J. D. & WOLFE, A. M. 2000a Possible Evidence for a variable fine structure constant from QSO absorption lines: motivations, analysis and results, astro-ph/0012419.

MURPHY, M.T., WEBB, J. K., FLAMBAUM, V. V., PROCHASKA, J. X. & WOLFE, A. M. 2000c Further constraints on variation of the fine structure constant from alkali doublet QSO absorption lines, astro-ph/0012421.

NACHMAN, P. & HOBBS, L. M. 1973, The doublet-ratio method and interstellar abundances., ApJ, 182, 481.

NAJITA, J., SILK, J. & WACHTER, K. W. 1990, Dust obscuration by an evolving galaxy population, ApJ, 348, 383.

O'MEARA, J. M., TYTLER, D., KIRKMAN, D., SUZUKI, N., PROCHASKA, J. X., LUBIN, D. & WOLFE, A. M. 2001, The Deuterium to Hydrogen Abundance Ratio toward a Fourth QSO: HS 0105+1619, ApJ, 552, 718.

ORTIZ-GIL, A., LANZETTA, K. M., WEBB, J. K., BARCONS, X. & FERNÁNDEZ-SOTO, A. 1999, The Gaseous Extent of Galaxies and the Origin of Lyα Absorption Systems. IV. Lyα Absorbers Arising in a Galaxy Group, ApJ, 523, 72.

OSTRIKER, J. P., BAJTLIK, S. & DUNCAN, R. C. 1988, Clustering and voids in the Lyα forest, ApJ, 327, L35.

OSTRIKER, J. P. & HEISLER, J. 1984, Are cosmologically distant objects obscured by dust? - A test using quasars, ApJ, 278, 1.

OSTRIKER, J. P. & VISHNIAC, E. T. 1986, Generation of microwave background fluctuations from nonlinear perturbations at the era of galaxy formation, ApJ, 306, L51.

OSTRIKER, J. P., VOGELEY, M. S. & YORK, D. G. 1990, Dust in QSO Absorption-Line Systems, ApJ, 364, 405.

OUTRAM, P. J., CARSWELL, R. F. & THEUNS, T. 2000, Non-Voigt Lyα Absorption Line Profiles, ApJ, 529, L73.

PARNELL, H. C. & CARSWELL, R. F. 1988, Effects of blending on estimates of the redshift evolution of QSO Ly-alpha absorbers MNRAS, 230, 491.

PEI, Y. C. & FALL, S. M. 1995, Cosmic Chemical Evolution ApJ, 454, 69.

PEI, Y. C., FALL, S. M. & BECHTOLD, J. 1991, Confirmation of dust in damped Lyα systems, ApJ, 378, 6.

PEI, Y. C., FALL, S. M. & HAUSER, M. G. 1999, Cosmic Histories of Stars, Gas, Heavy Elements, and Dust in Galaxies, ApJ, 522, 604.

PENTON, S. V., SHULL, J. M. & STOCKE, J. T. 2000, The Local Lyα Forest. II. Distribution of H I Absorbers,Doppler Widths, and Baryon Content, ApJ, 544, 150.

PENTON, S. V., STOCKE, J. T. & SHULL, J. M. 2000, The Local Lyα Forest. I. Observations with the GHRS/G160M on the Hubble Space Telescope, ApJS, 130, 121.

PETITJEAN, P. & CHARLOT, S. 1997 *Structure and Evolution of the Intergalactic Medium from QSO Absorption Line Systems, Proceedings of the 13th IAP Astrophysics Colloquium*, 1-5 July 1997, Institut d'Astrophysique, Paris, Editions Frontieres

PETITJEAN, P., SRIANAND, R. & LEDOUX, C. 2000, Molecular hydrogen and the nature of damped Lyman-alpha systems, A&A, 364, L26.

Petitjean, P., Webb, J. K., Rauch, M., Carswell, R. F. & Lanzetta, K. 1993, MNRAS, 262, 499.

Petry, C. E., Impey, C. D. & Foltz, C. B. 1998, Small-Scale Structure in the Lyα Forest at High Redshift, ApJ, 494, 60.

Pettini, M. 1999, Element Abundances at High Redshifts, in *Chemical Evolution from Zero to High Redshift*, Edited by Jeremy R. Walsh, Michael R. Rosa, Berlin: Springer-Verlag, p. 233.

Pettini, M. & Bowen, D. V. 1997, Zinc and chromium abundances in a third damped Lyα system at intermediate redshift, A&A, 327, 22.

Pettini, M., Ellison, S. L., Steidel, C. C. & Bowen, D. V. 1999, Metal Abundances at $z < 1.5$: Fresh Clues to the Chemical Enrichment History of Damped Lyα Systems, ApJ, 510, 576.

Pettini, M., Ellison, S. L., Steidel, C. C., Shapley, A. E. & Bowen, D. V. 2000, Si and Mn Abundances in Damped Lyα: Systems with Low Dust Content, ApJ, 532, 65.

Pettini, M., King, D. L., Smith, L. J. & Hunstead, R. W. 1997, Dust in High-Redshift Galaxies, ApJ, 478, 536.

Pettini, M., Lipman, K. & Hunstead, R. W. 1995, Element Abundances at High Redshifts: The N/O Ratio in a Primeval Galaxy, ApJ, 451, 100.

Pettini, M., Shapley, A. E., Steidel, C. C., Cuby, J., Dickinson, M., Moorwood, A. F. M., Adelberger, K. L. & Giavalisco, M. 2001, The Rest-Frame Optical Spectra of Lyman Break Galaxies: Star Formation, Extinction, Abundances, and Kinematics, ApJ, 554, 981.

Pettini, M., Smith, L. J., Hunstead, R. W. & King, D. L. 1994, Metal enrichment, dust, and star formation in galaxies at high redshifts : Zn and CR abundances for 17 damped Lyman-alpha systems, ApJ, 426, 79.

Pettini, M., Smith, L. J., King, D. L. & Hunstead, R. W. 1997, The Metallicity of High-Redshift Galaxies: The Abundance of Zinc in 34 Damped Lyα Systems from $z = 0.7$ to 3.4, ApJ, 486, 665.

Pilachowski, C., Dekker, H., Hinkle, K., Tull, R., Vogt, S., Walker, D. D., Diego, F. & Angel, R. 1995, High-Resolution Spectrographs for Large Telescopes, PASP, 107, 983.

Press, W. H. & Rybicki, G. B. 1993, Properties of High-Redshift Lyman-Alpha Clouds. II. Statistical Properties of the Clouds, ApJ, 418, 585.

Press, W. H., Rybicki, G. B. & Schneider, D. P. 1993, Properties of high-redshift Lyman-alpha clouds. I - Statistical analysis of the Schneider-Schmidt-Gunn quasars, ApJ, 414, 64.

Prochaska, J. X. & Wolfe, A. M. 1996, A Keck HIRES Investigation of the Metal Abundances and Kinematics of the Z = 2.46 Damped Lyα System toward Q0201+365, ApJ, 470, 403.

Prochaska, J. X. & Wolfe, A. M. 1997a, A Keck HIRES Investigation of the Metal Abundances and Kinematics of Three Damped Lyα Systems toward Q2206-199, ApJ, 474, 140.

Prochaska, J. X. & Wolfe, A. M. 1997b, On the Kinematics of the Damped Lyα Protogalaxies, ApJ, 487, 73.

Prochaska, J. X. & Wolfe, A. M. 1998, Protogalactic Disk Models of Damped Lyalpha Kinematics, ApJ, 507, 113.

Prochaska, J. X. & Wolfe, A. M. 1999, Chemical Abundances of the Damped Lyα Systems at $z > 1.5$, ApJS, 121, 369.

Rao, S. M. & Turnshek, D. A. 2000, The Incidence of Damped Lyα Systems in the Redshift Interval $0 < z < 4$, ApJS, 130, 1.

Rao, S. M., Turnshek, D. A. & Briggs, F. H. 1995, Discovery of Damped Lyα Systems at Redshifts Less than 1.65 and Results on Their Incidence and Cosmological Mass Density, ApJ, 449, 488.

Rauch, M. 1998, *The Lyman Alpha Forest in the Spectra of QSOs*, ARA&A, 36, 267.

RAUCH, M. et al. 1997, The Opacity of the Lyα Forest and Implications for Ω_B and the Ionizing Background, ApJ, 489, 7.

RAUCH, M., CARSWELL, R. F., CHAFFEE, F. H., FOLTZ, C. B., WEBB, J. K., WEYMANN, R. J., BECHTOLD, J. & GREEN, R. F. 1992, The Lyman forest of 0014 + 813, ApJ, 390, 387.

RAUCH, M. & HAEHNELT, M. G. 1995, $\Omega_{\rm baryon}$ and the geometry of intermediate-redshift Lymanα absorption systems MNRAS, 275, L76.

RAUCH, M., SARGENT, W. L. W. & BARLOW, T. A. 1999, Small-Scale Structure at High Redshift. I. Glimpses of the Interstellar Medium at Redshift 3.5, ApJ, 515, 500.

REES, M. J. & SETTI, G. 1970, Absorption and Scattering of Ultraviolet and X-Ray Photons by Intergalactic Gas, A&A, 8, 410.

REEVES, H., AUDOUZE, J., FOWLER, W. A. & SCHRAMM, D. N. 1973, On the Origin of Light Elements, ApJ, 179, 909.

REIMERS, D., BAADE, R., HAGEN, H.-J. & LOPEZ, S. 2001, High-resolution O VI absorption line observations at $1.2 \leq z \leq 1.7$ in the bright QSO HE 0515-4414, A&A, 374, 871.

REIMERS, D., KOHLER, S., WISOTZKI, L., GROOTE, D., RODRIGUEZ-PASCUAL, P. & WAMSTEKER, W. 1997, Patchy intergalactic He II absorption in HE 2347-4342. II. The possible discovery of the epoch of He-reionization, A&A, 327, 890.

RICOTTI, M., GNEDIN, N. Y. & SHULL, J. M. 2000, The Evolution of the Effective Equation of State of the Intergalactic Medium ApJ, 534, 41.

ROBERTS, M. S., BROWN, R. L., BRUNDAGE, W. D., ROTS, A. H., HAYNES, M. P. & WOLFE, A. M. 1976, Detection at z=0.5 of a 21 CM absorption line in AO0235+164 : the first coincidence of large radio and optical redshifts, AJ, 81, 293.

ROTH, K. C. & BAUER, J. M. 1999, The Z = 1.6748 C I Absorber toward PKS 1756+237, ApJ, 515, L57.

RUGERS, M. & HOGAN, C. J. 1996a, Confirmation of High Deuterium Abundance in Quasar Absorbers, ApJ, 459, L1.

RUGERS, M. & HOGAN, C. J. 1996b, High Deuterium Abundance in a New Quasar Absorber, AJ, 111, 2135.

RYBICKI, G. & LIGHTMAN, A. 1979 *Radiative Processes in Astrophysics* New York: Wiley.

SANDAGE, A. 1965 The Existance of a Major New Constituent of the Universe: The Quasi-stellar Galaxies, ApJ, 141, 1560.

SARAZIN, C. L., FLANNERY, B. P. & RYBICKI, G. B. 1979, On the distance from quasars to absorbing clouds, ApJ, 227, L113.

SARGENT, W. L. W., STEIDEL, C. C. & BOKSENBERG, A. 1988a, Mg II absorption in the spectra of high and low redshift QSOs, ApJ, 334, 22.

SARGENT, W. L. W., STEIDEL, C. C. & BOKSENBERG, A. 1988b, C IV absorption in a new sample of 55 QSOs - Evolution and clustering of the heavy-element absorption redshifts, ApJS, 68, 539.

SARGENT, W. L. W., STEIDEL, C. C. & BOKSENBERG, A. 1989, A survey of Lyman-limit absorption in the spectra of 59 high-redshift QSOs, ApJS, 69, 703.

SARGENT, W. L. W., YOUNG, P. J., BOKSENBERG, A., CARSWELL, R. F. & WHELAN, J. A. J. 1979, A high-resolution study of the absorption spectra of the QSOs Q0002-422 and Q0453-423 ApJ, 230, 49.

SARGENT, W. L. W., YOUNG, P. J., BOKSENBERG, A. & TYTLER, D. 1980 The Distribution of Lyman-alpha Absorption Lines in the Spectra of Six QSOs: Evidence for an Intergalactic Origin, ApJS, 42, 41.

SARGENT, W. L. W., YOUNG, P. & SCHNEIDER, D. P. 1982, Intergalactic Lyman-alpha absorption lines in a close pair of high-redshift QSOs, ApJ, 256, 374.

SAVAGE, B. D. et al. 1993, The Hubble Space Telescope Quasar Absorption Line Key Project. III - First observational results on Milky Way gas ApJ, 413, 116.

SAVAGE, B. D. et al. 2000, The Hubble Space Telescope Quasar Absorption Line Key Project.

XV. Milky Way Absorption Lines, ApJS, 129, 563.

SAVAGE, B. D. & SEMBACH, K. R. 1991, The analysis of apparent optical depth profiles for interstellar absorption lines, ApJ, 379, 245.

SAVEDOFF, M. P. 1956, Nature, 178, 688.

SCHAYE, J., RAUCH, M., SARGENT, W. L. W. & KIM, T. 2000, The Detection of Oxygen in the Low-Density Intergalactic Medium, ApJ, 541, L1.

SCHAYE, J., THEUNS, T., LEONARD, A. & EFSTATHIOU, G. 1999, Measuring the equation of state of the intergalactic medium, MNRAS, 310, 57.

SCHAYE, J., THEUNS, T., RAUCH, M., EFSTATHIOU, G. & SARGENT, W. L. W. 2000, The thermal history of the intergalactic medium, MNRAS, 318, 817.

SCHNEIDER, D. P. et al. 1993, The Hubble Space Telescope Quasar Absorption Line Key Project. II. Data calibration and absorption-line selection ApJS, 87, 45.

SCHRAMM, D. N. & TURNER, M. S. 1998, Big-bang nucleosynthesis enters the precision era, Reviews of Modern Physics, 70, 303.

SCHROEDER, D. J. 2000 *Astronomical Optics* San Diego: Academic Press.

SCOTT, J., BECHTOLD, J. & DOBRZYCKI, A. 2000a, A Uniform Analysis of the Lyα Forest at Z=0-5. I. The Sample and Distribution of Clouds at Z$>$ 1.7, ApJS, 130, 37.

SCOTT, J., BECHTOLD, J., DOBRZYCKI, A. & KULKARNI, V. P. 2000b, A Uniform Analysis of the Lyα Forest at $z = 0-5$. II. Measuring the Mean Intensity of the Extragalactic Ionizing Background Using the Proximity Effect, ApJS, 130, 67.

SCOTT, J., BECHTOLD, J., MORITA, M., DOBRZYCKI, A. & KULKARNI, V. P. 2001, A Uniform Analysis of the Lyman Alpha Forest at z = 0 - 5: V. The extragalactic ionizing background at low redshift ApJS, submitted.

SEMBACH, K. R., STEIDEL, C. C., MACKE, R. J. & MEYER, D. M. 1995, A critical analysis of interstellar Zn and CR as galactic abundance benchmarks for quasar absorbers, ApJ, 445, L27.

SHULL, J. M., PENTON, S. V., STOCKE, J. T., GIROUX, M. L., VAN GORKOM, J. H., LEE, Y. H. & CARILLI, C. 1998, A Cluster of Low-Redshift Lyα Clouds toward PKS 2155-304: I. Limits on Metals and D/H, AJ, 116, 2094.

SHULL, J. M., PENTON, S. & STOCKE, J. T. 1999, The Low-Redshift Intergalactic Medium, PASA, 16, 95.

SHULL, J. M., STOCKE, J. T. & PENTON, S. 1996, Intergalactic Hydrogen Clouds at Low Redshift: Connections to Voids and Dwarf Galaxies, AJ, 111, 72.

SMETTE, A., HEAP, S. R., WILLIGER, G. M., TRIPP, T. M., JENKINS, E. B. & SONGAILA, A. 2001, ASP Conf. Ser. 240: Gas and Galaxy Evolution, 17.

SMETTE, A., ROBERTSON, J. G., SHAVER, P. A., REIMERS, D., WISOTZKI, L. & KOEHLER, T. 1995, The gravitational lens candidate HE 1104-1805 and the size of absorption systems, A&A, 113, 199.

SMETTE, A., SURDEJ, J., SHAVER, P. A., FOLTZ, C. B., CHAFFEE, F. H., WEYMANN, R. J., WILLIAMS, R. E. & MAGAIN, P. 1992, A spectroscopic study of UM 673 A and B - On the size of Lyα clouds, ApJ, 389, 39.

SONGAILA, A. 1998, The Redshift Evolution of the Metagalactic Ionizing Flux Inferred from Metal Line Ratios in the Lyman Forest AJ, 115, 2184.

SONGAILA, A. et al. 1994, Measurement of the Microwave Background Temperature at a Redshift of 1.776, Nature, 371, 43.

SONGAILA, A. & COWIE, L. L. 1996, Metal enrichment and Ionization Balance in the Lyman Alpha Forest at Z = 3, AJ, 112, 335.

SONGAILA, A., COWIE, L. L., HOGAN, C. J. & RUGERS, M. 1994, Deuterium Abundance and Background Radiation Temperature in High Redshift Primordial Clouds, Nature, 368, 599.

SONGAILA, A., HU, E. M., COWIE, L. L. & MCMAHON, R. G. 1999, Limits on the Gunn-Peterson Effect at Z = 5, ApJ, 525, L5.

SONGAILA, A., WAMPLER, E. J. & COWIE, L. L. 1997, A high deuterium abundance in the early Universe, Nature, 385, 137.

SPITZER, L. 1978, *Interstellar Matter*, New York: Wiley.

SPITZER, L., DRAKE, J. F., JENKINS, E. B., MORTON, D. C., ROGERSON, J. B. & YORK, D. G. 1973, Spectrophotometric Results from the Copernicus Satellite: IV. Molecular Hydrogen in Interstellar Space, ApJ, 181, L116.

STEIDEL, C. C. 1995, QSO Absorption Lines, Proceedings of the ESO Workshop Held at Garching, Germany, 21 - 24 November 1994, edited by Georges Meylan. Springer-Verlag Berlin Heidelberg New York. Also ESO Astrophysics Symposia p.139, 139.

STEIDEL, C. C., BOWEN, D. V., BLADES, J. C. & DICKENSON, M. 1995, The z = 0.8596 damped Lyα absorbing galaxy toward PKS 0454+039, ApJ, 440, L45.

STEIDEL, C. C., DICKINSON, M., MEYER, D. M., ADELBERGER, K. L. & SEMBACH, K. R. 1997, Quasar Absorbing Galaxies at $z \leq 1$. I. Deep Imaging and Spectroscopy in the Field of 3C 336, ApJ, 480, 568.

STEIDEL, C. C., DICKINSON, M. & PERSSON, S. E. 1994, Field galaxy evolution since Z approximately 1 from a sample of QSO absorption-selected galaxies, ApJ, 437, L75.

STEIDEL, C. C., PETTINI, M. & ADELBERGER, K. L. 2001, Lyman-Continuum Emission from Galaxies at Z ~ 3.4, ApJ, 546, 665.

STEIDEL, C. C., PETTINI, M., DICKINSON, M. & PERSSON, S. E. 1994, Imaging of two damped Lyα absorbers at intermediate redshifts, AJ, 108, 2046.

STEIDEL, C. C. & SARGENT, W. L. W. 1987, A new upper limit on the density of generally distributed intergalactic neutral hydrogen, ApJ, 318, L11.

STEIDEL, C. C. & SARGENT, W. L. W. 1992, MG II absorption in the spectra of 103 QSOs - Implications for the evolution of gas in high-redshift galaxies, ApJS, 80, 1.

STEINMETZ, M. 2001, ASP Conf. Ser. 230: *Galaxy Disks and Disk Galaxies*, p. 633

STENGLER-LARREA, E. A. *et al.* 1995, The Hubble Space Telescope Quasar Absorption Line Key Project. V: Redshift Evolution of Lyman limit absorption in the spectra of a large sample of quasars, ApJ, 444, 64.

STOCKE, J. T., SHULL, J. M., PENTON, S., DONAHUE, M. & CARILLI, C. 1995, The Local Lyα Forest: Association of Clouds with Superclusters and Voids, ApJ, 451, 24.

STORRIE-LOMBARDI, L. J. & WOLFE, A. M. 2000, Surveys for $z > 3$ Damped Lyα Absorption Systems: The Evolution of Neutral Gas, ApJ, 543, 552.

SWIHART, T. 1976 in Introductory Theoretical Astrophysics, ed. R. Weymann Tucson: Pachart Press.

TARAMOPOULOS, A., GARWOOD, R., BRIGGS, F. H. & WOLFE, A. M. 1995, A search for H I 21 cm absorption and emission from a damped Lyα absorption line system at redshift z ~3.06 in the direction of PKS 0336-017, AJ, 109, 480.

THEUNS, T., LEONARD, A., EFSTATHIOU, G., PEARCE, F. R. & THOMAS, P. A. 1998, P^{3}M-SPH simulations of the Lyα forest, MNRAS, 301, 478.

THOMPSON, R. I. 1975, The determination of the electron to proton inertial mass ratio via molecular transitions, ApL, 16, 3.

TRIMBLE, V. 1995, QSO Absorption Lines, Proceedings of the ESO Workshop Held at Garching, Germany, 21 - 24 November 1994, edited by Georges Meylan. Springer-Verlag Berlin Heidelberg New York. Also ESO Astrophysics Symposia, 1995, p.453, 453

TRIPP, T. M., GREEN, R. F. & BECHTOLD, J. 1990, IUE observations of PG 1115 + 080 - The He I Gunn-Peterson test and a search for the lensing galaxy, ApJ, 364, L29.

TRIPP, T. M., LU, L. & SAVAGE, B. D. 1998, The Relationship between Galaxies and Low-Redshift Weak Lyα Absorbers in the Directions of H1821+643 and PG 1116+215, ApJ, 508, 200.

TRIPP, T. M. & SAVAGE, B. D. 2000, O VI and Multicomponent H I Absorption Associated with a Galaxy Group in the Direction of PG 0953+415: Physical Conditions and Baryonic Content, ApJ, 542, 42.

TRIPP, T. M., SAVAGE, B. D. & JENKINS, E. B. 2000, Intervening O VI Quasar Absorption

Systems at Low Redshift: A Significant Baryon Reservoir, ApJ, 534, L1.

TURNSHEK, D. A. 1984, Properties of the Broad Lines QSOs ApJ, 280, 51.

TURNSHEK, D. A. 1995, The Covering Factors, Ionization Structure, and Chemical Composition of QSO Broad Absorption Line Region Gas, in QSO Absorption Lines, Proceedings of the ESO Workshop Held at Garching, Germany, 21 - 24 November 1994, edited by Georges Meylan. *Ibid.*, p. 223.

TURNSHEK, D. A. & BOHLIN, R. C. 1993, Damped Lyman-alpha absorption in Q0957+561A, B, ApJ, 407, 60.

TURNSHEK, D. A., WOLFE, A. M., LANZETTA, K. M., BRIGGS, F. H., COHEN, R. D., FOLTZ, C. B., SMITH, H. E. & WILKES, B. J. 1989, Damped Lyα absorption by disk galaxies with large redshifts. III - Intermediate-resolution spectroscopy, ApJ, 344, 567.

TYTLER, D. 1982, QSO Lyman limit absorption, Nature, 298, 427.

TYTLER, D. 1987 The Distribution of QSO Absorption System Column Densities: Evidence for a Single Population, ApJ 321, 49.

TYTLER, D., BURLES, S., LU, L., FAN, X., WOLFE, A. & SAVAGE, B. D. 1999, The Deuterium Abundance at z=0.701Z = 0.701 toward QSO 1718+4807, AJ, 117, 63.

TYTLER, D., FAN, X.-M., & BURLES, S. 1996, Cosmological baryon density derived from the deuterium abundance at redshift $z = 3.57$, Nature, 381, 207.

TYTLER, D., FAN, X.-M., BURLES, S., COTTRELL, L., DAVIS, C., KIRKMAN, D. & ZUO, L. 1995, QSO Absorption Lines, Proceedings of the ESO Workshop Held at Garching, Germany, 21 - 24 November 1994, edited by Georges Meylan. Springer-Verlag Berlin Heidelberg New York. Also ESO Astrophysics Symposia, p.289.

TYTLER, D., O'MEARA, J. M., SUZUKI, N. & LUBIN, D. 2000, Deuterium and the baryonic density of the Universe, Physics Reports, 333, 409.

ULMER, A. 1996, Strong Clustering in the Low-Redshift Lyα Forest, ApJ, 473, 110.

VAN GORKOM, J. H., CARILLI, C. L., STOCKE, J. T., PERLMAN, E. S. & SHULL, J. M. 1996, The HI Environment of Nearby Lyα Absorbers, AJ, 112, 1397.

VARSHALOVICH, D. A. & POTEKHIN, A. Y. 1995, Cosmological Variability of Fundamental Physical Constants, Space Science Reviews, 74, 259.

VARSHALOVICH, D. A. & POTEKHIN, A. Y. 1996, Have the masses of molecules changed during the lifetime of the Universe?, Astronomy Letters, 22, 1.

VERNER, D. A., BARTHEL, P. D. & TYTLER, D. 1994 Atomic data for absorption lines from the ground level at wavelengths greater than 228A, A&A, 108, 287.

VERNER, D. A., TYTLER, D. & BARTHEL, P. D. 1994, Far-ultraviolet absorption spectra of quasars: How to find missing hot gas and metals, ApJ, 430, 186.

VOGT, S. S. *et al.* 1994, HIRES: the High-resolution Echelle Spectrometer on the Keck 10-m Telescope, Proc. SPIE, 2198, 362.

WADSLEY, J. & BOND, J. R. 1996, Probing the High Redshift IGM: SPH+P(3) MG Simulations of the Lyman-alpha Forest, A&AS, 189, 110402.

WAMPLER, E. J., WILLIGER, G. M., BALDWIN, J. A., CARSWELL, R. F., HAZARD, C. & MCMAHON, R. G. 1996, High resolution observations of the QSO BR 1202-0725: deuterium and ionic abundances at redshifts above z=4, A&A, 316, 33.

WARREN, S. J. & MOLLER, P. 1996, Further spectroscopy of emission from a damped Lyα absorber at z=2.81, A&A, 311, 25.

WARREN, S. J., MØLLER, P., FALL, S. M. & JAKOBSEN, P. 2001, NICMOS imaging search for high-redshift damped Lyα galaxies, MNRAS, 326, 759.

WEBB, J. K., BARCONS, X., CARSWELL, R. F. & PARNELL, H. C. 1992, The Gunn-Peterson effect and the H I column density distribution of Lyα forest clouds at $z = 4$, MNRAS, 255, 319.

WEBB, J. K., CARSWELL, R. F., IRWIN, M. J. & PENSTON, M. V. 1991, On measuring the deuterium abundance in QSO absorption systems, MNRAS, 250, 657.

WEBB, J. K., CARSWELL, R. F., LANZETTA, K. M., FERLET, R., LEMOINE, M., VIDAL-

MADJAR, A. & BOWEN, D. V. 1997, A high deuterium abundance at redshift z=0.7, Nature, 388, 250.

WEBB, J. K., MURPHY, M.T., FLAMBAUM, V.V., DZUBA, V. A., BARROW, J. D., CHURCHILL, C. W., PROCHASKA, J. X., WOLFE, A. M. 2000, Further Evidence for Cosmological Evolution of the Fine Structure Constant, astro-ph/0012539.

WEINBERG, D. H., CROFT, R. A. C., HERNQUIST, L., KATZ, N., & PETTINI, M. 1999, Closing In on Ω_M: The Amplitude of Mass Fluctuations from Galaxy Clusters and the Lyα Forest, ApJ, 522, 563.

WEINBERG, D. H., HERNQUIST, L. & KATZ, N. 1997, Photoionization, Numerical Resolution, and Galaxy Formation, ApJ, 477, 8.

WEISHEIT, J. C. 1978, On the use of line shapes in the analysis of QSO absorption spectra, ApJ, 219, 829.

WEYMANN, R. J. et al. 1998, The Hubble Space Telescope Quasar Absorption Line Key Project. XIV: The Evolution of Lyα Absorption Lines in the Redshift Interval Z = 0-1.5, ApJ, 506, 1.

WEYMANN, R. J., CARSWELL, R. F., SMITH, M. G. 1981 Absorption Lines in the Spectra of Quasistellar Objects, ARA&A, 19, 41.

WEYMANN, R. J. & FOLTZ, C. B. 1983, Common Lyman-alpha absorption lines in the triple QSO PG 1115+08, ApJ, 272, L1.

WIKLIND, T. & COMBES, F. 1994a, First detection of CO(J=0-¿1) absorption at cosmological distances (z=0.247), A&A, 286, L9.

WIKLIND, T. & COMBES, F. 1994b, A search for millimeterwave CO emission in damped Lyα systems, A&A, 288, L41.

WIKLIND, T. & COMBES, F. 1995, CO, HCO+ and HCN absorption in the gravitational lens candidate B0218+357 at z =0.685, A&A, 299, 382.

WIKLIND, T. & COMBES, F. 1996a, The redshift of the gravitational lens of PKS 1830-211 determined from molecular absorption lines, Nature, 379, 139.

WIKLIND, T. & COMBES, F. 1996b, Molecular absorptions towards the AGN 1504+377, A&A, 315, 86.

WIKLIND, T. & COMBES, F. 1997, Molecular absorption lines at high redshift: PKS 1413+135 (z=0.247), A&A, 328, 48.

WIKLIND, T. & COMBES, F. 1998, The Complex Molecular Absorption Line System at z = 0.886 toward PKS 1830-211, ApJ, 500, 129.

WILLIGER, G. M., BALDWIN, J. A., CARSWELL, R. F., COOKE, A. J., HAZARD, C., IRWIN, M. J., MCMAHON, R. G., & STORRIE-LOMBARDI, L. J. 1994, Lyα absorption in the spectrum of the z = 4.5 QSO BR 1033-0327, ApJ, 428, 574.

WILSON, T. L. & MATTEUCCI, F. 1992, Abundances in the interstellar medium, A.Ap.Rev., 4, 1.

WOLFE, A. M. & BRIGGS, F. H. 1981, Detection of 21-CENTIMETER Absorption at $z \sim 1.94$ in the QSO PKS1157+014, ApJ, 248, 460.

WOLFE, A. M., BRIGGS, F. H., TURNSHEK, D. A., DAVIS, M. M., SMITH, H. E. & COHEN, R. D. 1985, Detection of 21 centimeter absorption at z = 2.04 in the QSO PKS 0458-02 ApJ, 294, L67.

WOLFE, A. M., BROWN, R. L., & ROBERTS, M. S. 1976, Limits on the variation of fundamental atomic quantities over cosmic time scales, Physical Review Letters, 37, 179.

WOLFE, A. M. & DAVIS, M. M. 1979, Detection of 21-cm absorption at $z \sim 1.8$ in the quasi-stellar object 1331+170, AJ, 84, 699.

WOLFE, A. M., LANZETTA, K. M., TURNSHEK, D. A. & OKE, J. B. 1992, Lyα emission from the damped Lyα system toward H0836 + 113, ApJ, 385, 151.

WOLFE, A. M., TURNSHEK, D. A., LANZETTA, K. M., & LU, L. 1993, Damped Lyα absorption by disk galaxies with large redshifts. IV - More intermediate-resolution spectroscopy, ApJ, 404, 480.

WOLFE, A. M., TURNSHEK, D. A., SMITH, H. E. & COHEN, R. D. 1986, Damped Lyα absorption by disk galaxies with large redshifts. I - The Lick survey, ApJS, 61, 249.

WOMBLE, D. S., SARGENT, W. L. W., & LYONS, R. S. 1996, Heavy Elements in the Lyα Forest: Abundances and Clustering at z = 3, ASSL Vol. 206: *Cold Gas at High Redshift*, p.249.

WRIGHT, E. L. 1986, Effect of intervening galaxies on quasar counts and colors, ApJ, 311, 156.

WRIGHT, E. L. 1990, Are high-redshift quasars hidden by dusty galaxies?, ApJ, 353, 411.

YORK, D. G., DOPITA, M., GREEN, R., & BECHTOLD, J. 1986, On the origin of some QSO absorption lines, ApJ, 311, 610.

YORK, D. G., YANNY, B., CROTTS, A., CARILLI, C., GARRISON, E., & MATHESON, L. 1991, An inhomogeneous reference catalogue of identified intervening heavy element systems in spectra of QSOs, MNRAS, 250, 24.

YOUNG, P. J., SARGENT, W. L. W. & BOKSENBERG, A. 1982a, C IV Absorption in an unbiased sample of 33 QSOs – Evidence for the intervening galaxy hypothesis ApJS, 48, 455.

YOUNG, P., SARGENT, W. L. W. & BOKSENBERG, A. 1982b, A high-resolution study of the absorption spectra of three QSOs - Evidence for cosmological evolution in the Lyman-alpha lines ApJ 252, 10.

YOUNG, P. J., SARGENT, W. L. W., BOKSENBERG, A., CARSWELL, R. F. & WHELAN, J. A. J. 1979 A high-resolution study of the absorption spectrum of PKS 2126-158, ApJ, 229, 891.

YOUNG, P., SARGENT, W. L. W., OKE, J. B., & BOKSENBERG, A. 1981, The origin of a new absorption system discovered in both components of the double QSO Q0957+561, ApJ, 249, 415.

ZHANG, Y., ANNINOS, P., & NORMAN, M. L. 1995, A Multispecies Model for Hydrogen and Helium Absorbers in Lyα Forest Clouds, ApJ, 453, L57.

ZHANG, Y., ANNINOS, P., NORMAN, M. L., & MEIKSIN, A. 1997, Spectral Analysis of the Lyα Forest in a Cold Dark Matter Cosmology, ApJ, 485, 496.

ZHANG, Y., MEIKSIN, A., ANNINOS, P., & NORMAN, M. L. 1998, Physical Properties of the Ly alpha Forest in a Cold Dark Matter Cosmology, ApJ, 495, 63.

ZUO, L. 1992a, Fluctuations in the ionizing background, MNRAS, 258, 36.

ZUO, L. 1992b, Intensity correlation of ionizing background at high redshifts, MNRAS, 258, 45.

ZUO, L., BEAVER, E. A., BURBIDGE, E. M., COHEN, R. D., JUNKKARINEN, V. T., & LYONS, R. W. 1997, The Dust-to-Gas Ratio in the Damped Lyα Clouds toward the Gravitationally Lensed QSO 0957+561, ApJ, 477, 568.

ZWAAN, M. A., BRIGGS, F. H., SPRAYBERRY, D., & SORAR, E. 1997, The H I Mass Function of Galaxies from a Deep Survey in the 21 Centimeter Line, ApJ, 490, 173.

Stellar Population Synthesis Models at Low and High Redshift

By GUSTAVO BRUZUAL A.

Centro de Investigaciones de Astronomía (CIDA), A.P. 264, Mérida, Venezuela

The basic assumptions behind Population Synthesis and Spectral Evolution models are reviewed. The numerical problems encountered by the standard population synthesis technique when applied to models with truncated star formation rates are described. The Isochrone Synthesis algorithm is introduced as a means to circumvent these problems. A summary of results from the application of this algorithm to model galaxy spectra by Bruzual & Charlot (1993, 2000) follows. I present a comparison of these population synthesis model predictions with observed spectra and color magnitude diagrams for stellar systems of various ages and metallicities. It is argued that models built using different ingredients differ in the resulting values of some basic quantities (e.g. M/L_V), without need to invoking violations of physical principles. The range of allowed colors in the observer frame is explored for several galaxy redshifts.

1. Introduction

The number distribution of the stellar populations present in a galaxy is a function of time. Thus, the number of stars of a given spectral type, luminosity class, and metallicity content changes as the galaxy ages. In early-type galaxies (E/S0) most of the stars were formed during, or very early after the initial collapse of the galaxy and the stellar population ages as times goes by. The chemical abundance in these systems must have reached the value measured in the stars very quickly during the formation process since most E/S0 galaxies show little evidence of recent major events of star formation. In late-type galaxies the stellar population also ages, but there is a significant number of new stars being formed. Depending on the star formation rate, $\Psi(t)$, the mean age of the stars in a galaxy may even decrease as the galaxy gets older. In general, in late-type systems the metal content of the stars and the interstellar medium is an increasing function of time. $\Psi(t)$ can also increase above its typical value due to interactions between two or more galaxies or with the environment.

As a consequence of the aging of the stellar population, or its renewal in the case of galaxies with high recent $\Psi(t)$, the rest-frame spectral distribution of the light emitted by a galaxy is a function of its proper time. Observational properties such as photometric magnitude and colors, line strength indices, metal content of gas and young stars, depend on the epoch at which we observe a galaxy on its reference frame. This *intrinsic* evolution should not be mistaken with the *apparent* evolution produced by the cosmological redshift z. For distant galaxies both effects may be equally significant.

In order to search for spectral evolution in galaxy samples we must *(a)* quantify the amount of evolution expected between cosmological epoch t_1 and t_2, and *(b)* design observational tests that will reveal this amount of evolution, if present. Evolutionary population synthesis models predict the amount of evolution expected under different scenarios and allow us to judge the feasibility of measuring it. In the ideal case (Aragón-Salamanca et al. 1993; Stanford et al. 1995, 1998; Bender et al. 1996) spectral evolution is measured simply by comparing spectra of galaxies obtained in such a way that the same rest frame wavelength region is sampled in all galaxies, irrespective of z. In this case there is no need to apply uncertain K-corrections to transform all the spectra to a common wavelength scale. Alternatively, if this approach cannot be applied, e.g. when studying large samples of faint galaxies, one can use models which allow for different degrees of evolution (including none) to derive indirectly the amount of evolution consistent with the data (Pozzetti et al. 1996; Metcalfe et al. 1996). Clearly, the first approach is to be preferred whenever possible. In all cases we must rule out possible deviations from the natural or passive evolution of the stellar population in some of the galaxies under study induced, for example, by interactions with other galaxies or cluster environment, etc.

The large amount of astrophysical data that has become available in the last few years has made possible to build several complete sets of stellar population synthesis models. The predictions of these models have been used to study many types of stellar systems, from local normal galaxies to the most distant galaxies discovered so far (approaching z of 4 to 5), from globular clusters in our galaxy to proto-globular clusters forming in different environments in distant, interacting galaxies. In this paper I present an overview of results from population synthesis models directly applicable to the interpretation of galaxy spectra.

2. The Population Synthesis Problem

Stellar evolution theory provides us with the functions $T_{eff}(m, Z, t)$ and $L(m, Z, t)$ which describe the behavior in time t of the effective temperature T_{eff} and luminosity L of a star of mass m and metal abundance Z. For fixed m and Z, $L(t)$ and $T_{eff}(t)$ describe parametrically in the H-R diagram the evolutionary track for stars of this mass and metallicity. The initial mass function (IMF), $\phi(m)$, indicates the number of stars of mass m born per unit mass when a stellar population is formed. The star formation rate (SFR), $\Psi(t)$, gives the amount of mass transformed into stars per unit time according to $\phi(m)$. The metal enrichment function (MEF), $Z(t)$, also follows from the theory of stellar evolution. The population synthesis problem can then be stated as follows. Given a complete set of evolutionary tracks and the functions $\phi(m)$ and $\Psi(t)$, compute the number of stars present at each evolutionary stage in the H-R diagram as a function of time. To solve this problem exactly we need additional knowledge about the MEF, $Z(t)$, which gives the time evolution of the chemical abundance of the gas from which the successive generations of stars are formed. For simplicity, it is commonly assumed that $\phi(m)$ and $\Psi(t)$ are decoupled from $Z(t)$, even though it is recognized that in real stellar systems these three quantities are most likely closely interrelated. The spectral evolution problem can be solved trivially once the population synthesis problem is solved, provided that we know the spectral energy distribution (SED) at each point in the H-R diagram representing an evolutionary stage in our set of tracks. The discussion that follows is based in the work of Charlot & Bruzual (1991, hereafter CB91) and Bruzual & Charlot (1993, 2000, hereafter BC93 and BC2000). See also Bruzual (1998, 1999, 2000). Unless otherwise indicated, I will ignore chemical evolution and assume that at all epochs stars form with a single metallicity, $Z(t) = $ constant, and the same $\phi(m)$.

Let N_i^o be the number of stars of mass M_i born when an instantaneous burst of star formation occurs at $t = 0$. This kind of burst population has been called simple stellar population (SSP, Renzini 1981). When we look at this population at later times, we will see the stars traveling along the corresponding evolutionary track. If the stars live in their k^{th} evolutionary stage from time $t_{i,k-1}$ to time $t_{i,k}$, then at time t the number of stars of this mass populating the kth stage is simply

$$N_{i,k}(t) = \begin{cases} N_i^o, & \text{if } t_{i,k-1} \leq t < t_{i,k}; \\ 0, & \text{otherwise.} \end{cases} \quad (2.1)$$

For an arbitrary SFR, $\Psi(t)$, we compute the number of stars $\eta_{i,k}(t)$ of mass M_i at the kth evolutionary stage from the following convolution integral

$$\eta_{i,k}(t) = \int_0^t \Psi(t-t') N_{i,k}(t') dt', \qquad (2.2)$$

which in view of (2.1) can be written as

$$\eta_{i,k}(t) = N_i^o \int_{t_{i,k-1}}^{min(t,t_{i,k})} \Psi(t-t') dt'. \qquad (2.3)$$

From (2.3) we see that the commonly heard statement that the number of stars expected in a stellar population at a given position in the H-R diagram is proportional to the time spent by the stars at this position, i.e.

$$\eta_{i,k} \propto N_i^o (t_{i,k} - t_{i,k-1}), \qquad (2.4)$$

is accurate only for a constant $\Psi(t)$. For non-constant SFRs, the integral in (2.3) assigns more weight to the epochs of higher star formation. For instance, for an exponentially decaying SFR with e-folding time τ, $\Psi(t) = \exp(-t/\tau)$, we have

$$\eta_{i,k}(t) \propto N_i^o \{\exp[-(t-t_{i,k})/\tau] - \exp[-(t-t_{i,k-1})/\tau]\}. \qquad (2.5)$$

$\Psi(t)$ was stronger at time $(t - t_{i,k})$ than at time $(t - t_{i,k-1})$, which is clearly taken into account in (2.5).

A prerequisite for building trustworthy population synthesis and spectral evolution models is an adequate algorithm to follow the evolution of consecutive generations of stars in the H-R diagram. This goal is accomplished by the standard technique described above provided that the function $\Psi(t)$ extends from $t = 0$ to $t = \infty$. Special caution is required if $\Psi(t)$ becomes 0 at a finite age. As an illustration, let us consider the case of a burst of star formation which lasts for a finite length of time τ,

$$\Psi(t) = \begin{cases} \Psi_o, & \text{if } 0 \leq t \leq \tau; \\ 0, & \text{otherwise,} \end{cases} \qquad (2.6)$$

or equivalently,

$$\Psi(t-t') = \begin{cases} \Psi_o, & \text{if } t-\tau \leq t' \leq t; \\ 0, & \text{otherwise.} \end{cases} \qquad (2.7)$$

In this case, equation (2.2) reduces to

$$\eta_{i,k}(t) = \begin{cases} > 0, & \text{if } [t_{i,k-1}, t_{i,k}] \cap [t-\tau, t] \neq \emptyset; \\ 0, & \text{otherwise.} \end{cases} \qquad (2.8)$$

From (2.8) we see that $\eta_{i,k}(t)$ can be $= 0$ depending on the value of τ and on the value of t chosen to sample the stellar population. The most extreme case is that of the SSP ($\tau = 0$), for which $\Psi(t) = \delta(t)$, and $\eta_{i,k}(t)$ in (2.8) is identical to $N_{i,k}(t)$ in (2.1). For a typical set of evolutionary tracks, there is no grid of values t_j of the time variable t for which all the stellar evolutionary stages included in the tracks can be adequately sampled for arbitrarily chosen values of τ. This is obviously an undesirable property of population synthesis models. The models should be capable of representing galaxy properties in a continuous and well behaved form, independent of the sampling time scale t. Consequently, standard population synthesis models for truncated $\Psi(t)$ as given by (2.6) will reflect the coarseness of the set of evolutionary tracks, and may miss stellar evolutionary phases depending on our choice of t. This results in unwanted and unrealistic numerical noise in the predicted properties of the stellar systems studied with the synthesis code (see example in CB91).

A solution to this problem is to build a set of evolutionary tracks with a resolution in mass which is high enough to guarantee that all evolutionary stages, i.e. all values of k in (2.2), can be populated for any choice of t or τ in (2.6). In other words, there must be tracks for so many stellar masses in this ideal library that for any model age, we have at hand the position in the H-R diagram of the star which at that age is in the kth evolutionary stage. The realization of this library is not possible with present day computers. This limitation can be circumvented by careful interpolation in a relatively complete set of evolutionary tracks.

3. The Isochrone Synthesis Algorithm

The isochrone synthesis algorithm described in this section allows us to compute continuous isochrones of any age from a carefully selected set of evolutionary tracks. The isochrones are then used to build population synthesis models for arbitrary SFRs $\Psi(t)$, without encountering any of the problems mentioned above. This algorithm has been used by CB91, BC93, and BC2000. The details of the algorithm follow.

A relationship is built between the main sequence (MS) mass of a star M and the age t of the star during the kth evolutionary stage (CB91). Ignoring mass loss is justified because M is used only to label the tracks. For the set of tracks used by BC93 (described in the next section) 311 different relationships $\log M$ vs. $\log t$ are built, which can be visualized as 311 different curves in the $(\log t, \log M)$ plane. We derive by linear interpolation the MS mass $m_k(t')$ of the star which will be at the kth evolutionary stage at age t', given by

$$\log m_k(t') = A_{k,i} \log(M_i) + (1 - A_{k,i}) \log(M_{i+1}), \quad (3.9)$$

where

$$A_{k,i} = \frac{\log t_{i+1,k} - \log t'}{\log t_{i+1,k} - \log t_{i,k}}. \quad (3.10)$$

$t_{i,k}$ represents the age of the star of mass M_i at the kth evolutionary stage, and

$$t_{i,k} \leq t' < t_{i+1,k},$$

and

$$M_{i+1} \leq m_k(t') < M_i.$$

The procedure is performed for all the curves that intersect the $\log t = \log t'$ line. We thus obtain a series of values of m_k which must now be assigned values of $\log L$ and $\log T_{eff}$ in order to define the isochrone corresponding to age t'.

To compute the integrated properties of the stellar population, we must specify the number of stars of mass m_k. This number is determined from the IMF, which we can write generically as

$$n(m_k) = \phi(m_k^-, m_k^+, 1 + x). \quad (3.11)$$

Of this number of stars,

$$N_{i,k} = A_{k,i} n(m_k) \quad (3.12)$$

stars are assigned the observational properties of the star of mass M_i at the kth evolutionary stage, and

$$N_{i+1,k} = (1 - A_{k,i}) n(m_k) \quad (3.13)$$

stars, the observational properties of the star of mass M_{i+1} at the same k^{th} stage. This procedure is equivalent to interpolating the tracks for the stars of mass M_i and M_{i+1} to derive the values of log L and log T_{eff} to be assigned to the star of mass m_k, but this intermediate step is unnecessary. In (3.11) m_k^- and m_k^+ are computed from

$$m_k^- = (m_{k-1} m_k)^{1/2}, \qquad m_k^+ = (m_k m_{k+1})^{1/2}. \tag{3.14}$$

In this case m_{k-1}, m_k, and m_{k+1} represent the masses obtained by interpolation at the given age in the segments representing the $(k-1)^{th}$, k^{th}, and $(k+1)^{th}$ stages, respectively. The IMF has been assumed to be a power law of the form (Salpeter 1955)

$$\phi(m_1, m_2, 1+x) = c \int_{m_1}^{m_2} m^{-(1+x)} dm. \tag{3.15}$$

For a given $\Psi(t)$ equation (2.2) gives the number $\eta_{i,k}(t)$ of stars of mass M_i at the kth evolutionary stage. If $f_{i,k}(\lambda)$ represents the SED corresponding to the star of mass M_i during the kth stage, then the contribution of the $\eta_{i,k}(t)$ stars to the integrated evolving SED of the stellar population is simply given by

$$F_{i,k}(\lambda, t) = \eta_{i,k}(t) f_{i,k}(\lambda). \tag{3.16}$$

The resulting SED for the stellar population is given by

$$F(\lambda, t) = \sum_{i,k} F_{i,k}(\lambda, t). \tag{3.17}$$

4. Evolutionary Population Synthesis Models

A number of groups has developed in recent years different population synthesis models which provide a sound framework to investigate the problem of spectral evolution of galaxies. Some of the most commonly used models are Arimoto & Yoshii (1987), Guiderdoni & Rocca-Volmerange (1987), Buzzoni (1989), Bressan, Chiosi & Fagotto (1994), Fritze-v.Alvensleben & Gerhard (1994), Worthey (1994), Bruzual & Charlot (1993, 2000). The basic astrophysical ingredients used in these models are: *(1)* Stellar evolutionary tracks of one or more metallicities; *(2)* Spectral libraries, either empirical or theoretical model atmospheres; *(3)* Sets of rules, or calibration tables, to transform the theoretical HR diagram to observational quantities (e.g. $B - V$ vs. T_{eff}, $V - K$ vs. T_{eff}, $B.C.$ vs. T_{eff}, etc.). These rules are not necessary when theoretical model atmosphere libraries are used which are already parameterized according to T_{eff}, log g, and [Fe/H]; *(4)* Additional information, such as analytical fitting functions, required to compute various line strength indices (Worthey et al. 1994). Regardless of the specific computational algorithm used, all evolutionary synthesis models depend on three adjustable parametric functions: *(1)* the stellar initial mass function, $f(m)$, or IMF; *(2)* the star formation rate, $\Psi(t)$; and *(3)* the chemical enrichment law, $Z(t)$. For a given choice of $f(m)$, $\Psi(t)$, and $Z(t)$, a particular set of evolutionary synthesis models provides: *(1)* Galaxy spectral energy distribution *vs.* time, $F_\lambda(\lambda, Z(t), t)$; *(2)* Galaxy colors and magnitude *vs.* time; *(3)* Line strength and other spectral indices *vs.* time. Some authors (e.g. Bressan et al. 1994; Fritze-v.Alvensleben & Gerhard 1994) consider that $Z(t)$ can be derived self-consistently from their models. In other instances $Z(t)$ is introduced as an external piece of information. I discuss below some results from work still in progress in collaboration with S. Charlot.

5. Stellar Ingredients

BC2000 have extended the BC93 evolutionary population synthesis models to provide the evolution in time of the spectrophotometric properties of SSPs for a wide range of

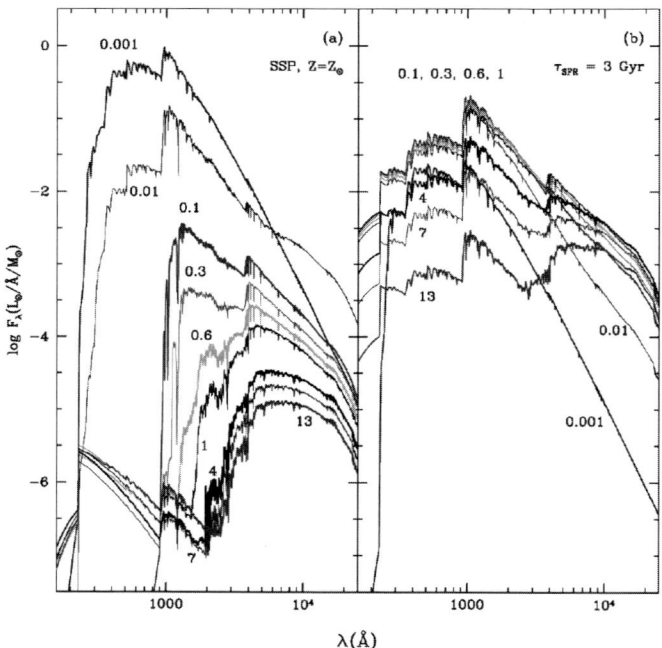

FIGURE 1. Evolving spectral energy distributions. (a) Evolution in time of the SED of a SSP computed for the Salpeter IMF ($m_L = 0.1$, $m_U = 125$ $M_\odot$). The age in Gyr is indicated next to each spectrum. (b) Same as (a) but for a composite population in which stars form according to $\Psi(t) = \exp(-t/\tau)$ for $\tau = 3$ Gyr. The total mass of each model galaxy is 1 $M_\odot$. F_λ in frame (b) has been multiplied by 100 to use a common vertical scale.

stellar metallicity. In an SSP all the stars form at $t = 0$ and evolve passively afterward. The BC2000 models are based on the stellar evolutionary tracks computed by Alongi et al. (1993), Bressan et al. (1993), Fagotto et al. (1994a, b, c), and Girardi et al. (1996), which use the radiative opacities of Iglesias et al. (1992). This library includes tracks for stars with initial chemical composition $Z = 0.0001, 0.0004, 0.004, 0.008, 0.02, 0.05$, and 0.10 (Table 1), with $Y = 2.5Z + 0.23$, and initial mass $0.6 \leq m/M_\odot \leq 120$ for all metallicities, except $Z = 0.0001$ ($0.6 \leq m/M_\odot \leq 100$) and $Z = 0.1$ ($0.6 \leq m/M_\odot \leq 9$). This set of tracks will be referred to as the Padova or P tracks hereafter. A similar set of tracks for slightly different values of Z has been published by Girardi et al. (2000). A comparison of the predictions of models built with both sets of Padova tracks will be shown elsewhere (but see Fig. 7 and Bruzual 2000).

The published tracks go through all phases of stellar evolution from the zero-age main sequence to the beginning of the thermally pulsing regime of the asymptotic giant branch (AGB, for low- and intermediate-mass stars) and core-carbon ignition (for massive stars), and include mild overshooting in the convective core of stars more massive than 1 $M_\odot$. The Post-AGB evolutionary phases for low- and intermediate-mass stars were added to the tracks by BC2000 from different sources (see BC2000 for details).

BC2000 use as well a parallel set of tracks for solar metallicity computed by the Geneva

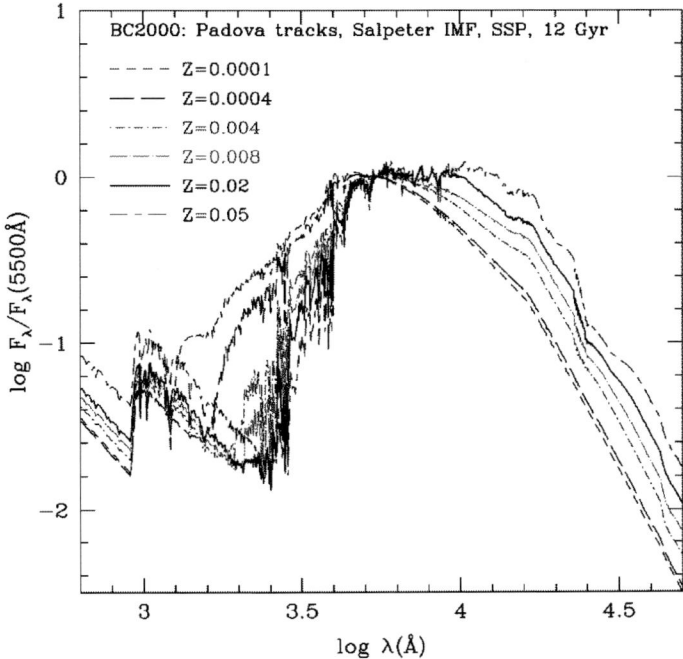

FIGURE 2. Chemically homogeneous BC2000 SSP model galaxy SEDs at age = 12 Gyr. Each line pattern represents a different metallicity, as indicated inside the frame. All the models shown were computed for the Salpeter IMF ($m_L = 0.1$, $m_U = 125\ M_\odot$). The total mass of each model galaxy is 1 $M_\odot$. The SEDs have been normalized at $\lambda = 5500$Å.

group (Geneva or G tracks hereafter), which provides a framework for comparing models computed with two different sets of tracks.

The BC2000 models use the library of synthetic stellar spectra compiled by Lejeune et al. (1997, 1998, LCB97 and LCB98 hereafter) for all the metallicities in Table 1. This library consists of Kurucz (1995) spectra for the hotter stars (O-K), Bessell et al. (1989, 1991) and Fluks et al. (1994) spectra for M giants, and Allard & Hauschildt (1995) spectra for M dwarfs. For $Z = Z_\odot$, BC2000 also use the Pickles (1998) stellar atlas, assembled from empirical stellar data.

6. Spectral Evolution at Fixed Metallicity

Fig. 1a shows the evolution in time of the SED for the SSP model. In an SSP all the stars form at $t = 0$ and evolve passively afterward. In all the examples shown in this paper I assume that stars form according to the Salpeter (1955) IMF in the range from $m_L = 0.1$ to $m_U = 125\ M_\odot$. The total mass of the model galaxy is 1 $M_\odot$. The evolution is fast and is dominated by massive stars during the first Gyr in the life of the SSP (6 top SEDs). The flux seen around 2000 Å at 4 and 7 Gyr is produced by the turn-off stars. The UV-rising branch (Burstein et al. 1988, Greggio & Renzini 1990) seen after 10 Gyr is produced by the PAGB stars. These stars are also responsible of the decrease in the amplitude of the 912 Å discontinuity observed after 4 Gyr. The SSP

TABLE 1. Model chemical composition

Z	X	Y	[Fe/H]
0.0001	0.7696	0.2303	-2.2490
0.0004	0.7686	0.2310	-1.6464
0.0040	0.7560	0.2400	-0.6392
0.0080	0.7420	0.2500	-0.3300
0.0200	0.7000	0.2800	0.0932
0.0500	0.5980	0.3520	0.5595
0.1000	0.4250	0.4750	1.0089

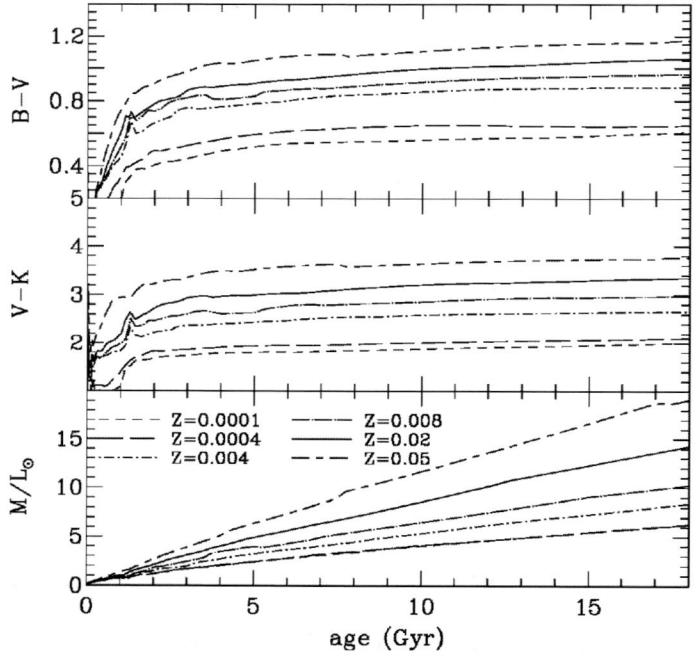

FIGURE 3. Evolution in time of the $B-V$, and $V-K$ colors, and the M/L_V ratio for the BC2000 SSP models shown in Fig. 2. Each line represents a different metallicity, as indicated in the bottom panel.

model is the basic ingredient which, together with the convolution integral (2.2), is used to compute models with arbitrary SFRs and equal IMF. For illustration I show in Fig. 1b the evolution of a model with $\Psi(t) = \exp(-t/\tau)$ for $\tau = 3$ Gyr. The UV to optical spectrum remains roughly constant during the main episode of star formation because of the continuous input of young massive stars, but the near-infrared light rises as evolved stars accumulate. When star formation drops, the spectral characteristics at various wavelengths are determined by stars in advanced stages of stellar evolution.

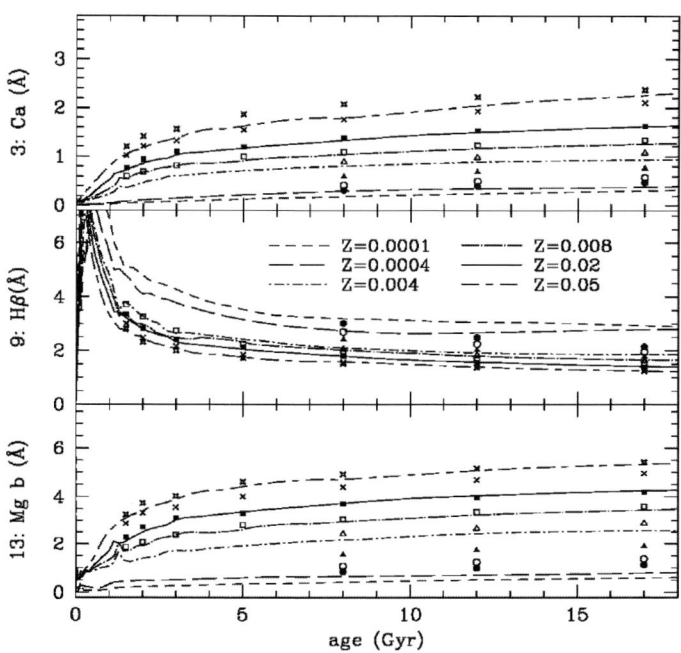

FIGURE 4. Evolution in time of the Mg_b, H_β, and Ca spectral indices as defined by Worthey (1994) for the BC2000 models shown in Fig. 2. The different symbols represent the values of the indices computed by Worthey for the same range in Z. Each line represents a different metallicity, as indicated in the middle panel.

7. Dependence of Galaxy Properties on Stellar Metallicity

Fig. 2 shows the predicted SEDs at $t = 12$ Gyr for chemically homogeneous SSPs of the indicated metallicity. The SEDs shown in Fig. 2 have been normalized at $\lambda = 5500$Å to make the comparison more clear. Fig. 3. shows the evolution in time of the $B - V$ and $V - K$ colors, and the M/L_V ratio predicted by BC2000 for the same SSPs shown in Fig. 2.

From Figs. 2 and 3 it is apparent that there is a uniform tendency for galaxies to become redder in $B - V$ as the metallicity increases from $Z = 0.0001$ ($\frac{1}{200}Z_\odot$) to $Z = 0.05$ ($2.5 \times Z_\odot$). The $V - K$ color and the M/L_V ratio show the expected tendency with metallicity, i.e. $V - K$ becomes redder and M/L_V becomes higher with increasing Z.

Fig. 4 shows the evolution in time of the Mg_b, H_β, and Ca spectral indices as defined by Worthey (1994) for the same BC2000 SSP models shown in Figs. 2 and 3. Again, the models show the expected tendency with Z and match the values computed by Worthey (1994). It should be remarked that the time behavior of the line strength indices at constant Z is due to the change in the number of stars at different positions in the HR diagram produced by stellar evolution and is not related to chemical evolution. The indices change also in chemically homogeneous populations. The H_β index is less sensitive to the stellar metallicity than the Mg_b and Ca index. Instead, the H_β index is high when there is a large fraction of MS A-type stars ($t < 1$) Gyr.

In Fig. 5 I compare the behavior of the SSP models in the $(U - B)$ vs. $(B - V)$ color

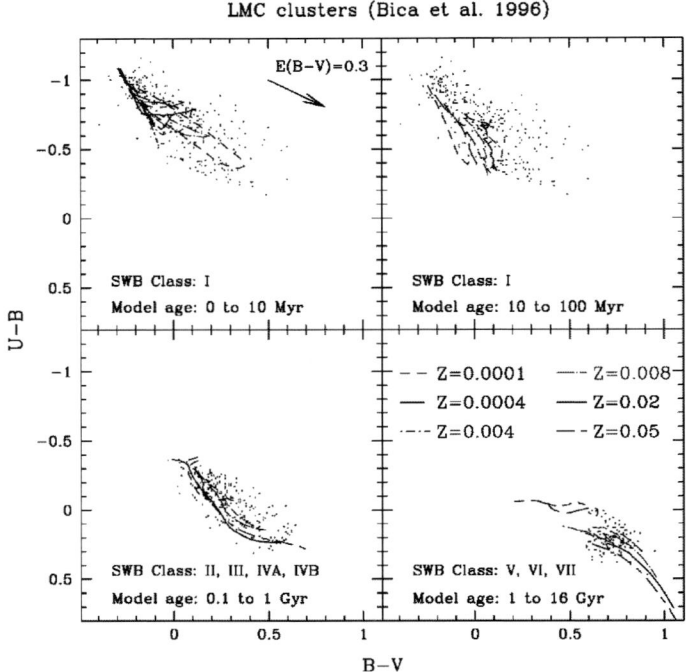

FIGURE 5. Points: LMC cluster $(U-B)$ vs. $(B-V)$ colors from Bica et al. (1996b), discriminated by SWB class. The lines represent SSP models at various age ranges.

plane with the LMC cluster data from Bica et al. (1996b), discriminated by SWB class and model age. In each panel the models (lines) are shown in the range of age for which the predicted colors overlap the observed colors for the class. It is apparent from the figure that the models reproduce quite well the observed colors.

8. Calibration of the Models in the C-M Diagram

Population synthesis models, frequently used to study composite stellar populations in distant galaxies, are rarely confronted with observations of local simple stellar populations, such as globular star clusters, whose age and metal content have been constrained considerably in later years. If the models cannot reproduce the CMDs and SEDs of these objects for the correct choice of parameters, their predictive power becomes weaker and their usage to study complex galaxies may not be justified.

It is important to compare the properties of the population synthesis models to observations of stellar systems whose age and metallicity is well determined. This is a means to test to what extent the adopted relationships between stellar color and magnitude and effective temperature and luminosity (or surface gravity) introduce systematic shifts between the predicted and observed isochrones in the C-M diagram.

For each value of the metallicity Z listed in Table 1, and a particular choice of the IMF $\phi(m)$, a BC2000 SSP model consists of a set of 221 evolving integrated SEDs spanning from 0 to 20 Gyr. The isochrone synthesis algorithm used to build these models renders it straightforward to compute the loci described by the stellar population in the CMD

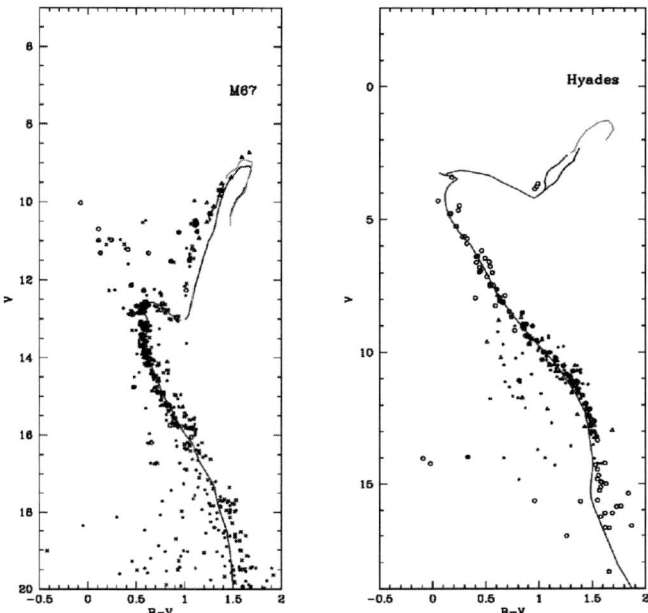

FIGURE 6. CMD of M67 and the Hyades compared with isochrones derived from the Padova tracks for solar metallicity. See text for details.

at any time step and in any photometric band. We can extract the model SED that best reproduces a given observed SED and assign an age to the program object, and then examine how well the isochrone computed at this age fits the most significant features in the CMD of this object, if available.

Fig. 6 compares the $(B, B - V)$ observations of the clusters M67 and the Hyades with the isochrones obtained from the BC2000 models. These two clusters have nearly solar metallicity: M67 ([Fe/H]$\approx$ 0.01), the Hyades ([Fe/H]$\approx$ 0.15). For M67 we adopted a distance modulus of 9.5 mag and a color excess $E(B - V) = 0.06$ mag (Janes 1985). For the Hyades a distance modulus of 3.4 mag (Peterson & Solensky 1988) and $E(B-V) \approx 0$. Estimates of the ages of these clusters vary from 4 to 4.3 Gyr for M67 and from 0.5 to 0.8 Gyr for the Hyades. The isochrones are shown at 4 Gyr (M67) and 0.6 Gyr (Hyades). The data points for M67 are from Eggen & Sandage (1964, *open circles*), Racine (1971, *filled circles*), Janes & Smith (1984, *triangles*), and Gilliland et al. (1991; *crosses*). For the Hyades the observations are from Upgren (1974, *triangles*), Upgren & Weis (1977, *filled circles*), and Micela et al. (1988, *open circles*).

Fig. 7 shows a comparison of the excellent HST CMD diagram of NGC 6397 assembled from various sources by D'Antona (1999) with isochrones computed from the Padova-1994 tracks for $Z = 0.0004$ and the Padova-2000 tracks for $Z = 0.001$ at ages 10 to 16 Gyr. The *original* version of the model atmospheres in the LCB atlas was used to derive the colors in Fig. 7. The *corrected* version of these models produces considerably worse agreement with the observations, mainly in the MS from the turn-off down. The redder cluster RGB most likely reflects a slightly higher metallicity than $Z = 0.0004$, close to

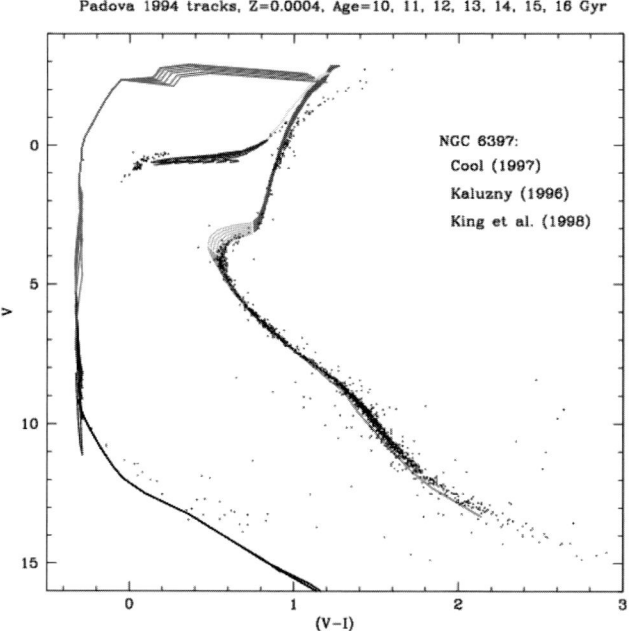

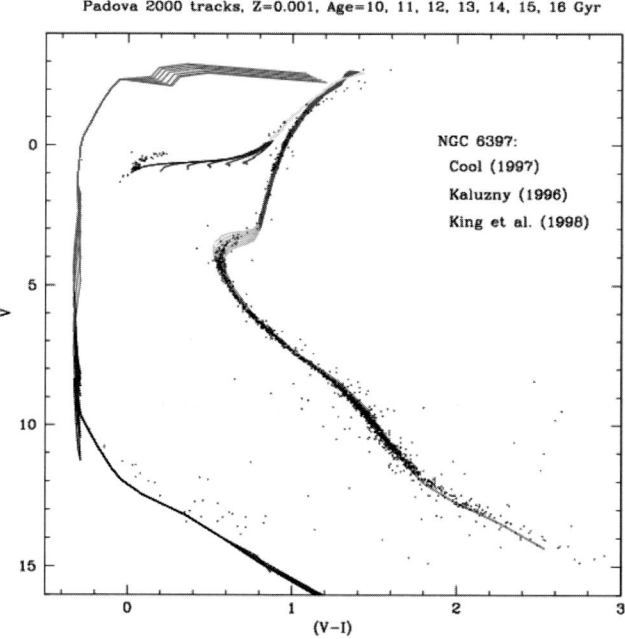

FIGURE 7. CMD of NGC 6397 compared with isochrones derived from the Padova-1994 tracks for $Z = 0.0004$ (top), and the Padova-2000 tracks for $Z = 0.001$ (bottom). The observations were assembled by D'Antona (1999) from the sources indicated in the figure.

TABLE 2. $Z_\odot$ model fits to SED of NGC 6528

Model	Spectral Library	Stellar Tracks	Best-fitting age (Gyr)	Σ^2_{min}
1	Pickles	P	11.75	2.22
2	"	G	10.25	1.34
3	LCB97-C	P	10.00	1.73
4	"	G	8.50	1.38
5	LCB97-O	P	11.50	3.35
6	"	G	9.00	1.95

$Z = 0.001$. Despite the discrepancies seen in Figs. 6 and 7 (mainly in the RGB of M67 and NGC 6397) the agreement between the predicted isochrones and the loci in the CMD of these clusters may be regarded as satisfactory, and is excellent in some parts of the diagram.

9. Observed Color-Magnitude Diagrams and Integrated Spectra

The high quality and depth of the HST VI-photometry NGC 6528 ($Z \lesssim Z_\odot$), as well as the high signal-to-noise ratio of the integrated SED of NGC 6528 currently available, provide an excellent framework for testing the range of validity of population synthesis models (see Bruzual et al. 1997 for details).

The integrated spectrum of NGC 6528 in the wavelength range $\lambda = 3500 - 9800$ Å was obtained by combining the visible, near-infrared, and near-ultraviolet spectra of Bica & Alloin (1986, 1987) and Bica et al. (1994), respectively. We applied a reddening correction of E(B-V) = 0.66, adopted in Bica et al. (1994 and references therein). The SED of NGC 6528 is typical of old stellar populations of high metal content (Santos et al. 1995). It may happen that NGC 6528, similarly to NGC 6553, has [Fe/H] < 0.0 (Barbuy et al. 1997), whereas the [α-elements/Fe] are enhanced, resulting in [$Z/Z_\odot$] $\approx$ 0.0, or possibly slightly below solar. Since we have evolutionary tracks for $Z = 0.02$ and $Z = 0.008$, we adopt $Z = 0.02$.

The data available for this cluster provide a unique opportunity to examine the 6 different options for the $Z = Z_\odot$ models considered by BC2000. Thus, we can study objectively which of the basic building blocks used by BC2000 for $Z = Z_\odot$: P or G tracks; Pickles, LCB97-O (original version), or LCB97-C (corrected version) stellar libraries, is most successful at reproducing the data. In Table 2 I show the age at which Σ^2, defined as the sum of squared residuals $[\log F_\lambda(observed) - \log F_\lambda(model)]^2$, is minimum for various $Z = Z_\odot$ models. The values of Σ^2_{min} given in Table 2 indicate the goodness-of-fit. According to this criterion, models 2 and 4 provide the best fit to the integrated SED of NGC 6528 in the wavelength range λ 3500-9800 Å. This fit is shown in Fig. 8. Except for the differences in the best-fitting age, model 3 provides a comparable, although somewhat poorer, fit to the SED of this cluster. The residuals for models 1, 5 and 6 are considerably larger. Spectral evolution is slow at these ages, and the minimum in the function Σ^2 vs. age is quite broad. Reducing or increasing the model age by 1 or 2 Gyr produces fits of comparable quality to the one at which Σ^2 is minimum. For

198 Gustavo Bruzual A.: *Population Synthesis at low and high z*

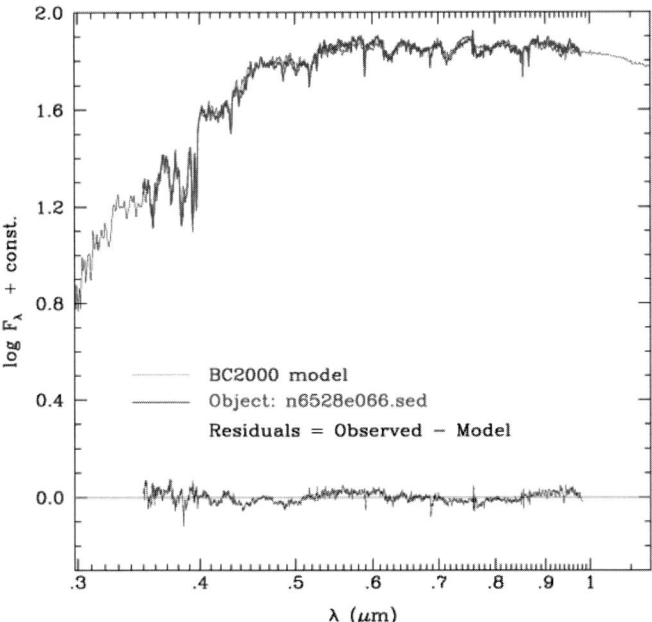

FIGURE 8. Best fit to the integrated spectrum of NGC 6528 (heavy line) in the range $\lambda\lambda$ 3500 - 9800 Å for model 2 in Table 2 (thin line extending over the full wavelength range). The best fit occurs at 10.25 Gyr. The residuals of the fit, $\log F_\lambda(observed) - \log F_\lambda(model)$, are shown as a function of wavelength.

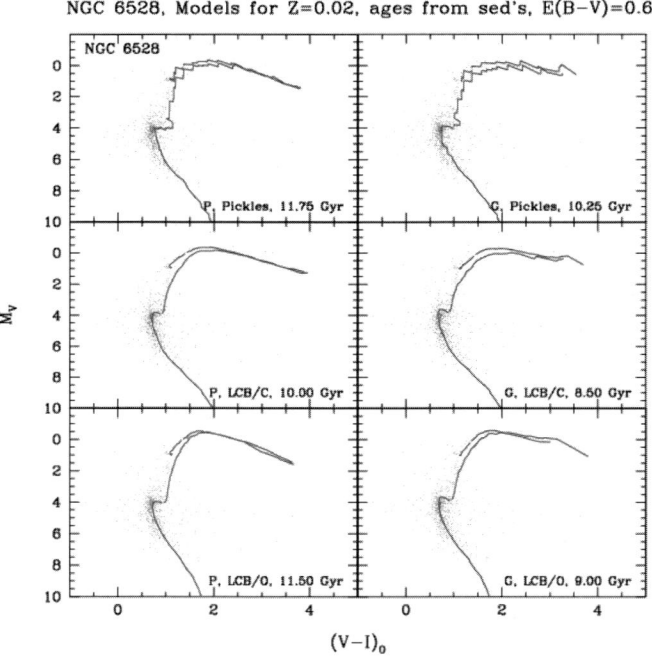

FIGURE 9. Intrinsic M_V vs. $(V-I)_0$ CMD of NGC 6528 shown together with theoretical isochrones for the $Z = Z_\odot$ models listed in Table 2.

instance, for model 2, Σ^2(12 Gyr) = 1.88, and Σ^2(8 Gyr) = 2, which are still better fits according to the Σ^2 criterion than the best fits provided by some models in Table 2.

Fig. 9 shows the intrinsic HST VI CMD of NGC 6528 together with the isochrones corresponding to the models in Table 2. It is apparent from this figure that all the isochrones shown provide a good representation of the cluster population in this CMD, especially the position of the turn-off and the base of the asymptotic giant branch (AGB). We note that NGC 6528 shows a double turnoff, the upper one being due to contamination from the field star main sequence. The appropriate TO location would be around that indicated by the 10, 11 and 12 Gyr isochrones. Despite the fact that models 2 and 4 provide better fits to the SED of this cluster than models 1 and 3, the isochrones from model 3 reproduce more closely the CMD diagram than models 1 and 2. The noisy nature of the isochrones computed with models 1 and 2 is due to the lack of some stellar types in the Pickles stellar library.

From these results we conclude that:

(1) The $\lambda = 3500 - 9800$ Å SED for E(B-V)=0.66 and the VI CMD of NGC 6528 are well reproduced by $Z = Z_\odot$ models at an age from 9 to 12 Gyr.

(2) The age derived from the fit to the observed SED of NGC 6528 is extremely sensitive to the assumed E(B-V). Using E(B-V) = 0.59 instead of 0.66, increases the best-fitting ages by 2 to 3 Gyr, since the observed SED is then intrinsically redder. On the other hand, if E(B-V) = 0.69, the observed SED is intrinsically bluer and the derived ages are 2 to 3 Gyr younger than for E(B-V) = 0.66. However, inspecting the isochrones in the CMD, the ages derived for E(B-V) = 0.66 listed in Table 2, seem appropriate.

(3) The SED and CMD of this cluster are consistent with those expected for a $Z = Z_\odot$ population at an age of $\approx$ 9-12 Gyr, if overshooting occurs in the convective core of stars down to 1 $M_\odot$ (P tracks, models 1, 3, and 5 in Table 2). If overshooting stops at 1.5 $M_\odot$, as in the G tracks, this age is reduced to $\approx$ 8-10 Gyr (models 2, 4, and 6 in Table 2).

(4) For the same spectral library, the ages derived from the P tracks are older than the ones derived from the G tracks. This is due to the fact that the P tracks include overshooting in the convective core of stars more massive than 1 $M_\odot$ whereas the G tracks stop overshooting at 1.5 $M_\odot$. Thus, stars in this mass range require more time in the P tracks to leave the main sequence than in the G tracks.

(5) For the same set of evolutionary tracks, the corrected LCB97 library seems to provide a better fit to the CMD than the Pickles atlas. Interpolation in the finer LCB97 grid of models produces smoother isochrones than in the coarser Pickles atlas. We attribute this to the fact that M stars are very sparse in the Pickles atlas. Furthermore, the temperature scale becomes problematic for these stars in the Pickles atlas. In the LCB97 library, the temperature scale for giants relies on measurements of angular diameters and fluxes, which enter directly in the definition of effective temperature. For dwarfs, the temperature scale is more difficult to define, as discussed in LCB98.

(6) Noticeable differences exist in the isochrones computed for both sets of LCB97 libraries. The differences are more pronounced for the M giants of $(V - I)_0 > 1.6$, and $(J - K)_0 > 1$, corresponding to a temperature of $T_e \leq 4000K$, and for the cool dwarfs of $(V - I)_0 > 1$, corresponding to a temperature of $T_e \leq 4700K$. There are very few of these stars in the CMD of NGC 6528 to favor a particular choice of library. However, the corrected library produces better fits to the observed SED. We attribute this fact to the relative importance of the luminosity of M giants.

(7) In general the LCB97 corrections redden the stellar SEDs in the optical range, producing redder SSP models at an earlier age. As a consequence, the ages derived in Table 1 for the LCB97-C library are younger than the ones derived from the LCB97-O library for the same set of tracks.

(8) We have adopted $Z = Z_\odot$ for this cluster. However, for a slightly lower value of Z, the derived age would be older.

10. Comparison of Model and Observed Spectra

10.1. Solar metallicity

Fig. 10 shows a model fit to the average spectrum of an E galaxy (kindly provided by M. Rieke). The model SED is the line extending over the complete wavelength range shown in the figure. The observed SED covers the range from 3300 Å to 2.75 μm. The residuals (observed - model) are shown at the bottom of the figure in the same vertical scale. The model corresponds to a 10 Gyr $Z = Z_\odot$ SSP computed for the Salpeter (1955) IMF ($m_L = 0.1$, $m_U = 125\ M_\odot$) using the P tracks and the Pickles (1998) stellar atlas. The fit is excellent over most of the spectral range. A minor discrepancy remains in the region from 1.1 to 1.7 μm. The source of this discrepancy is not understood at the moment. In Fig. 11 I show the same model and E galaxy SED as in Fig. 10 but in different units. In addition, in Fig. 11 I include the broad band fluxes representing the average of many E galaxies in the Coma cluster (solid squares) from A. Stanford (private communication). The observed SED is the one with the lowest spectral resolution. Fig. 12 shows a closer look at the same data in an enlarged scale. Again the agreement is excellent for the 3 data sets. The discrepant line in Fig. 12 corresponds to the same model shown in Figs. 10 and 11 but I used the LCB97 synthetic stellar atlas instead of the empirical stellar SEDs. Fig. 12 shows clearly that the models based on empirical stellar SEDs are to be preferred over the ones based on theoretical model atmospheres. Unfortunately, complete libraries of empirical stellar SEDs are available only for solar metallicity.

10.2. Non-solar metallicity

Figs. 13 and 14 show the results of a comparison of SSP models built for various metallicities using the LCB97 atlas, all for the Salpeter IMF, with several of the average spectra compiled by Bica et al. (1996a). The name and the metal content of the observed spectra indicated in each panel is as given by Bica et al. The quoted age is derived from the best fit of our model spectra to the corresponding observations. The residuals (observed - model) are shown in the same vertical scale. See the description to Fig. 10 above for more details. Even though, in detail, the fits for non-solar metallicity stellar populations are not as good as the ones for solar metallicity, over all the models reproduce the observations quite well over a wide range of $[Z/Z_\odot]$, and provide a reliable tool to study these stellar systems. The discrepancy can be due both to uncertainties in the synthetic stellar atlas or the evolutionary tracks at these $[Z/Z_\odot]$. I have used SSPs in all the fits, neglecting possibly composite stellar populations, as well as any interstellar reddening.

11. Different Sources of Uncertainties in Population Synthesis Models

11.1. Uncertainties in the astrophysics of stellar evolution

There are significant differences in the fractional contribution to the integrated light by red giant branch (RGB) and asymptotic giant branch (AGB) stars in SSPs computed for different sets of evolutionary tracks. Fig. 15 shows the contribution of stars in various evolutionary stages to the bolometric light, and to the broad-band $UBVRIKL$ fluxes for a $Z = Z_\odot$ model SSP computed for the Salpeter IMF ($m_L = 0.1$, $m_U = 125\ M_\odot$) using

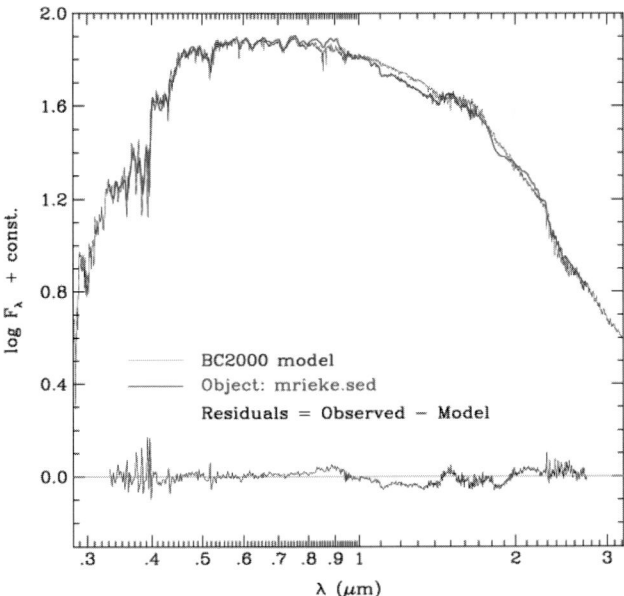

FIGURE 10. Best fit to an average Elliptical galaxy SED (heavy line) in the range $\lambda\lambda$ 3300 - 27500 Å. The model is the thin line extending over the full wavelength range. The best fit occurs at 10 Gyr for this model SED. The residuals of the fit, $\log F_\lambda(observed) - \log F_\lambda(model)$, are shown as a function of wavelength.

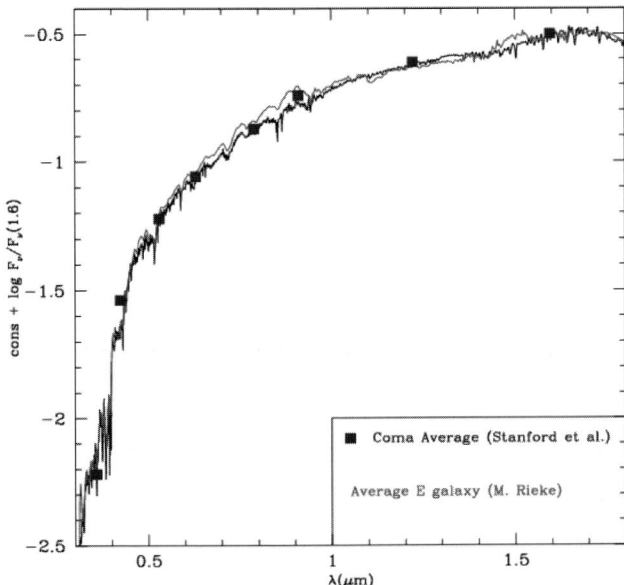

FIGURE 11. Best fit to an average Elliptical galaxy SED (same model and E galaxy SED as in Fig. 10 but in different units). The broad band fluxes representing the average of many E galaxies in the Coma cluster (solid squares) from A. Stanford (private communication) are shown. The observed SED is the one with the lowest spectral resolution.

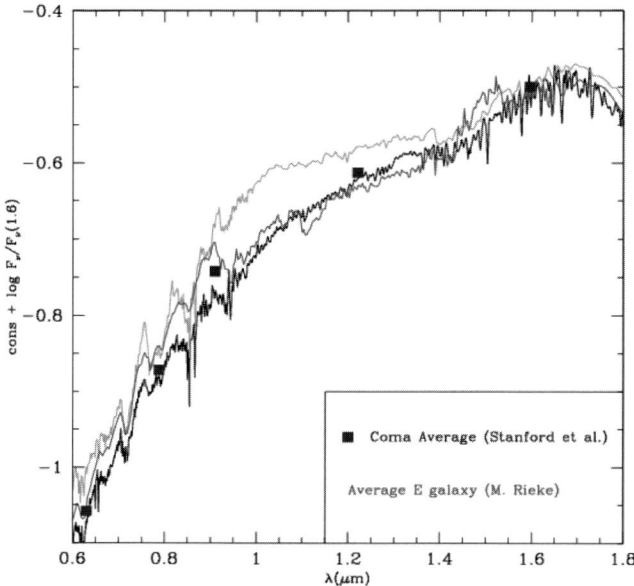

FIGURE 12. Fits to an average Elliptical galaxy SED. This figure shows a closer look at the same data of Fig. 11 in an enlarged scale. The discrepant line corresponds to the same model shown in Figs. 10 and 11 but using the LCB97 synthetic stellar atlas instead of the empirical stellar SEDs.

the P tracks and the Pickles (1998) stellar atlas. The meaning of each line is indicated in the top central frame. Fig. 16 shows the corresponding plot for an equivalent model computed according to the G tracks. The contribution of the RGB stars is higher in the P track model than in the G track model. Correspondingly, the AGB stars contribute less in the P track model than in the G track model. For instance, for $t > 1$ Gyr, RGB and AGB stars contribute 40% and 10%, respectively, to the bolometric light in the P track model (Fig. 15). These fractions change to 30% and 20% in the G track model (Fig. 16). These differences are seeing more clearly in Fig. 17 which shows the ratio of the fractional contribution by different stellar groups in the G track model to that in the P track model. According to the fuel consumption theorem (Renzini 1981), these numbers reflect relatively large differences in the amount of fuel used up in the RGB and AGB phases by stars of the same mass and initial chemical composition depending on the stellar evolutionary code.

Fig. 20 shows the difference in B magnitude and $B - V$ and $V - K$ color between a G track model SSP and a P track model SSP as seen both in the rest frame of the galaxy (vs. galaxy age in the left hand side panels) and the observer frame (vs. redshift in the right hand side panels). These differences reach quite substantial values. The observer frame quantities include both the k and the evolutionary corrections. Here and elsewhere in this paper I assume $H_0 = 65$ km s^{-1} Mpc^{-1}, $\Omega = 0.10$, and the age of galaxies to be $tg = 12$ Gyr.

11.2. *On the energetics of model stellar populations*

Buzzoni (1999) has argued that most population synthesis models violate basic prescriptions from the fuel consumption theorem (FCT). Fig. 18 should be compared to Fig. 2 of Buzzoni (1999). The line with square dots along it is reproduced from Buzzoni's Fig.

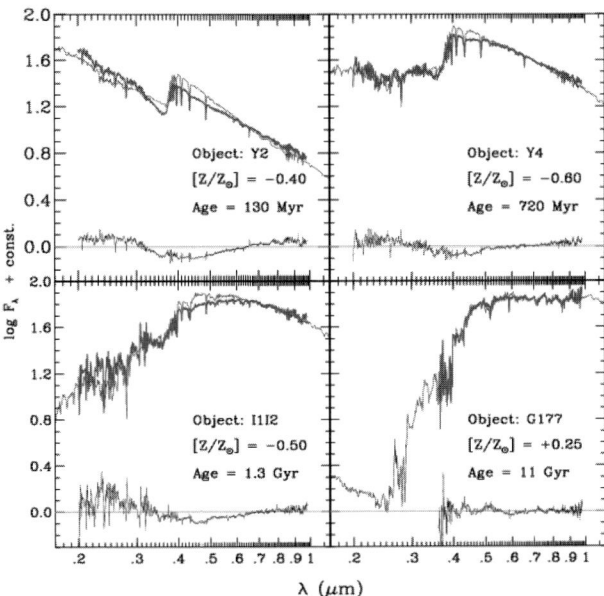

FIGURE 13. Best fit to average optical SED (heavy line) of star clusters of various metallicities compiled by Bica et al. (1996a). The SSP model is the thin line extending over the full wavelength range. The residuals of the fit, $\log F_\lambda(observed) - \log F_\lambda(model)$, are shown as a function of wavelength. The name and the metal content of the observed spectra indicated in each panel is as given by Bica et al. The quoted age is derived from the best fit of our model spectra for the indicated metallicity to the corresponding observations.

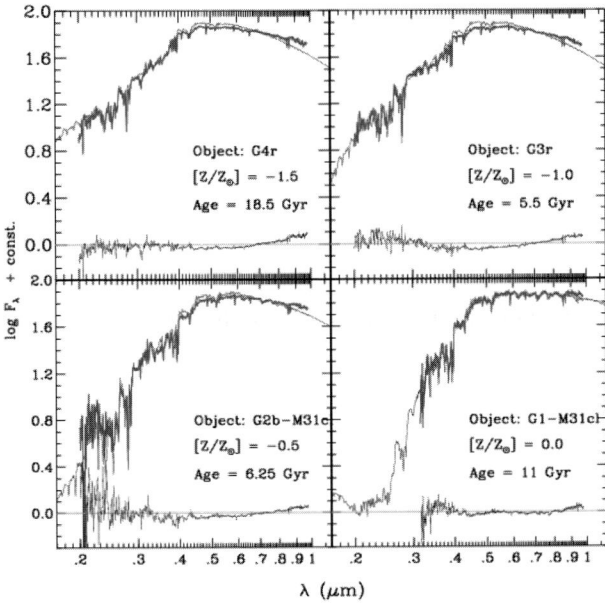

FIGURE 14. Same as Fig. 13, but for four different star clusters.

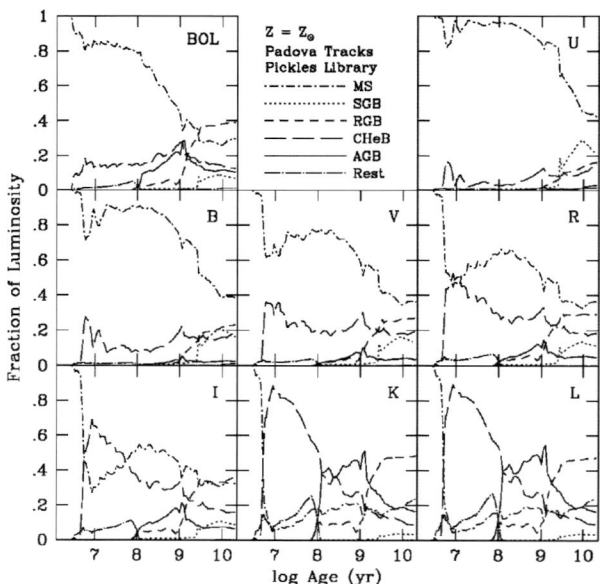

FIGURE 15. Contribution of stars in various evolutionary stages to the bolometric light, and to the broad-band $UBVRIKL$ fluxes for a $Z = Z_\odot$ model SSP computed for the Salpeter IMF ($m_L = 0.1$, $m_U = 125$ $M_\odot$) using the P tracks and the Pickles (1998) stellar atlas.

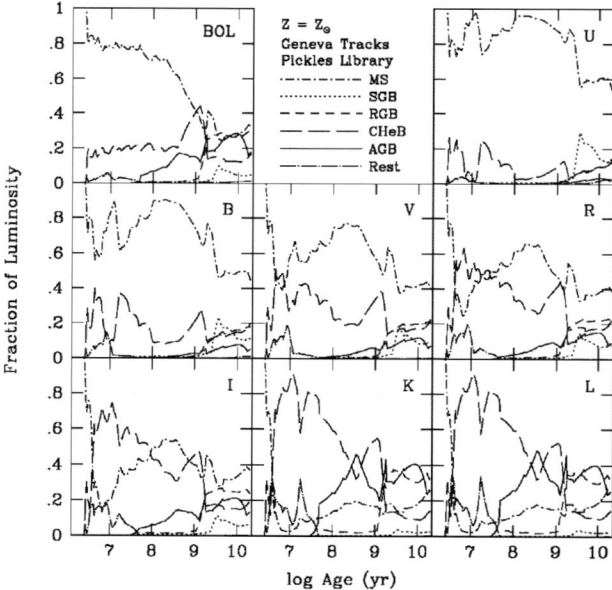

FIGURE 16. Contribution of stars in various evolutionary stages to the bolometric light, and to the broad-band $UBVRIKL$ fluxes for a $Z = Z_\odot$ model SSP computed for the Salpeter IMF ($m_L = 0.1$, $m_U = 125$ $M_\odot$) using the G tracks and the Pickles (1998) stellar atlas.

2. The other lines show the dependence of the ratio of the Post-MS to MS contribution to the bolometric flux for different models. The heavy lines correspond to the Salpeter IMF models. The thin lines to the Scalo IMF models. The solid lines correspond to P track models, whereas the dashed lines correspond to the G track models. The G track model for the Salpeter IMF (heavy dashed line) is in quite good agreement with Buzzoni's model for $t > 5$ Gyr.

Fig. 19 (after Buzzoni's Fig. 1) plots the M/L_V ratio vs. the Post-MS to MS contribution in the V band. The open dots correspond to the models shown in Buzzoni's Fig. 1. The solid dot is Buzzoni's model marked B in his Fig. 1. The solid triangles correspond to our $Z_\odot$ SSP models for various stellar atlas using the P tracks and the Salpeter IMF. The open triangles are for the same models but for the Scalo IMF. The solid pentagons represent the G track models for the Salpeter IMF and the open pentagons the same models but for the Scalo IMF. The three solid squares joined by a line represent sub-solar metallicity models for the P tracks and the Salpeter IMF. The three open squares joined by a line are for identical models using the Scalo IMF. Fig. 19 shows clearly that the position of points representing various models in this diagram is a strong function of the stellar IMF, the set of evolutionary tracks, and the chemical composition of the stellar population. It may be too simplistic to attribute the dispersion of the points to a violation of the FCT (Buzzoni 1999).

11.3. *Uncertainties in the stellar IMF*

It is constructive to compare models computed for identical ingredients except for the stellar IMF. Fig. 21 shows the results of such a comparison. Brightness and color differences with respect to the Salpeter IMF model SSP are shown vs. galaxy age in the galaxy rest frame (LHS panels) and vs. redshift in the observer frame (RHS panels) for SSP P track models computed for the following IMFs: Scalo (1986, *solid line*), Miller & Scalo (1979, *short dashed line*), and Kroupa et al. (1993, *long dashed line*).

Fig. 22 compares in the same format as before the results of using different solar metallicity stellar libraries for a P track SSP model. Brightness and color differences with respect to the SSP model computed with the Pickles (1998) stellar atlas are shown vs. galaxy age in the galaxy rest frame (LHS panels) and vs. redshift in the observer frame (RHS panels) for SSP P track models computed for the following stellar libraries: Extended Gunn & Stryker (1983) atlas used by BC93 (*solid line*), LCB97 uncorrected atlas (*short dashed line*), and LCB97 corrected atlas (*long dashed line*).

11.4. *Different chemical composition*

In this section we explore the differences between SSP models for non-solar composition and the solar case. Fig. 23 shows the brightness and color differences with respect to the SSP model for $Z = Z_\odot$. All models shown in this figure are for the P tracks and use the LCB97 stellar atlas. The lines in this figure have the following meaning: $Z = 0.0001$ (*solid line*), $Z = 0.0004$ (*short dashed line*), $Z = 0.004$ (*long dashed line*), $Z = 0.008$ (*dot - short dashed line*), $Z = 0.05$ (*short dash - long dashed line*), and $Z = 0.1$ (*dot - long dashed line*).

11.5. *Different history of chemical evolution*

Fig. 24 shows three possible chemical evolutionary histories, $Z(t)$, quick (Z_Q, *short dashed line*), linear (Z_L, *solid line*), and slow (Z_S, *long dashed line*), that reach $Z = 0.1 = 5 \times Z_\odot$ at 15 Gyr. The dotted lines indicate $Z_\odot$ and $t_g = 12$ Gyr. I have computed models for a SFR $\Psi(t) \propto \exp(-t/\tau)$, with $\tau = 5$ Gyr, which evolve chemically accordingly to the lines shown in Fig. 24. The difference in brightness and color of these three models with

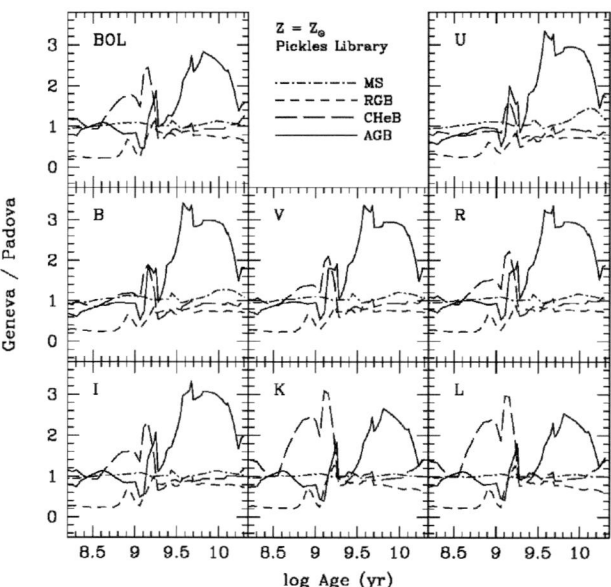

FIGURE 17. Ratio of the fractional contribution by different stellar groups in the G track model to that in the P track model.

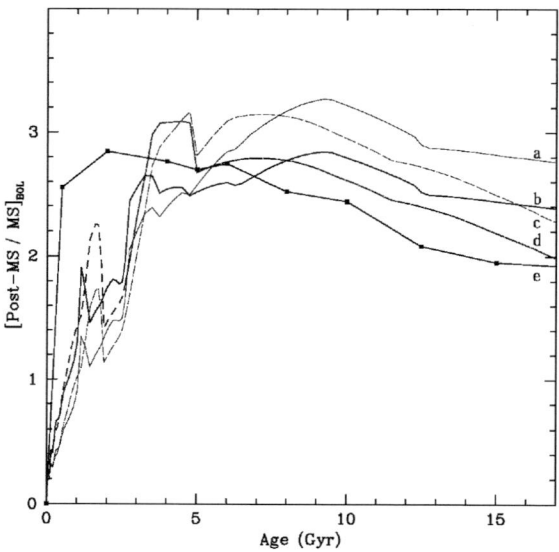

FIGURE 18. Ratio of the Post-MS to MS contribution to the bolometric flux vs. age for different models. This figure should be compared to Fig. 2 of Buzzoni (1999). The line with square dots along it is reproduced from Buzzoni's Fig. 2. The heavy lines correspond to the Salpeter IMF models. The thin lines to the Scalo IMF models. The solid lines correspond to P track models, whereas the dashed lines correspond to the G track models. The G track model for the Salpeter IMF (heavy dashed line) is in quite good agreement with Buzzoni's model for $t > 5$ Gyr.

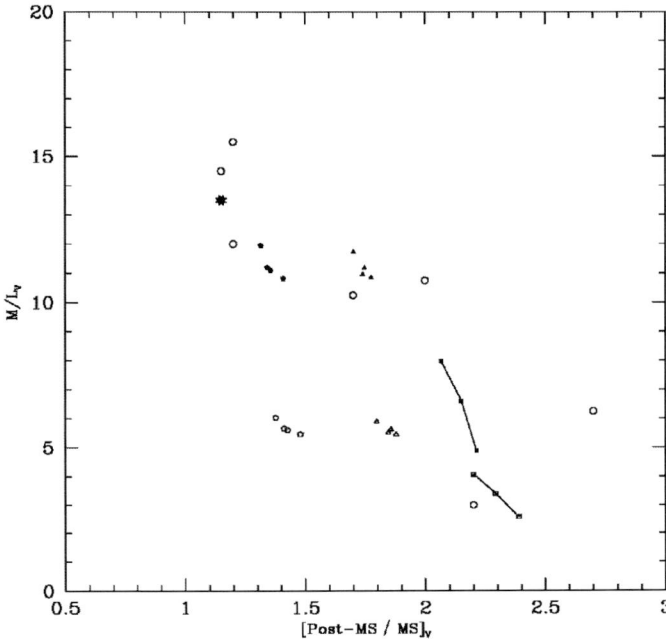

FIGURE 19. M/L_V ratio vs. the Post-MS to MS contribution in the V band. This figure should be compared to Fig. 1 of Buzzoni (1999). The open dots correspond to the models shown in Buzzoni's Fig. 1. The solid dot is Buzzoni's model marked B in his Fig. 1. The solid triangles correspond to our $Z_\odot$ SSP models for various stellar atlas using the P tracks and the Salpeter IMF. The open triangles are for the same models but for the Scalo IMF. The solid pentagons represent the G track models for the Salpeter IMF and the open pentagons the same models but for the Scalo IMF. The three solid squares joined by a line represent sub-solar metallicity models for the P tracks and the Salpeter IMF. The three open squares joined by a line are for identical models using the Scalo IMF.

respect to a $Z = Z_\odot$ model for the same SFR are shown in Fig. 25. The meaning of the lines is as follows: $Z(t) = Z_Q$ (short dashed line), $Z(t) = Z_L$ (solid line), and $Z(t) = Z_S$ (long dashed line).

11.6. Evolution in the observer frame at various cosmological epochs

In Figs. 26 to 31, I summarize the range of values expected in the measured $(V - R)$ and $(V - K)$ colors in the observer frame at various redshifts z as a function of galaxy age. In these figures, the panel marked $TRACKS$ shows the range of colors obtained for solar metallicity SSP models computed using the Pickles empirical stellar atlas with the Salpeter IMF for the P and the G tracks. In the panel marked IMF I show $Z = Z_\odot$ SSP models computed for the P tracks, the Pickles stellar atlas, and the Salpeter, the Scalo, and the Miller-Scalo IMFs. The panel marked $SEDs$ shows the evolution of $Z = Z_\odot$ SSP models computed with the P tracks and the Salpeter IMF, using the empirical Gunn-Stryker and Pickles stellar libraries, as well as the original and corrected versions of the LCB atlas for $Z = Z_\odot$. The panel marked SFR shows the evolution of an SSP model together with a model in which stars form at a constant rate during the first Gyr

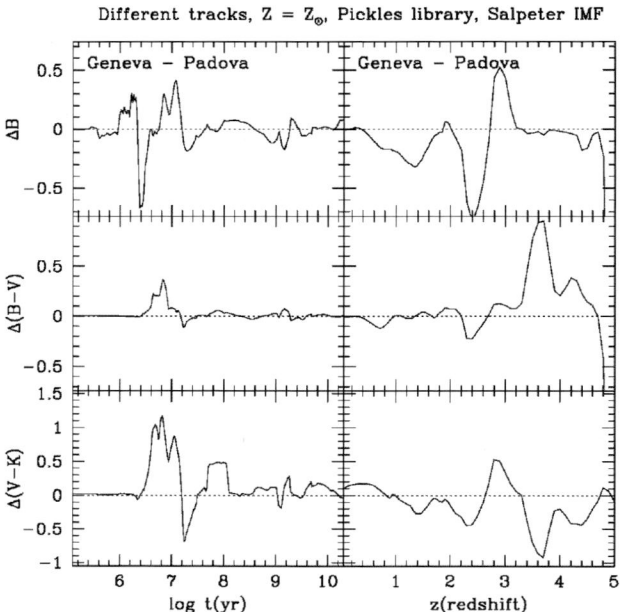

FIGURE 20. Difference in B magnitude and $B-V$ and $V-K$ color between a G track model SSP and a P track model SSP, vs. age in the galaxy rest frame (LHS panels) and vs. redshift in the observer frame (RHS panels).

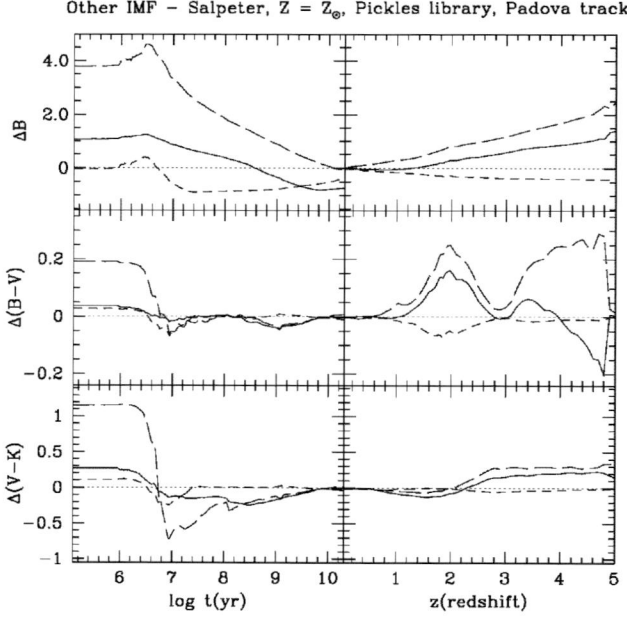

FIGURE 21. Difference in B magnitude and $B-V$ and $V-K$ color for various IMF, P track model SSP's, with respect to the Salpeter IMF model, vs. galaxy age in the galaxy rest frame (LHS panels) and vs. redshift in the observer frame (RHS panels). See §11.3 for details.

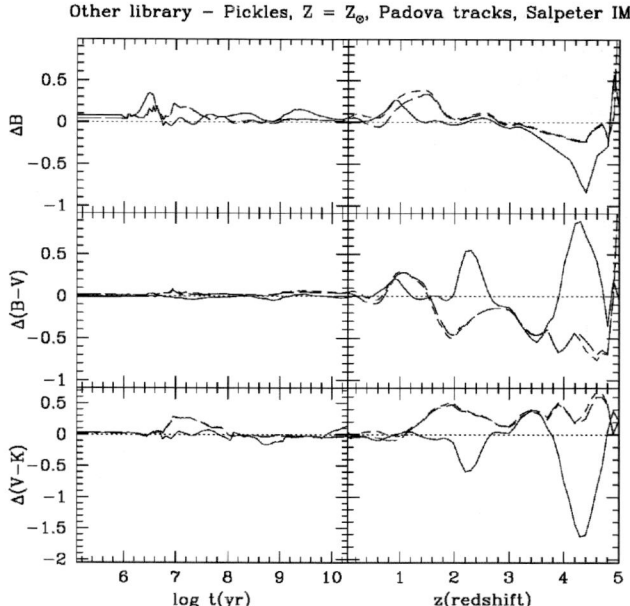

FIGURE 22. Difference in B magnitude and $B-V$ and $V-K$ color for P track model SSP's computed for various $Z = Z_\odot$ stellar libraries, with respect to the SSP model computed with the Pickles (1998) stellar atlas, vs. galaxy age in the galaxy rest frame (LHS panels) and vs. redshift in the observer frame (RHS panels). See §11.3 for details.

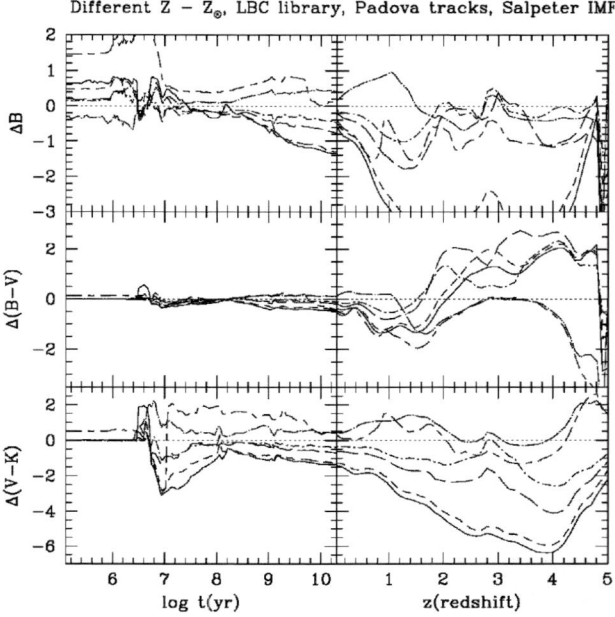

FIGURE 23. Differences between SSP models for non-solar composition and the solar case, vs. galaxy age in the galaxy rest frame (LHS panels) and vs. redshift in the observer frame (RHS panels). See §11.4 for details.

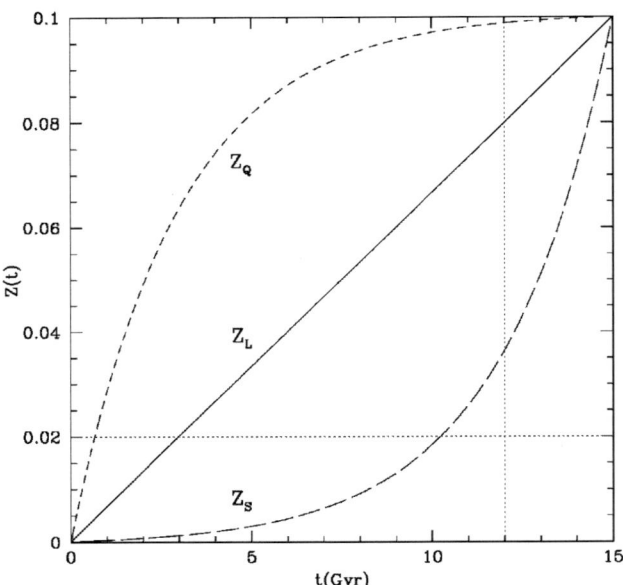

FIGURE 24. Three possible chemical evolutionary histories, $Z(t)$, quick (Z_Q, *short dashed line*), linear (Z_L, *solid line*), and slow (Z_S, *long dashed line*), that reach $Z = 0.1 = 5 \times Z_\odot$ at 15 Gyr.

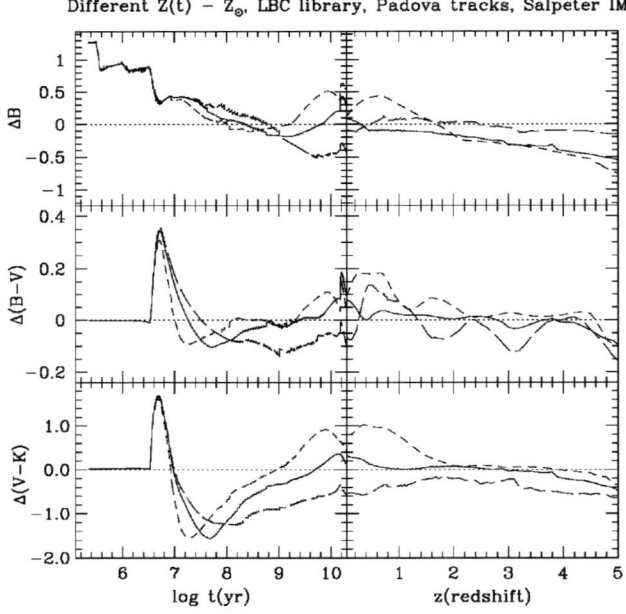

FIGURE 25. Difference in brightness and color of three models that evolve chemically according to Fig. 24 with respect to a $Z = Z_\odot$ model for the same SFR, vs. galaxy age in the galaxy rest frame (LHS panels) and vs. redshift in the observer frame (RHS panels). See §11.5 for details.

in the life of the galaxy (1 Gyr model), both computed with the P tracks, Salpeter IMF, and the Pickles stellar library. The panel marked Z shows the range of colors covered by SSP models of metallicity $Z = 0.004, 0.008$ and 0.02 (solar), computed with the P tracks and the Salpeter IMF. In the solar case, I repeat the models shown in the panel marked $SEDs$. The panel marked ALL summarizes the results of the previous panels. The reddest color obtained at any age in the previous 5 panels is shown as the top solid line. The bluest color is shown as a dotted line. The average color is indicated by the solid line between these two extremes. The 1 Gyr model is shown as a dashed line to show the dominant effects of star formation in galaxy colors.

Figs. 26 and 27 show the range of values expected in $(V-R)$ and $(V-K)$ in the observer frame at $z = 0$ as a function of galaxy age. The maximum age allowed is the age of the universe, $t_u = 13.5$ Gyr at $z = 0$ using $H_0 = 65$ km s^{-1} Mpc^{-1}, $\Omega = 0.10$. Figs. 28 and 29 show the same quantities but for galaxies seen at $z = 1.552$. The age of the universe for this z in this cosmology is $t_u = 4.6$ Gyr. Figs. 30 and 31 correspond to $z = 3$, in this case $t_u = 2.7$ Gyr. From Figs. 26 to 31 I conclude that metallicity Z and the SFR are the most dominant factors determining the range of allowed colors.

The horizontal lines shown across the panels in Figs. 28 and 29 indicate the color $\pm \sigma$ of the galaxy LBDS 53W091 observed by Spinrad et al. (1997). Our models reproduce the colors of this galaxy at an age close to 1.5 Gyr.

Figs. 32 to 35 are based on the panel marked ALL of Figs. 26 to 29, and similar figures for $(V-U)$, $(V-B)$, $(V-I)$, and $(V-J)$ not shown in this work. To build these figures I have subtracted from each line in the previous figures, the color of the $Z = Z_\odot$ SSP model computed with the P tracks, the Salpeter IMF, and the Pickles stellar library. We conclude that the evolution of $(V-R)$ is less model dependent than for any other color shown in these figures.

Figs. 36 to 39 are also based on the panel marked ALL of Figs. 26 to 29 and similar figures for other values of z not shown in this work. Figs. 36 to 39 show the evolution in time of $(V-U)$, $(V-B)$, $(V-R)$, and $(V-K)$ in the observer frame for several values of the redshift z. The color of the $Z = Z_\odot$ SSP model computed with the P tracks, the Salpeter IMF, and the Pickles stellar library has been subtracted from the lines in the previous figures. Again, $(V-R)$ shows less variations with model than the other colors.

12. Summary and Conclusions

Present population synthesis models show reasonable agreement with the observed spectrum of stellar populations of various ages and metal content. Differences in results from different codes can be understood in terms of the different ingredients used to build the models and do not necessarily represent violations of physical principles by some of these models. However, inspection of Fig. 20 shows that two different sets of evolutionary tracks for stars of the same metallicity produce models that at early ages differ in brightness and color from 0.5 to 1 mag, depending on the specific bands. The differences decrease at present ages in the rest frame, but are large in the observer frame at $z > 2$. Thus any attempt to date distant galaxies, for instance, based on fitting observed colors to these lines will produce ages that depend critically on the set of models which is used. Note that from z of 3 to 3.5 $(V-K)$ in the two models differs by more than 1 mag. This difference is produced by the corresponding difference between the models seen in the rest frame at 10 Myr. From Figs. 15 and 16 these differences can be understood in terms of the different contribution of the same stellar groups to the total V and K flux in the two models.

Even though at the present age models built with different IMFs show reasonably

212 Gustavo Bruzual A.: *Population Synthesis at low and high z*

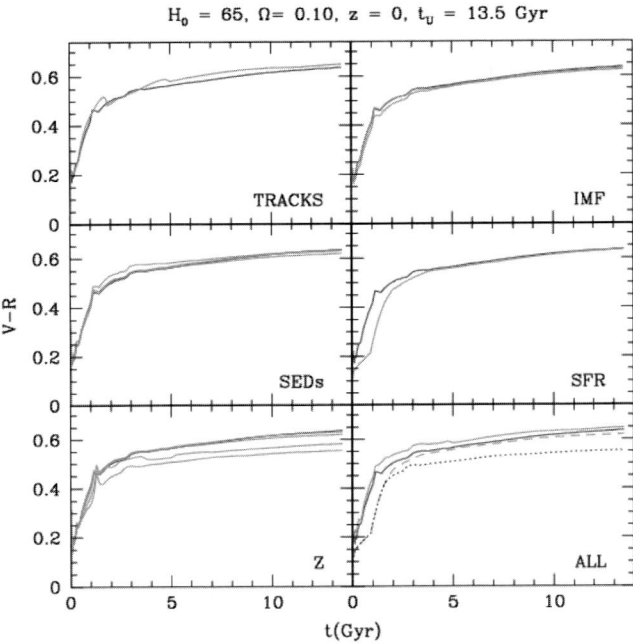

FIGURE 26. $(V - R)$ vs. time in the observer frame at $z = 0$. See §11.6 for details.

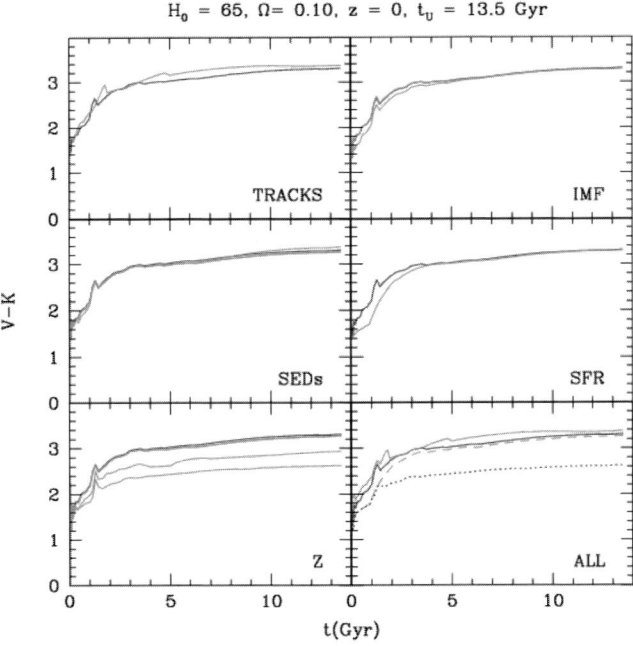

FIGURE 27. $(V - K)$ vs. time in the observer frame at $z = 0$. See §11.6 for details.

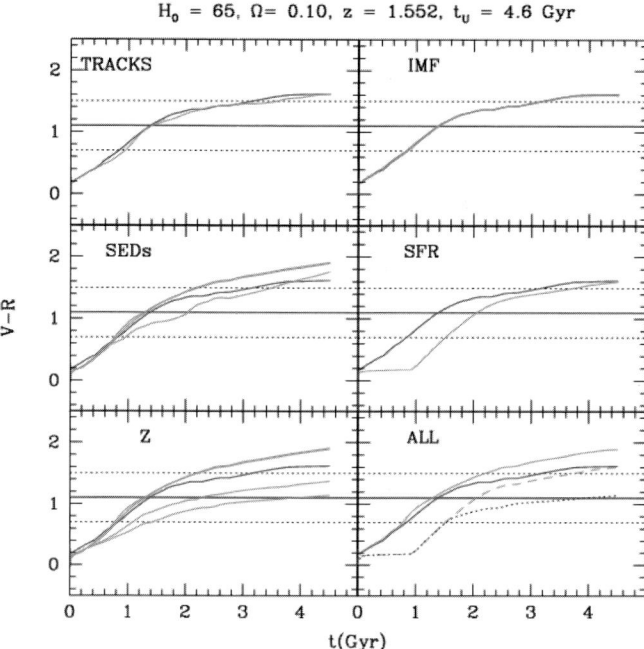

FIGURE 28. $(V - R)$ vs. time in the observer frame at $z = 1.552$. See §11.6 for details.

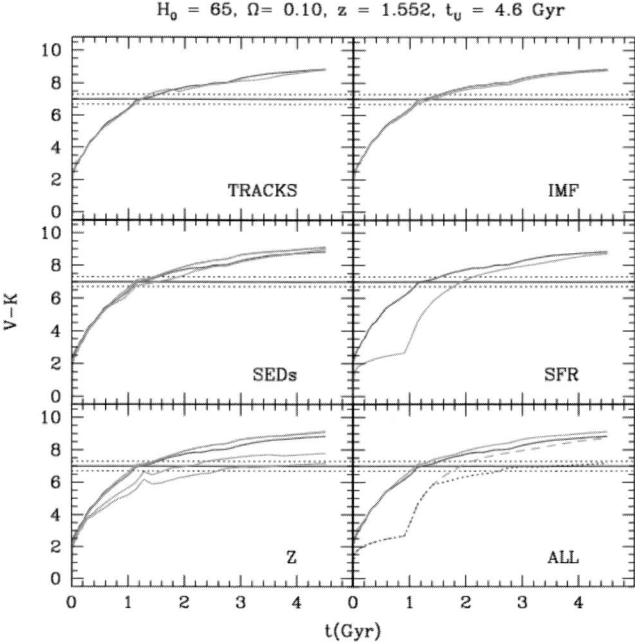

FIGURE 29. $(V - K)$ vs. time in the observer frame at $z = 1.552$. See §11.6 for details.

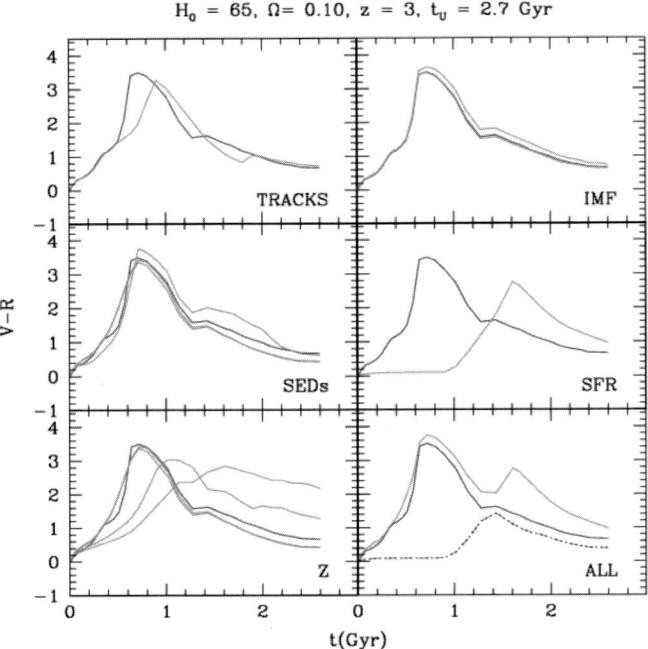

FIGURE 30. $(V - R)$ vs. time in the observer frame at $z = 3$. See §11.6 for details.

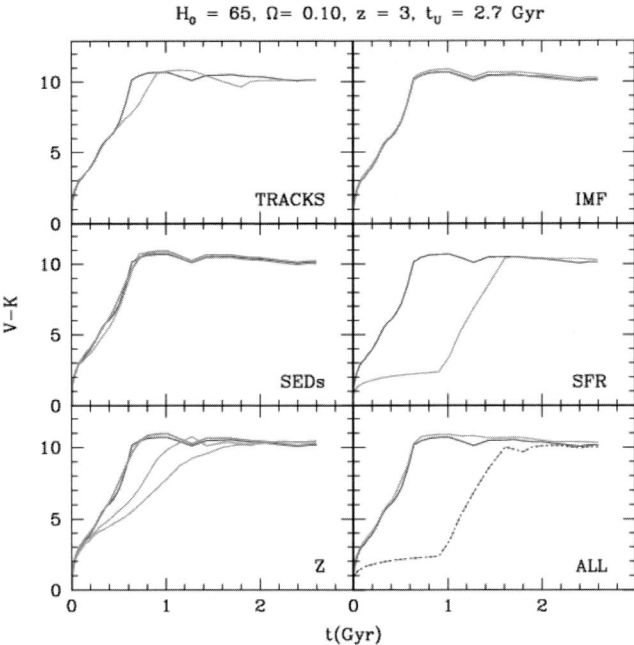

FIGURE 31. $(V - K)$ vs. time in the observer frame at $z = 3$. See §11.6 for details.

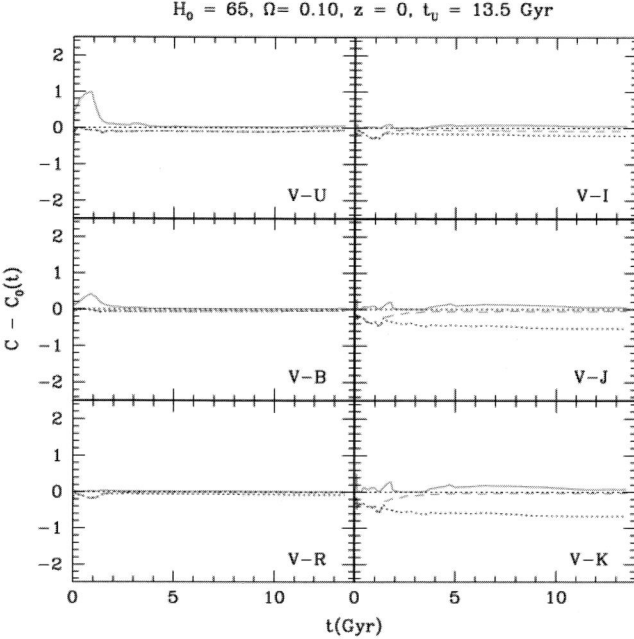

FIGURE 32. Color vs. time in the observer frame at $z = 0$. The color of the $Z = Z_\odot$ SSP model computed with the P tracks, the Salpeter IMF, and the Pickles stellar library have been subtracted from each line. See §11.6 for details.

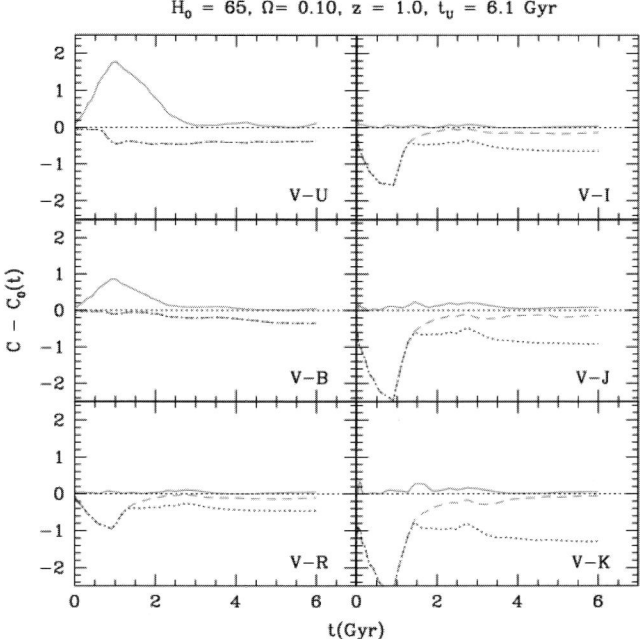

FIGURE 33. Color vs. time in the observer frame at $z = 1$. The color of the $Z = Z_\odot$ SSP model computed with the P tracks, the Salpeter IMF, and the Pickles stellar library have been subtracted from each line. See §11.6 for details.

216   Gustavo Bruzual A.: *Population Synthesis at low and high z*

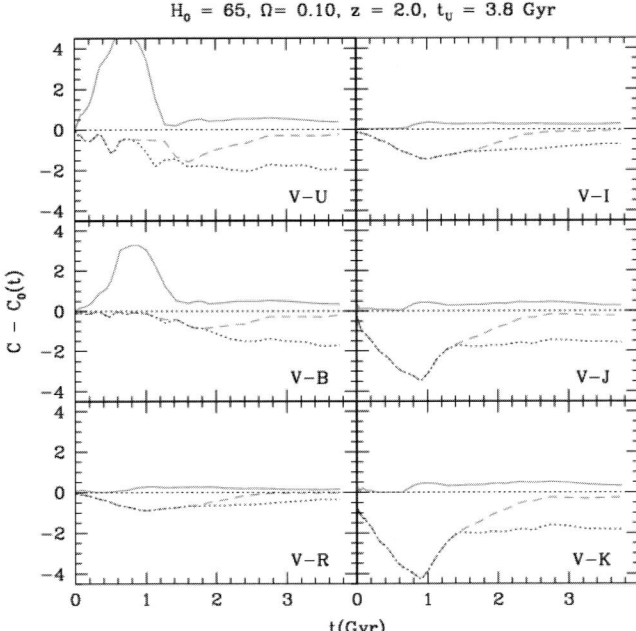

FIGURE 34. Color vs. time in the observer frame at $z = 2$. The color of the $Z = Z_\odot$ SSP model computed with the P tracks, the Salpeter IMF, and the Pickles stellar library have been subtracted from each line. See §11.6 for details.

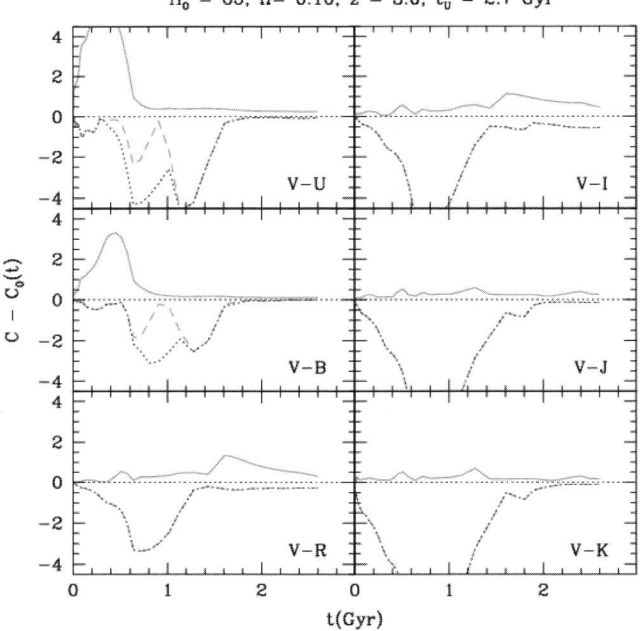

FIGURE 35. Color vs. time in the observer frame at $z = 3$. The color of the $Z = Z_\odot$ SSP model computed with the P tracks, the Salpeter IMF, and the Pickles stellar library have been subtracted from each line. See §11.6 for details.

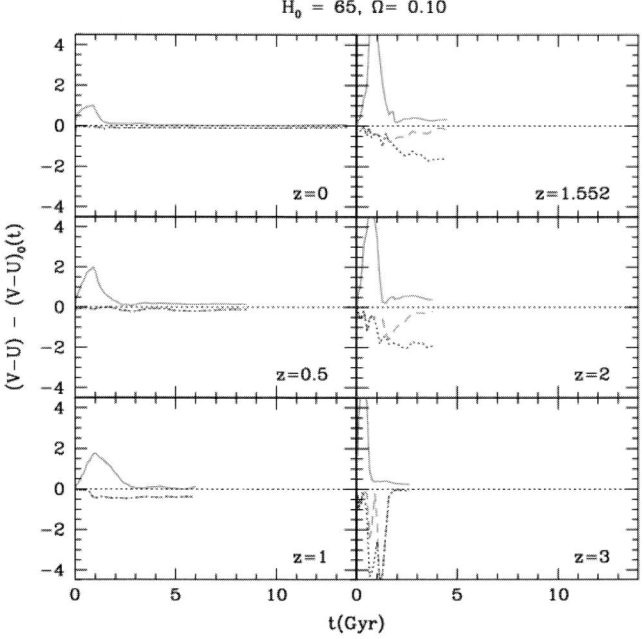

FIGURE 36. $(V - U)$ vs. time in the observer frame for various values of z. The color of the $Z = Z_\odot$ SSP model computed with the P tracks, the Salpeter IMF, and the Pickles stellar library have been subtracted from each line. See §11.6 for details.

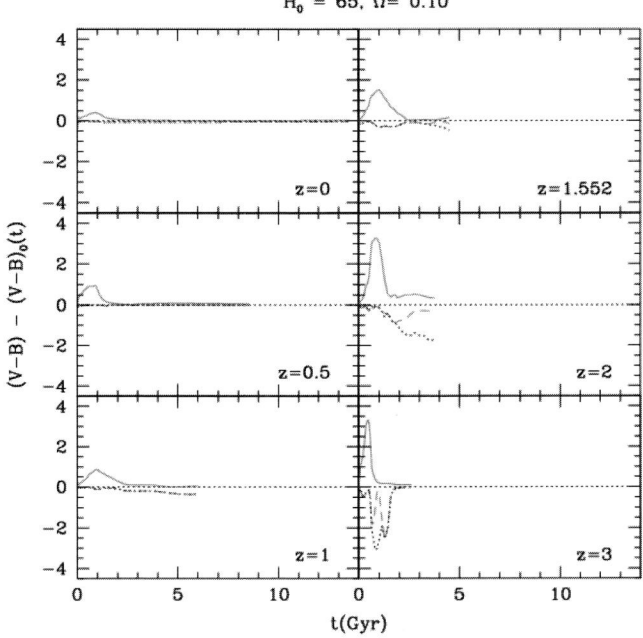

FIGURE 37. $(V - B)$ vs. time in the observer frame for various values of z. The color of the $Z = Z_\odot$ SSP model computed with the P tracks, the Salpeter IMF, and the Pickles stellar library have been subtracted from each line. See §11.6 for details.

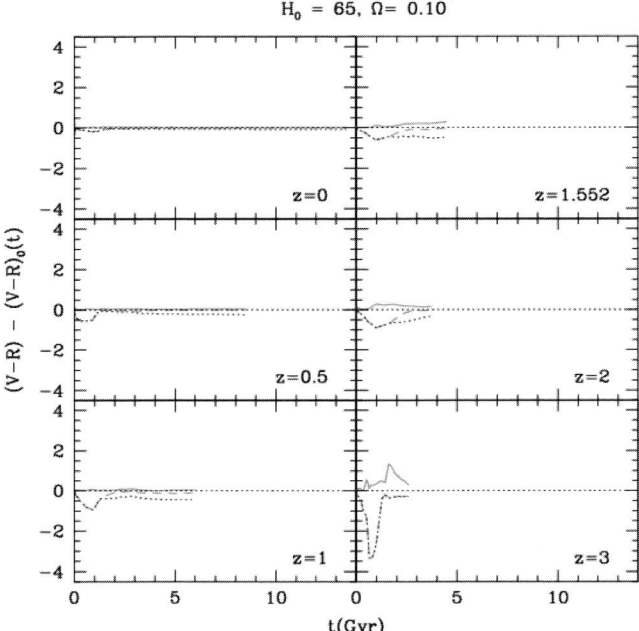

FIGURE 38. $(V - R)$ vs. time in the observer frame for various values of z. The color of the $Z = Z_\odot$ SSP model computed with the P tracks, the Salpeter IMF, and the Pickles stellar library have been subtracted from each line. See §11.6 for details.

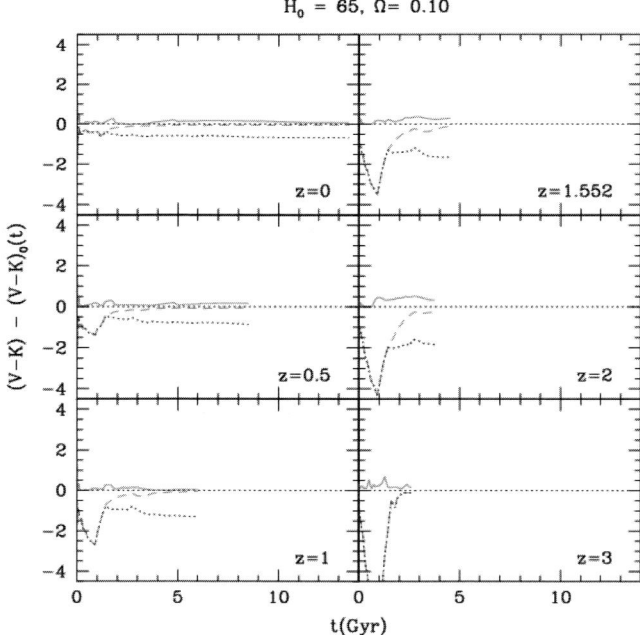

FIGURE 39. $(V - K)$ vs. time in the observer frame for various values of z. The color of the $Z = Z_\odot$ SSP model computed with the P tracks, the Salpeter IMF, and the Pickles stellar library have been subtracted from each line. See §11.6 for details.

similar colors and brightness, the early evolution of these models is quite different at early ages (Fig. 21), resulting in larger color differences in the observer frame at $z >$ 2. Thus, the more we know about the IMF, the better the model predictions can be constrained. The small color differences seen in the rest frame when different stellar libraries of the same metallicity are used, are magnified in the observer frame (Fig. 22). When the k correction brings opposing flux differences into each filter, the difference in the resulting color is enhanced. Fig. 23 shows the danger of interpreting data for one stellar system with models of the wrong metallicity. The color differences between these models, especially in the observer frame, are so large as to make any conclusion thus derived very uncertain.

It is common practice to use solar metallicity models when no information is available about the chemical abundance of a given stellar system. Galaxies evolving according the $Z(t)$ laws of Fig. 24 show color differences with respect to the $Z = Z_\odot$ model which are not larger than the differences introduced by the other sources of uncertainties discussed so far. Hence, the solar metallicity approximation may be justified in some instances. The color differences between the chemically inhomogeneous composite population and the purely solar case (Fig. 25), are much smaller than the ones shown in Fig. 23 for chemically homogeneous SSPs.

Figs. 26 to 39 indicate that some colors, especially $(V - R)$, when measured in the observer frame are less sensitive to model predictions than other colors. From Figs. 26 to 31, metallicity Z and the SFR are the most dominant factors determining the range of allowed colors.

I expect that through these simple examples the reader can get a feeling of the kind of uncertainties introduced by the many ingredients entering the stellar population synthesis problem, and that he or she will be motivated to try his or her own error estimates when using these models.

REFERENCES

ALLARD, F. & HAUSCHILDT, P.H. 1995, ApJ, 445, 433.
ALONGI, M., BERTELLI, G., BRESSAN, A., CHIOSI, C., FAGOTTO, F., GREGGIO, L. & NASI, E. 1993, A&AS, 97, 851.
ARAGÓN-SALAMANCA, A., ELLIS, R.S.E., COUCH, W. J., CARTER, D. 1993, MNRAS, 262, 764A.
ARIMOTO, N. & YOSHII, Y. 1987, A&A, 173, 23.
BARBUY, B., ORTOLANI, S., BICA, E., RENZINI, A. & GUARNIERI, M.D. 1997, in IAU Symp. 189, *Fundamental Stellar Parameters: Confrontation Between Observation and Theory*, eds. J. Davis, A. Booth & T. Bedding, Kluwer Acad. Pub., p. 203.
BENDER, R., ZIEGLER, B. & BRUZUAL A., G. 1996, ApJ Letters, 463, L51.
BESSELL, M.S., BRETT, J., SCHOLTZ, M. & WOOD, P. 1989, A&AS, 77, 1.
———. 1991, A&AS, 89, 335.
BICA, E. & ALLOIN, D. 1986, A&A, 162, 21.
———. 1987, A&A, 186, 49.
BICA, E., ALLOIN, D., BONATTO, C., PASTORIZA, M.G., JABLONKA, P., SCHMIDT, A. & SCHMITT, H.R. 1996a, in *A Data Base for Galaxy Evolution Modeling*, eds. C. Leitherer et al., PASP, 108, 996.
BICA, E., ALLOIN, D. & SCHMITT, H. 1994, A&A, 283, 805.
BICA, E., CLARIÁ, J.J., DOTTORI, H., SANTOS JR., J.F.C., PIATTI, A. E. 1996b, ApJS, 102, 57.
BRESSAN, A., CHIOSI, C. & FAGOTTO, F. 1994, ApJS, 94, 63.

Bressan, A., Fagotto, F., Bertelli, G. & Chiosi, C. 1993, A&AS, 100, 647.

Bruzual A. G. 1998, in *The Evolution of Galaxies on Cosmological Time scales*, eds. J.E. Beckman and T.J. Mahoney, ASP Conference Series, Vol. 187, p. 245.

———. 1999, in *The Hy-Redshift Universe: Galaxy Formation and Evolution at High Redshift*, eds. A. J. Bunker and W. J. M. van Breugel, ASP Conference Series, Vol. 193, p. 121.

———. 2000, in *Euroconference on The Evolution of Galaxies, I- Observational Clues*, eds. J.M. Vílchez, G. Stasinska, and E. Pérez, Kluwer Academic Publisher, in press.

Bruzual A. G., Barbuy, B., Ortolani, S., Bica, E., Cuisinier, F., Lejeune, T. & Schiavon, R. 1997, AJ, 114, 1531.

Bruzual A. G. & Charlot, S. 1993, ApJ, 405, 538 (BC93).

———. 2000, ApJ, in preparation (BC2000).

Burstein, D., Bertola, F., Buson, L.M., Faber, S.M. & Lauer, T.R. 1988, ApJ, 328, 440.

Buzzoni, A. 1989, ApJS, 71, 817.

———. 1999, in IAU Symposium No. 183 *Cosmological Parameters and the Evolution of the Universe*, ed. K. Sato, Dordrecht: Kluwer, p. 134.

Charlot, S. & Bruzual A. G. 1991, ApJ, 367, 126 (CB91).

D'Antona, F. 1999, in *The Galactic Halo: from Globular Clusters to Field Stars*, 35th Liege Int. Astroph. Colloquium, astro-ph/9910312.

Eggen, O. J. & Sandage, A.R. 1964, ApJ, 140, 130.

Fagotto, F., Bressan, A., Bertelli, G., & Chiosi, C. 1994a, A&AS, 100, 647.

———. 1994b, A&AS, 104, 365.

———. 1994c, A&AS, 105, 29.

Fluks, M. *et al.* 1994, A&AS, 105, 311.

Fritze-v.Alvensleben, U. & Gerhard, O.E. 1994, A&A, 285, 751.

Gilliland, R.L., Brown, T.M., Duncan, D.K., Suntzeff, N.B., Wesley Lockwood, G., Thompson, D.T., Schild, R.E., Jeffrey, W.A. & Penprase, B.E. 1991, AJ, 101, 541.

Girardi, L., Bressan, A., Bertelli, G. & Chiosi, C. 2000, A&AS, 141, 371.

Girardi, L., Bressan, A., Chiosi, C., Bertelli, G. & Nasi, E. 1996, A&AS, 117, 113.

Greggio, L. & Renzini, A. 1990, ApJ, 364, 35.

Guiderdoni, B. & Rocca-Volmerange, B. 1987, A&A, 186, 1.

Gunn, J.E. & Stryker, L.L. 1983, ApJS, 52, 121.

Iglesias, C.A., Rogers, F.J. & Wilson, B.G. 1992, ApJ, 397, 717.

Janes, K.A. 1985, in Calibration of Fundamental Stellar Quantities, IAU Symposium No. 111, D.S. Hayes, L.E. Pasinetti, and A.G. Davis Philip, (Dordrecht: Reidel), 361.

Janes, K.A. & Smith, G.H. 1984, AJ, 89, 487.

Kroupa, P., Tout, C.A. & Gilmore, G. 1993, MNRAS, 262, 545.

Kurucz, R. 1995, private communication.

Lejeune, T., Cuisinier, F. & Buser, R. 1997, A&AS, 125, 229 (LCB97).

———. 1998, A&AS, 130, 65 (LCB98).

Metcalfe, N., Shanks, T., Fong, R., Gardner, J., Roche, N. 1996, IAU Symp. 171, p. 225.

Micela, G., Sciortino, S., Vaiana, G.S., Schmitt, J.H.M.M., Stern, R.A., Harnden, F. R., Jr. & Rosner, R. 1988, ApJ, 325, 798.

Miller, G.E. & Scalo, J.M. 1979, ApJS, 41, 513.

Peterson, D.M. & Solensky, R. 1988, ApJ, 333, 256.

Pickles, A.J. 1998, PASP, 110, 863.

Pozzetti, L., Bruzual A., G., Zamorani, G. 1996, MNRAS, 281, 953.

Racine, R. 1971, ApJ, 168, 393.
Renzini, A. 1981, Ann. Phys. Fr., 6, 87.
Salpeter, E.E. 1955 ApJ, 121, 161.
Santos, J.F.C. Jr., Bica, E., Dottori, H., Ortolani, S. & Barbuy, B. 1995, A&A, 303, 753.
Scalo, J.M. 1986, Fund. Cosmic Phys, 11, 1.
Spinrad, H., Dey, A., Stern, D., Dunlop, J., Peacock, J., Jiménez, R., Windhorst, R. 1997, ApJ, 484, 581.
Stanford, S.A., Eisenhardt, P.R. & Dickinson, M. 1995, ApJ, 450, 512.
———. 1998, ApJ, 492, 461.
Upgren, A.R. 1974, ApJ, 193, 359.
Upgren, A.R. & Weis, E.W. 1977, AJ, 82, 978.
Worthey, G. 1994, ApJS, 95, 107.
Worthey, G., Faber, S.M., González, J.J. & Burstein, D. 1994, ApJS, 94, 687.

Elliptical Galaxies

By KENNETH C. FREEMAN

Mt. Stromlo Observatory, Canberra, Australia

Galaxies fall into two main classes: the disks, supported primarily by rotation, and the ellipticals, supported mainly by the random motions of their stars. Dissipation is an important element in the formation of the flat and orderly disk galaxies. On the other hand, it is widely believed that non-dissipative mergers are a major contributor to the formation of at least a large fraction of the giant elliptical galaxies. In these lectures, we will look at the clues from ellipticals at low redshift about when and how elliptical galaxies formed, and why they have the structural properties that are observed.

---

1. The Structure of Elliptical Galaxies

1.1. *The Radial Light Distribution*

Elliptical galaxies show some regularity in the radial light distributions. de Vaucouleurs (1948) proposed an empirical law known as the $r^{1/4}$ law, in which the surface brightness distribution is given by the equation

$$I(r) = I_e \exp\{-7.67[(r/r_e)^{1/4} - 1]\}.$$

Here r_e is the radius containing half of the total luminosity (the effective radius) and I_e is the corresponding effective surface brightness. This distribution works very well for many elliptical galaxies, particularly those of intermediate luminosity. Capaccioli *et al.* (1990) showed that the light distribution of the standard elliptical galaxy NGC 3379 is very well represented by the $r^{1/4}$ law over a surface brightness range of about 12 mag. See Figure 1.

The $r^{1/4}$ law is widely used as a representation of spheroidal systems like elliptical galaxies and the bulges of spirals (we will see later how well it works for some bulges). Dynamically, this law is important in our context because it is associated with violent relaxation processes. van Albada (1982) showed that the outcome of dissipationless collapse with irregular initial conditions and a large collapse factor is be well represented by the $r^{1/4}$ law. Similar results were found for the outcome of merging (e.g. Barnes 1988) and of tidal stripping (Aguilar & White 1986).

The $r^{1/4}$ law is not such a good fit to all ellipticals. Many authors have shown that there are systematic deviations from the $r^{1/4}$ law and that these deviations are similar for ellipticals of similar luminosities. See Caon *et al.* (1993) for references. A more general fitting function was introduced by Sersic (1968) of the form

$$I(r) = I_\circ \exp[-(r/r_\circ)^n]$$

which corresponds to the $r^{1/4}$ law for $n = 1/4$ and to an exponential law for $n = 1$. This function is now widely used to investigate how the structure of ellipticals changes with absolute magnitude. As n decreases, the surface brightness profile flattens and tends towards a power law of slope -2.

From the work of Caon *et al.* (1993) and Jerjen *et al.* (1999), for example, it seems clear now that the $r^{1/4}$ law is just a first approximation to the real situation. The fainter ellipticals have n closer to 1 (the exponential distribution) while the brighter ellipticals may have n even lower than 0.1. The Sersic scale length $r_\circ$ and the index

n vary fairly smoothly with absolute magnitude: see Figure 2. Jerjen et al. find that $\log n = 1.4 + 0.10 M_B$ where M_B is the absolute blue magnitude. Graham et al. (1996) also found a strong decrease of the index n with increasing r_e. (Note that some authors use $1/n$ in place of n).

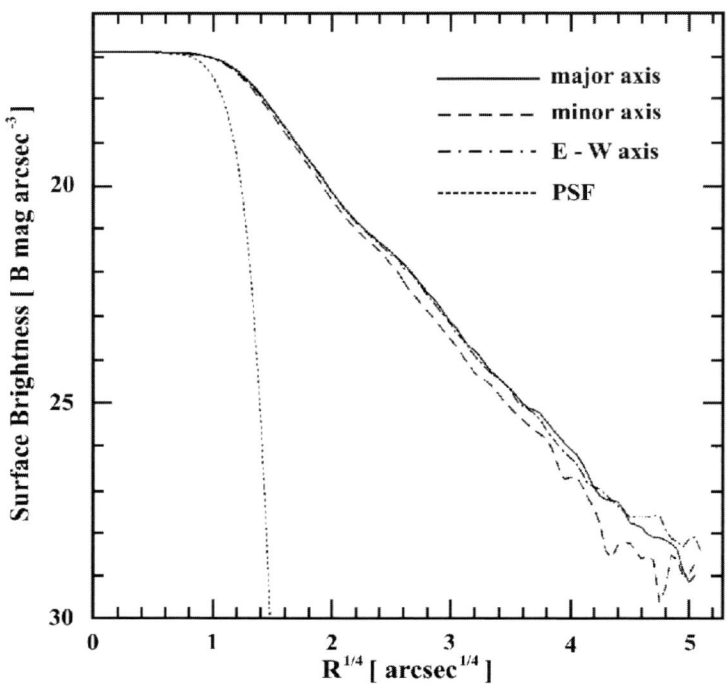

FIGURE 1. The surface brightness distribution of NGC 3379, showing how well the $r^{1/4}$ law works for this galaxy (from Capaccioli et al. 1990)

I should mention also a simple analytic *volume* density distribution due to Jaffe (1983) that gives a good representation of at least some of the brighter elliptical galaxies. It has the form

$$\rho(r) \propto [r(r + r_J)]^{-2}$$

where r_J is a scale length. For this simple distribution, it is possible to calculate the surface brightness distribution and dynamical quantities like the line-of-sight velocity dispersion, and also the (phase density) distribution function for different dynamical assumptions: see Binney & Tremaine (1987) for more discussion.

1.2. *Diskiness and Boxiness*

The isophotes of ellipticals are not always elliptical. Some ellipticals show a distinctly boxy stucture, while others have a more disky appearance. It turns out that the isophote shape is related to the absolute magnitude, to the structure of the innermost parts of the galaxy, and to its gross dynamical properties. See Figure 3.

Boxy galaxies are among the more luminous systems. Their inner regions shows cores with a shallow slope in the surface density distribution (see Faber et al. 1997; also section 9.1), and they are flattened primarily by the anisotropy of their velocity dispersion (see

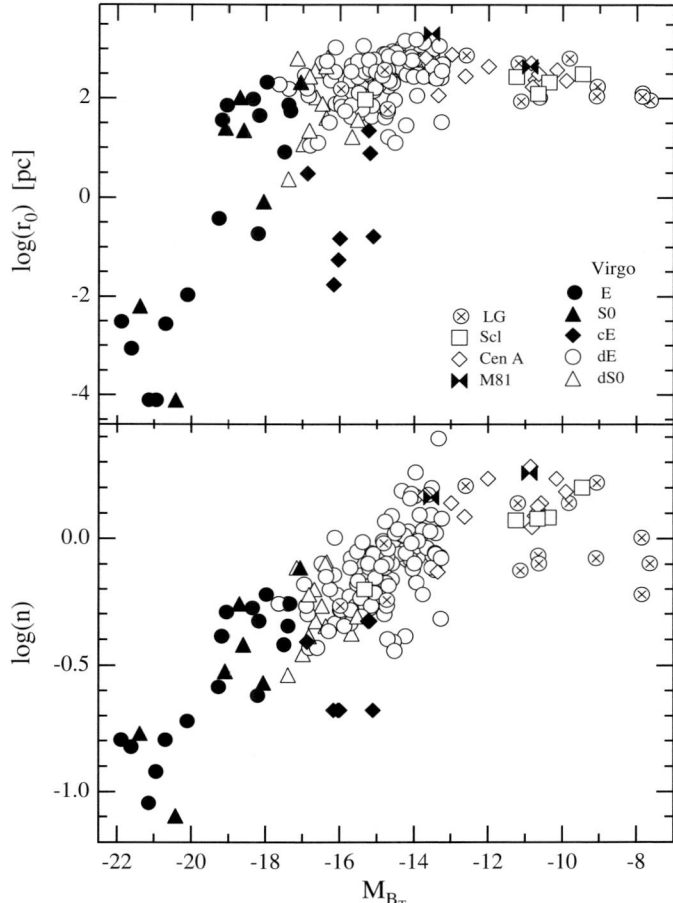

FIGURE 2. The best fitting Sersic parameters $\log r_\circ$ (upper) and $\log n$ (lower) as a function of the absorption-corrected absolute B magnitude for early-type galaxies. The $r^{1/4}$ law corresponds to $\log n = -0.60$ (from Jerjen et al. 1999).

section 3). The more disky systems are less luminous and typically have a small fraction of their light in faint disks: their spheroidal components are flattened by rotation, and they show a steep cusped density distribution at small radius. Their disks and their spheroidal components are well aligned.

For our purpose, we need to understand the origin of the disky/boxy dichotomy, because it probably relates to the formation processes of massive and less massive ellipticals. Naab et al. (1999) have shown that mergers of equal mass disks produce boxy merger products with slow rotation and flattened by anisotropic velocity dispersion, while the mergers of disks with a 3:1 mass ratio produce disky systems that are rotationally supported.

2. The Rotation of Ellipticals

The rotation of ellipticals is usefully displayed in a plot of V/σ vs ϵ, where V is the peak rotation of the galaxy, σ its central velocity dispersion and ϵ is a measure of the isophotal ellipticity $\epsilon = 1 - b/a$, where b/a is the isophotal major to minor axis ratio at some isophotal level (typically 25 mag arcsec^{-2}). Figure 4 shows a sample of elliptical galaxies from Davies et al. (1983). The curve represents the $V/\sigma - \epsilon$ relation for an oblate rotating stellar system with an isotropic velocity dispersion; these systems are flattened by their rotation. Figure 4 shows how the less luminous ellipticals lie close to the oblate isotropic curve, while the brighter ellipticals lie mostly well below it.

The slow rotation of the brighter ellipticals was a major discovery of the late 1970s. The location of the bright ellipticals in the $V/\sigma - \epsilon$ plane indicated that these galaxies were flattened by their anisotropic velocity dispersion rather than by rotation. It raised an important problem about the angular momentum of giant ellipticals that was clearly formulated by Fall (1983) in terms of the relationship between specific angular momentum J/M and mass M for spiral and elliptical galaxies. Figure 5 shows a version of the Fall diagram from Zurek et al. (1988). The specific angular momentum of the giant ellipticals appears to be about an order of magnitude lower than that of the spirals of similar mass. The different point symbols represent N-body cosmological simulations of the growth of dark halos, scaled to show luminous mass, for dense and less dense environments. Most of the simulated systems lay near the spiral locus, and Zurek et al. concluded that the specific angular momentum of the ellipticals had decreased by about an order of magnitude since their formation.

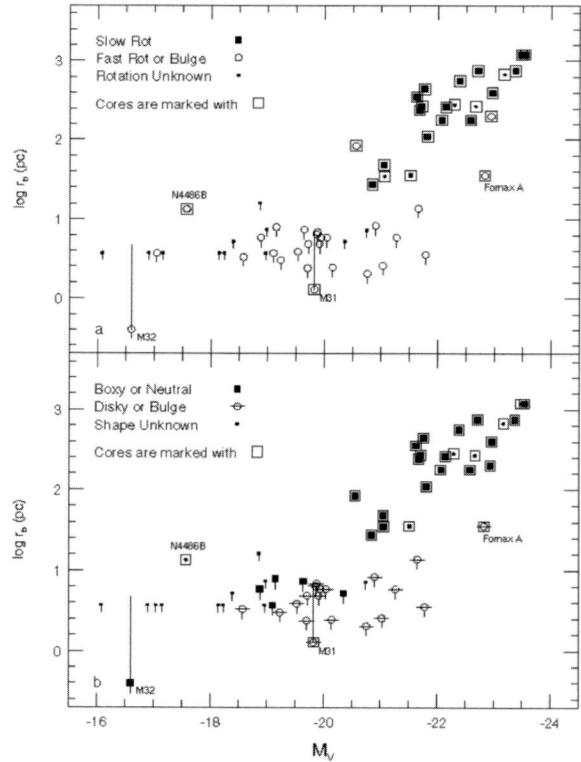

FIGURE 3. The brighter ellipticals have flatter cores, are slowly rotating and have boxy isophotes. The break radius r_b is the radius at which the core or cusp structure in the inner regions appears. From Faber et al. (1997).

The spectroscopy needed to measure rotation was possible only in the bright inner

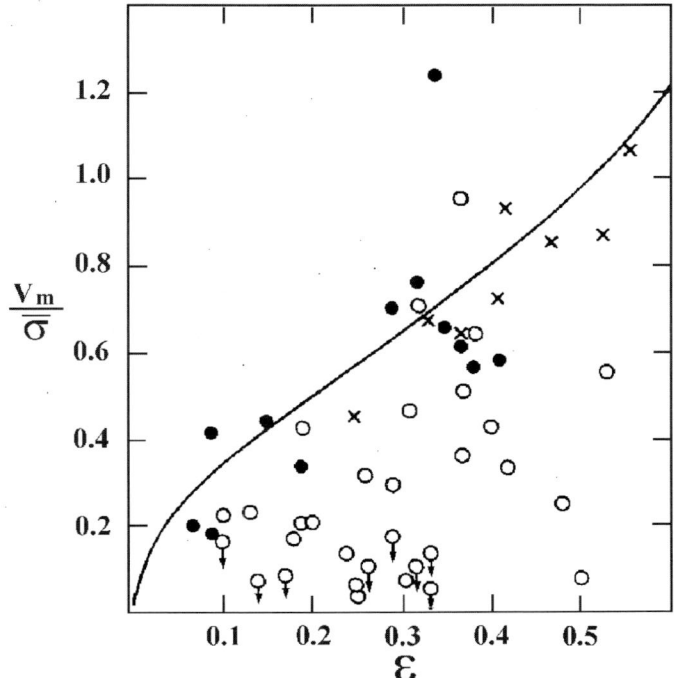

FIGURE 4. The $V/\sigma - \epsilon$ relation for brighter and fainter elliptical galaxies. The curve shows the locus of oblate rotating stellar systems with an isotropic velocity dispersion. This figure shows how the fainter ellipticals are flattened by rotation, while the brighter ellipticals are flattened by their anisotropic velocity dispersion. Open circles show brighter ellipticals ($M_B < 20.5$), filled circles show fainter ellipticals and crosses show bulges of disk galaxies. (From Davies et al. 1983).

parts of ellipticals, so the region outlined by the dashed lines in Figure 5 shows $J/M, M$ for the *inner* regions of giant ellipticals. Zurek et al. argued that angular momentum is transported outwards during the merger processes that build up their halos. Figure 6 shows the evolution of particles from five different bins of binding energy (defined at $z = 0$). The particles that were initially ($z = 5.25$) most bound become more bound and lose angular momentum. The least bound particles become slightly more bound but gain significant amounts of angular momentum through the torques that act during the merging process. In this picture, we would then expect to find a significant amount of angular momentum in the outer parts of these slowly rotating giant ellipticals. Observations of planetary nebulae in the outer regions of several giant ellipticals ($r > 4r_e$) show that this is indeed the situation. Rapid rotation was first detected in the outer parts of Cen A (Hui et al. 1995) and then for the more normal giants NGC 1399 and NGC 1316 (Arnaboldi et al. 1994, 1998). The total specific angular momentum of giant ellipticals appears to be similar to that of spirals of similar mass, although most of the angular momentum of the giant ellipticals appears to reside at large radius.

Planetary nebulae kinematics are not yet available for the outer regions of M87, but it seems likely that the angular momentum of M87 again lies mostly in its outer regions. Although the kinematics of globular cluster systems and the main body of the galaxy may be quite different, it is interesting that Cohen et al. (1997) find significant rotation in the M87 globular cluster system. The outer parts of M87 are clearly much more

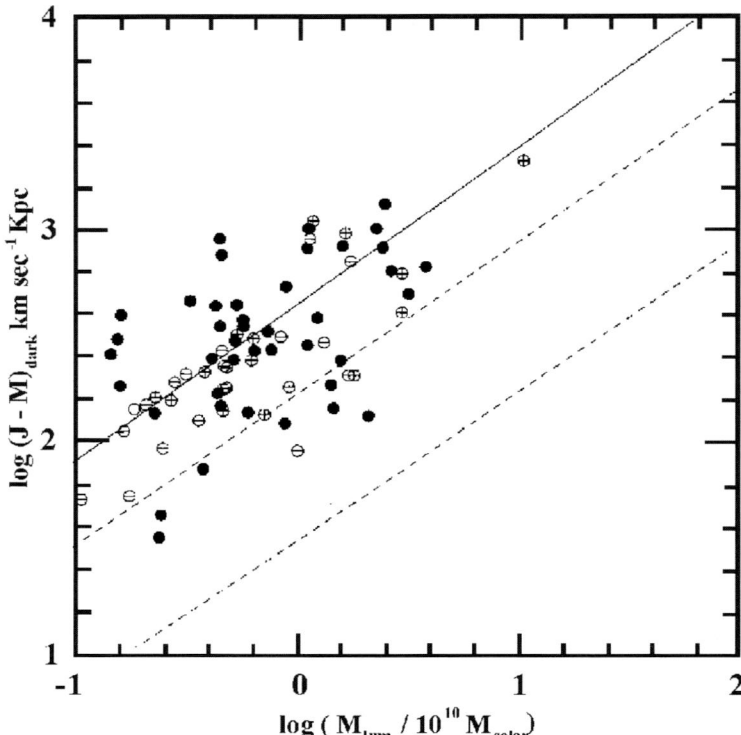

FIGURE 5. The specific angular momentum J/M vs the mass M for simulated halos, scaled to show luminous mass. The region of the giant ellipticals is represented by the dashed lines, and the locus for spirals is the solid line. The points show $J/M, M$ for simulated systems: see text. (From Zurek et al. 1988).

elliptical than the inner regions, consistent with a higher angular momentum content. Figure 7 compares an image of M87 from the DSS (limiting surface brightness ~ 24 B mag arcsec^{-2}) and a very deep image from Weil et al. (1997) (limiting surface brightness ~ 28 B mag arcsec^{-2}). This increase of isophotal ellipticity is common among the cluster giant ellipticals; see Porter et al. (1991). In summary, we would argue that, although the inner regions of giant ellipticals are flattened primarily by their anisotropic velocity dispersion, the outer regions are flattened mainly by their rotation.

3. The Fundamental Plane

The fundamental plane is a scaling relation between structural and kinematical parameters for elliptical galaxies. It started with the Faber-Jackson (1976) relation between luminosity and central velocity dispersion; a recent K-band derivation of this law gives $L_K \propto \sigma^{4.14 \pm 0.22}$ (Pahre et al. 1998). This law can easily be derived from the virial theorem, if the mean surface brightness and M/L ratio do not vary significantly from galaxy to galaxy. However, it soon became clear that the residuals from this relation correlated with the surface brightness, and the fundamental plane includes this correlation. It is usually presented in the form (e.g. Pahre et al. 1998)

$$r_e \propto \sigma^A I_e^B$$

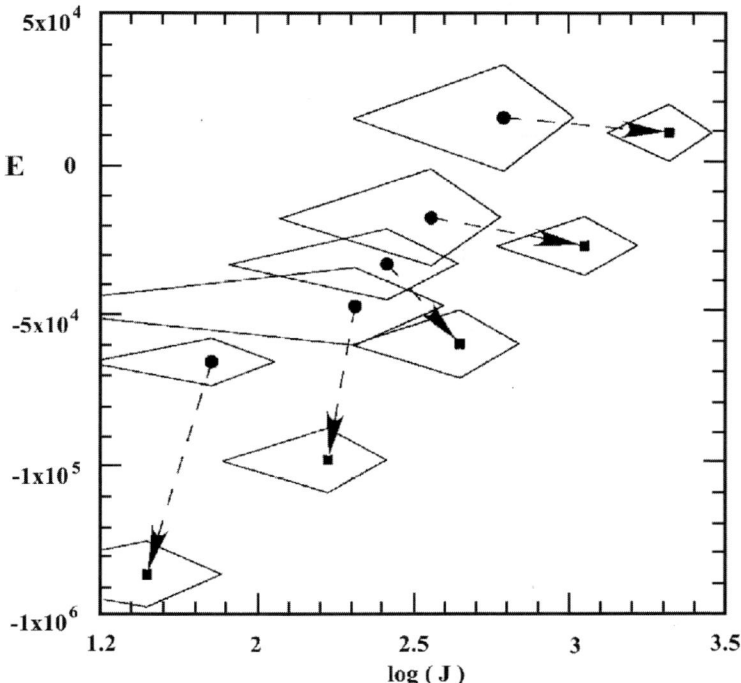

FIGURE 6. Evolution of a protohalo in the energy-angular momentum plane. At $z = 0$, the halo was partitioned into five bins of binding energy E. The energy and angular momenta of the particles in these bins at $z = 5.25$ (filled circles) were compared with their values at $z = 0$ (filled squares). During the merging processes, angular momentum is transported outwards from the inner regions of the system to the outer envelope (from Zurek et al. 1988).

where r_e is the effective radius, σ the central velocity dispersion and I_e the mean surface brightness within r_e. The virial theorem gives

$$R \propto V^2 I^{-1} (M/L)^{-1}$$

where R, V^2 and I are suitably defined means of radius, internal velocity and surface brightness. If M/L is constant, and the galaxies are homologous, i.e. $r_e \propto R$, $\sigma^2 \propto V^2$ and $I_e \propto I$, then the virial theorem corresponds to

$$r_e \propto \sigma^2 I_e^{-1}.$$

The parameter A depends strongly on the observed wavelength, changing from about 1.0 at U through 1.25 at B to 1.5 at K. In the K-band, where line blanketing effects on the M/L ratio are reduced, the observed fundamental plane for elliptical galaxies has the form (Pahre et al. 1998)

$$r_e \propto \sigma^{1.53 \pm 0.08} I_e^{-0.79 \pm 0.03}$$

with a scatter of about 0.096 in $\log r_e$. See Figure 8.

This departure from the virial theorem scaling (i.e. the *tilt* of the fundamental plane) can be produced by a systematic change of M/L with mass, due to stellar population changes with mass (e.g. mean age, metallicity, IMF), or to a change in the dark/luminous mass ratio within r_e: the K-band tilt corresponds to the relation $M/L \propto M^{0.16 \pm 0.01}$. Alternatively, this tilt of the fundamental plane could come from a systematic breakdown of the homology assumption, such as changes with mass in the shape of $I(r)$ and $\sigma(r)$, and

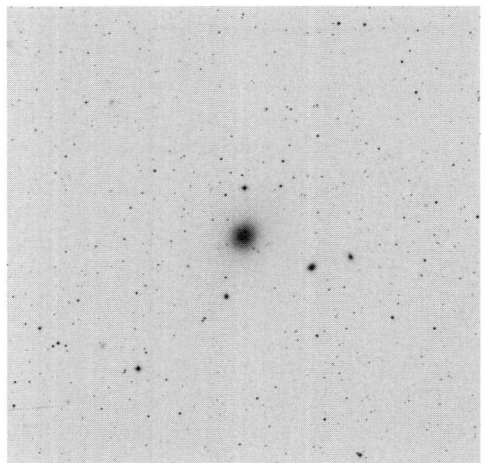 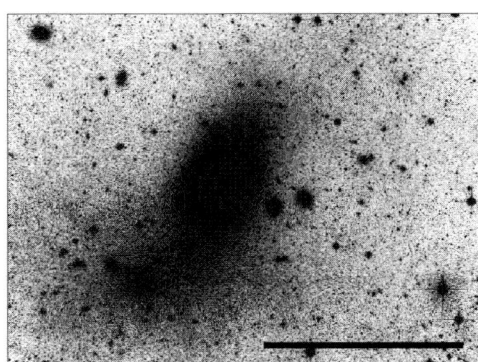

FIGURE 7. Two images of M87: in the shallow DSS image (left), M87 appears almost round, while in the deep image of Weil et al. (1997: right), it is much more elliptical. The two images cover similar regions of sky in the horizontal direction. The flattening at large radius is probably a consequence of the angular momentum of the outer regions.

in rotation and kinematic anisotropy. For example, we know already that the brighter ellipticals are more anisotropic and have shallower $I(r)$ distributions. Graham & Colless (1997) show how correcting for broken homology can bring the exponents of σ and I_e in the fundamental plane closer to the virial theorem values. Busarello et al. (1997) find that non-homology in the dynamical structure of ellipticals accounts for more than half of the tilt of the fundamental plane. The scatter about the fundamental plane cannot be accounted for by observational errors alone. Small variations in age, metallicity and non-homology at any point along the fundamental plane probably contribute to the observed scatter.

The globular clusters illustrate the homology problem nicely (Djorgovski 1995). Within the cores of the clusters, the structure and M/L ratios are fairly similar from cluster to cluster. The fundamental plane for the cores of globular clusters has the form

$$r_c \propto \sigma^{1.8 \pm 0.15} I_o^{-1.1 \pm 0.1}$$

which is close to the virial theorem scaling. On the other hand, at the half–light radius r_h, the M/L ratio is still fairly similar but the dissimilarities in structure from cluster to cluster are already well established: the corresponding fundamental plane is

$$r_h \propto \sigma^{1.45 \pm 0.2} I_h^{-0.85 \pm 0.1}$$

(here σ is still the central velocity dispersion), which is similar to the observed K-band fundamental plane for the elliptical galaxies.

Whatever the reason for the deviation of the fundamental plane from the virial theorem scaling, the intrinsic scatter about the fundamental plane for elliptical galaxies is small, despite the wide range in the rotation parameter (V/σ), and the different $I(r)$ and $\sigma(r)$ profiles, etc. So the departures from constant M/L and/or homology are well organised to keep the fundamental plane tight. This needs explaining. It may be an indication that elliptical galaxies can attain only a restricted subset of dynamical possibilities via their formation processes (violent relaxation, mergers ...)

Although the fundamental plane does not include metallicity as one of the variables, there is an important relationship between the metallicity and the luminosity, which is

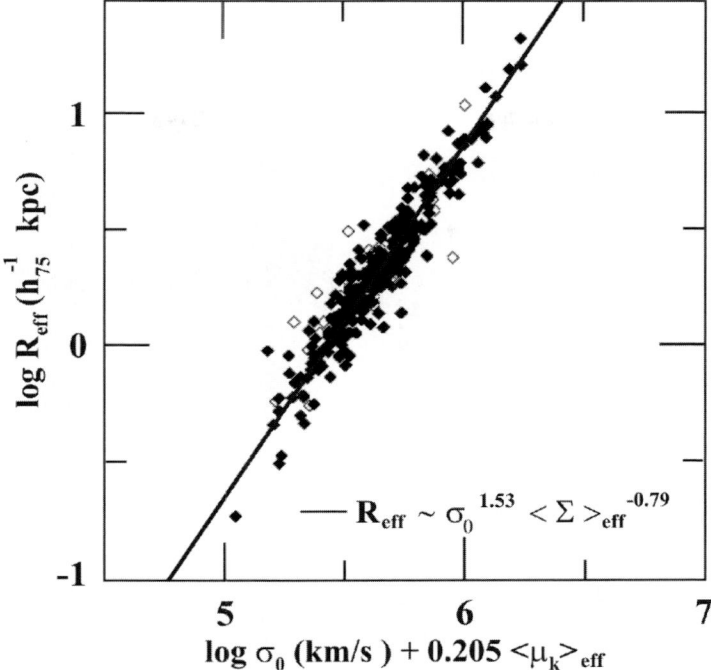

FIGURE 8. The K-band fundamental plane for 251 early-type galaxies, from Pahre et al.(1998), seen edge-on along its long side.

often expressed as an $Mg_2 - \sigma$ relation, because Mg_2 (a measure of the strength of the Mg b feature) and σ are distance-independent. In this form, it represents an increase of metallicity with deepening potential well: see Figure 9.

With the assumption of homology, the tilt of the fundamental plane can be interpreted as a relationship between the M/L ratio and mass or luminosity. For example, Mobasher et al. (1999) find that $M/L \propto M^{0.18\pm0.01}$ at K and $M/L \propto M^{0.23\pm0.01}$ at optical wavelengths. This is usually interpreted as coming from the mass-metallicity relation.

The fundamental plane at higher redshifts gives constraints on the evolution of M/L and the epoch of star formation in ellipticals, if the assumption of homology remains correct. Evolutionary effects on M/L have been derived out to redshifts of about 0.8. For example, van Dokkum et al. (1998) used fundamental plane arguments to derive the evolution of M/L from clusters with redshifts between 0.02 and 0.83; their M/L corresponds to a common rest wavelength near B. Relative to the Coma cluster, they find that $\Delta \log M/L = (-0.40 \pm 0.04)z$ if $\Omega_m = 0.3$. Comparison with population synthesis models gives constraints on the formation epoch and the value of Ω_m: their data favor high formation redshifts and low Ω_m (see Figure 10).

4. The Stellar Content of Elliptical Galaxies

In the previous section, we have already seen that the metallicity, as measured by the Mg2 index, increases with increasing luminosity or velocity dispersion. This is usually ascribed to the ability of galaxies with deeper potential wells to retain chemically enriched material from supernovae. The ratio of alpha elements like magnesium to iron also

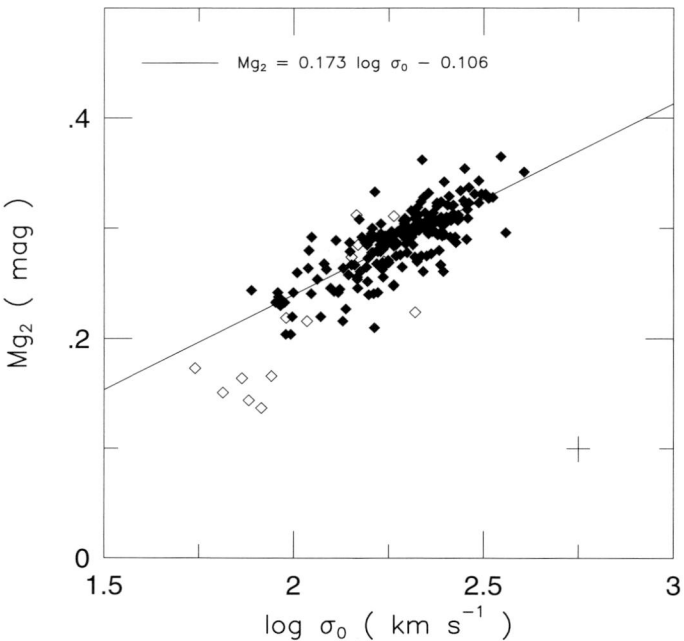

FIGURE 9. The $Mg_2 - \sigma$ relation for 182 galaxies, from Pahre et al. (1998).

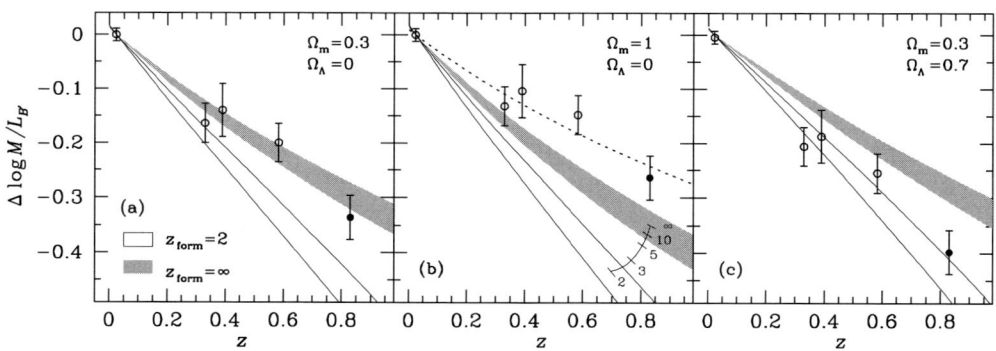

FIGURE 10. Evolution of M/L with redshift for three cosmological models. Model predictions for different formation redshifts of the stars are shown, assuming a Salpeter IMF ($x = 2.35$) and a range of metallicities. The dotted line in (b) indicates a model with $z_{form} = \infty$ and a steeper IMF ($x = 3.35$). The data favor high formation redshifts and low Ω_m (from van Dokkum et al. 1998).

increases with luminosity: see for example Fisher et al. (1995). For the most luminous ellipticals, [Mg/Fe] $\simeq 0.4$, while for the fainter ellipticals, it is close to the solar value. The usual interpretation is that SNII were more significant in the enrichment of the more luminous ellipticals, which in turn suggests that most of the star formation occurred rapidly (within about 1 Gyr) before enrichment by the Fe-producing SNIa could occur.

Fisher et al. (1995) and others noted that Hβ is stronger in galaxies with lower Mg/Fe ratios. This is a hint that the lower-luminosity ellipticals may be younger, but there is a

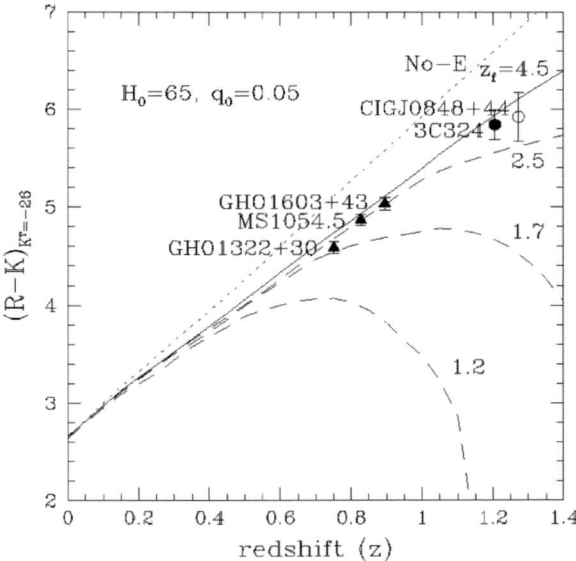

FIGURE 11. Evolution of the zero point of the C-M relation for distant clusters (Kodama et al. 1998). The dotted line shows their no-evolution model. The solid line shows the metallicity sequence model with $z_f = 4.5$ (age 12 Gyr). The dashed lines are models with $z_f = 2.5, 1.7$ and 1.2 (ages 11, 10 and 9 Gyr).

well-known degeneracy between the effects of age and metallicity on many spectral and color indices. Recently Kuntschner & Davies (1998) measured central line strengths for 11 ellipticals and 11 lenticular galaxies in the Fornax cluster, all brighter than $M_B = -17$. They used C4668 and Hγ indicies which at least partly separate the effects of age and metallicity. Comparing the indicies for their galaxies with those from stellar population models, they find that the ellipticals are all old and approximately coeval, with metallicities between solar and three times solar. On the other hand, the lenticular galaxies have luminosity-weighted ages that cover a much wider range, from about 2 to 12 Gyr. They interpret this age spread as an indication that a fraction of the spiral galaxies in the Fornax cluster have evolved to quiescence in the last few Gyr. This is consistent with the finding of Dressler et al. (1997) that the fraction of elliptical galaxies in clusters at $z \sim 0.5$ is similar to that in local clusters, while the fraction of S0 galaxies is 2-3 times lower than at $z \simeq 0$.

From fundamental plane arguments, and from studies of spectral indicies and the color-magnitude relations of ellipticals in clusters, the indications are that the stars in elliptical galaxies formed long ago, at redshifts > 2, and the subsequent evolution of their stellar content has been passive. For example, Kodama et al. (1998) studied the color-magnitude relations for 17 clusters at redshifts between 0.31 and 1.27. The color-magnitude diagrams are very tight, and are consistent with the ellipticals in each cluster being of similar age and having similar metallicity-luminosity relations from cluster to cluster, with the stars forming at a redshift $z > 2-4$. This does not necessarily mean that the ellipticals themselves formed so long ago. For example, van Dokkum et al. (1999) noted the presence of many ongoing mergers in the cluster MS1054-03 at a redshift of 0.83. The merger progenitors are mostly early-type disk galaxies or ellipticals. So these are primarily dissipationless mergers of systems whose stars formed much earlier, and the remnants will be luminous ellipticals ($\sim 2L_*$) whose *structure* formed relatively recently.

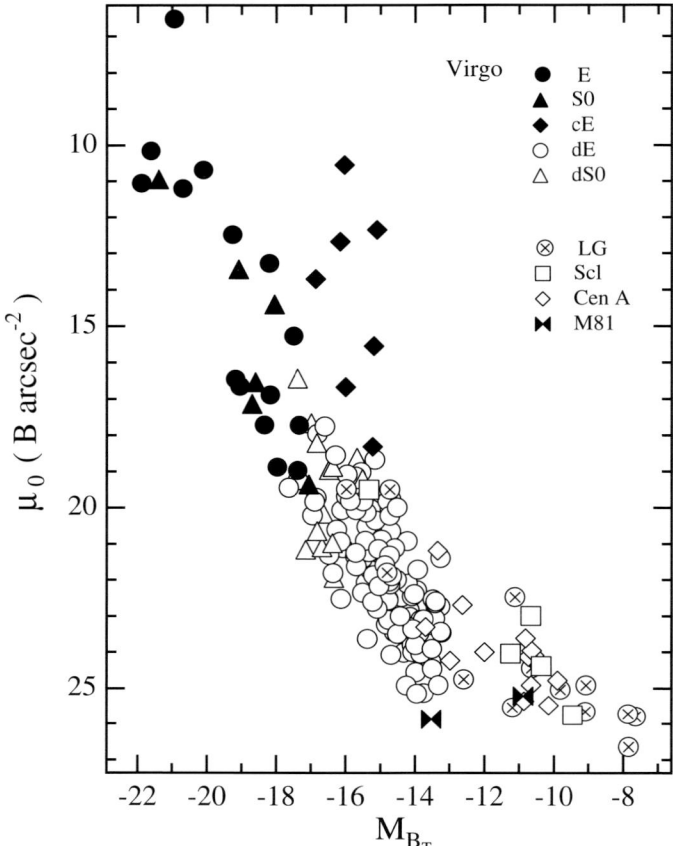

FIGURE 12. The central surface brightness μ_0 of the best-fitting Sersic model *vs.* the absolute absorption-corrected B magnitude. The galaxies are shown with different symbols according to their type (Virgo) and their group membership. Note the continuity between classical and dwarf early-type galaxies. Brighter galaxies have higher central surface brightnesses (from Jerjen *et al.* 1999).

See also Kauffmann (1996) and Bower *et al.* (1992) for other studies based on color-magnitude relations with similar conclusions.

5. Dwarf Ellipticals

Dwarf elliptical galaxies ($M_B > -18$) fall into two classes: compact systems of high surface brightness, like M32, and diffuse ellipticals of low surface brightness like the nearby dwarf spheroidal galaxies. The Sersic profile index n for a sample of giant and dwarf ellipticals is shown in Figure 2, with the dwarfs having surface brightness profiles that are close to exponential. The apparent continuity of the n - absolute magnitude relation is striking.

The Sersic profile fits by Jerjen *et al.* (1999) give a formal model value for the central surface brightness. Figure 12 shows how this central surface brightness becomes systematically fainter for the fainter ellipticals (except for the compact M32-like systems). Again, the continuity of the surface brightness - absolute magnitude relation is evident,

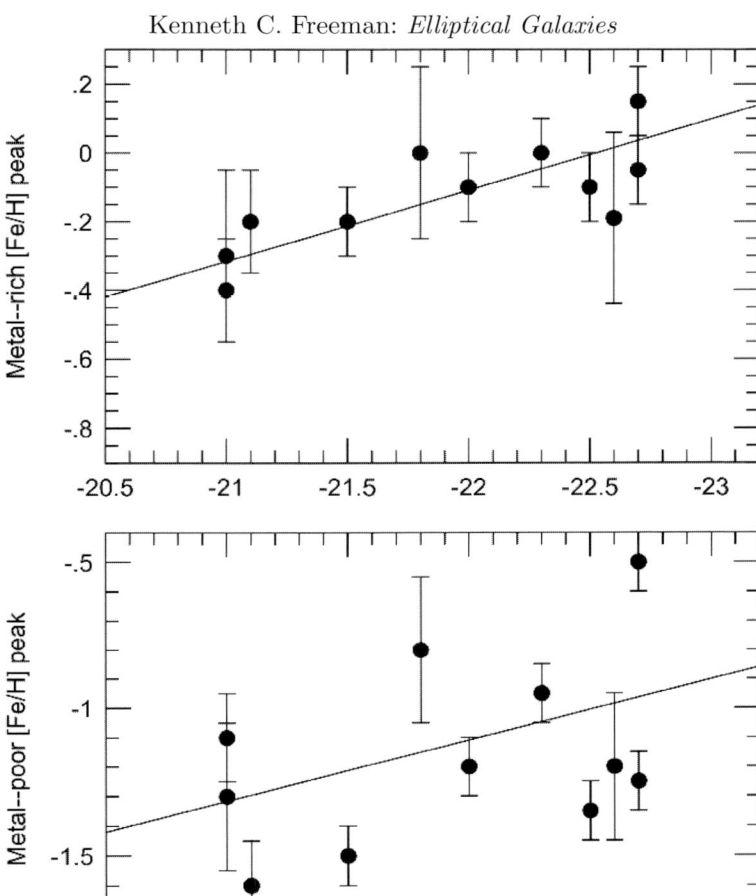

FIGURE 13. Mean metallicity *vs.* galaxy luminosity for the two metallicity modes of globular clusters in elliptical galaxies: the upper panel shows the metal-rich mode and the lower panel shows the metal-poor mode (from Forbes *et al.* 1997)

and indicates that these low surface brightness dwarfs are the low luminosity extension of the classical giant ellipticals.

Large numbers of dwarf ellipticals are found in nearby clusters like Virgo, Fornax and Centaurus (Sandage *et al.* 1985, Ferguson & Sandage 1988, Jerjen & Tammann 1997), with their luminosity function rising steeply towards fainter magnitudes. An interesting subclass of brighter dwarf ellipticals have sharp central nuclei. In both the Virgo and Fornax clusters, the nucleated and non-nucleated dwarf ellipticals show different radial distributions within the clusters. The nucleated dwarfs follow the strongly clustered distribution of the bright ellipticals, while the brighter non-nucleated dwarfs ($M_B < -14.2$) are more loosely distributed, like the spirals and dIrr galaxies (see Ferguson & Sandage 1989). The reason for the different distributions of nucleated and non-nucleated dwarfs in these clusters is not yet understood.

The origin of the dwarf ellipticals is also not well understood. Most of the faint nearby dwarf spheroidal galaxies have complex star formation histories (see for example Grebel

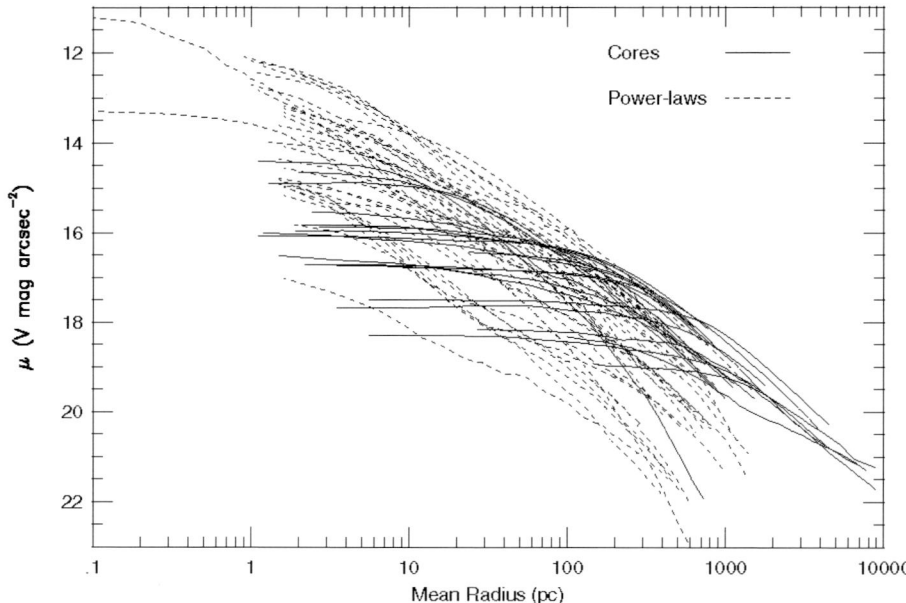

FIGURE 14. V-band surface brightness profiles of the inner regions of 55 early-type galaxies from HST observations. The dichotomy between the shallow inner power laws (core galaxies: solid lines) and the steeper distributions (power law galaxies: dashed lines) is evident. (from Faber et al. 1997).

1999). Some of these apparently structureless systems were forming stars until ~ 1 Gyr ago, and may well have looked like present-day dIrr or dIrr/dSph transition systems at that time. In clusters of galaxies, the harassment process is likely to be a significant source of dwarf ellipticals. This process transforms many of the lower surface brightness spiral galaxies, which have shallow extended potential wells, into dwarf ellipticals through the combined action of fast binary encounters and the tidal field of the cluster itself. Much of their mass is stripped and now resides in the intracluster medium. Moore et al. (1998) note the large numbers of small spirals in clusters at intermediate redshift, and their absence in nearby clusters where dwarf ellipticals comprise most of the faint end of the luminosity function. See also Moore et al. (1999).

6. The Formation of Ellipticals

The work of Toomre & Toomre (1972) first demonstrated the likely importance of mergers in forming ellipticals. Now this process is seen also in the context of hierarchical models for structure formation. While the stars of most elliptical galaxies appear to have formed at $z > 2$, the structures of the elliptical galaxies may have formed much later, as shown by the example of the cluster MS1054-03 at $z = 0.83$ in which many dissipationless mergers of early type galaxies are observed. It still seems likely that some ellipticals are formed through the later mergers of already-formed disk galaxies, as Toomre (1977) suggested.

At a more specific level, we recall the recent study by Naab et al. (1999) who showed how mergers of equal mass disks produce boxy slowly rotating systems flattened by anisotropy, while mergers of disks with 3:1 mass ratios produce disky systems that are

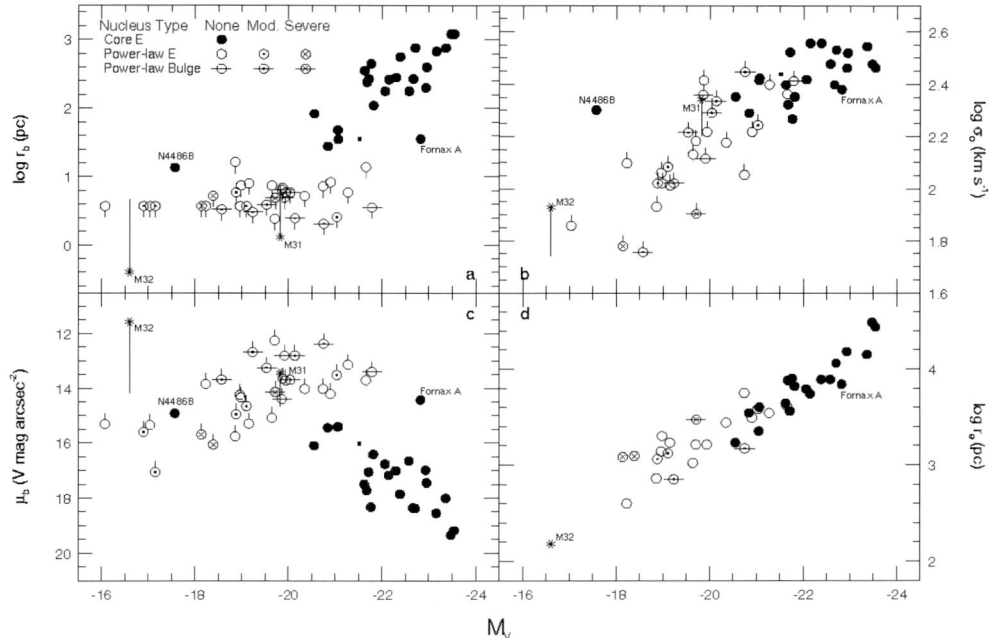

FIGURE 15. Central (left) and global parameters (right) against absolute magnitude for a sample of early-type galaxies. Profile type and degree of nucleation are shown at the top left. Bulges are distributed like the smaller ellipticals. For the galaxies with steep cusps (open symbols) the values of r_b and μ_b are limits only. M31 and M32 are plotted twice: asterisks are as observed, and the tails indicate their positions if they were located at the distance of the Virgo cluster. (From Faber et al. 1997)

rotationally flattened. We also recall that the surface density distribution of dissipationless merger products can be well represented by the $r^{1/4}$ law.

Some ellipticals appear to have been involved in relatively recent merger or accretion events. The shell systems like NGC 3923 (e.g. Quinn 1984) are evidence for a recent accretion of a small stellar system which was drawn in to the parent galaxy by dynamical friction and then tidally disrupted. Galaxies like Cen A with its minor axis lane of dust and gas and its complex shell system (Malin et al. 1983) are probably the product of a recent accretion of a more substantial gas-rich companion.

7. Globular Cluster Systems in Elliptical Galaxies

Giant elliptical galaxies have rich systems of globular clusters, which are believed to be very old. The properties of these cluster systems have been studied in detail and can give some insight into the formation processes of the ellipticals. Forbes et al. (1997) summarize many of the important properties of globular cluster systems in ellipticals. The specific frequency S_N is the number of clusters per unit absolute magnitude $M_V = -15$. For most elliptical galaxies S_N is typically about 5. However some of the brightest ellipticals have much higher S_N values of about 15. The origin of this high specific frequency is still not understood. Another important element is the distribution of metallicity within cluster systems, as inferred from the distribution of colors of the clusters. The metallicity distribution is typically bimodal, with the mean abundance of each mode increasing weakly with the luminosity of the parent galaxy. [Fe/H] $\simeq -0.2$ for the metal-rich mode and the metal-poor mode has [Fe/H] $\simeq -1.2$. See Figure 13. Some of the cluster giants, like M87 and NGC 1399, show more complex multimodal metallicity distributions.

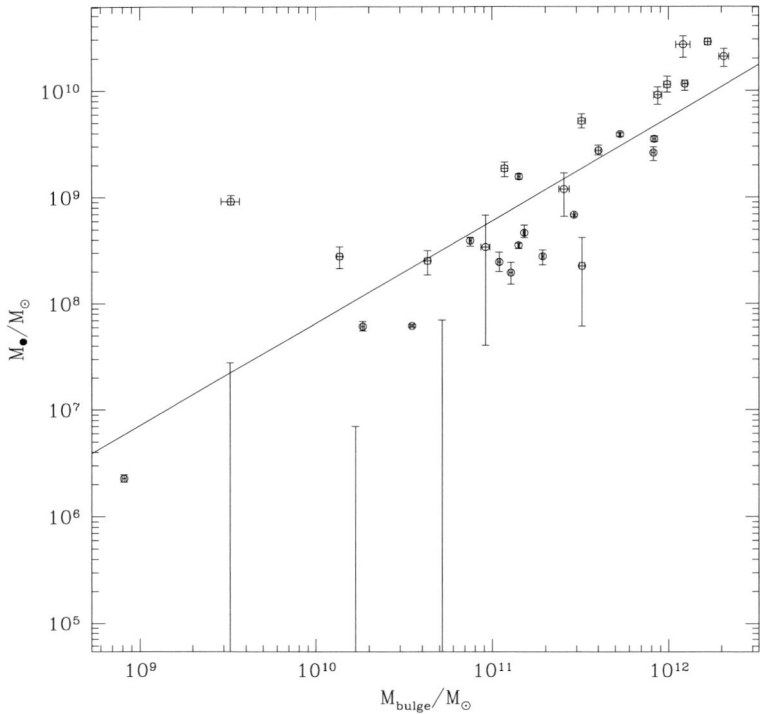

FIGURE 16. Correlation between MDO mass $M_\bullet$ and galaxy mass M_{bulge}. The aberrant point is NGC 4486B, a low luminosity galaxy which has the velocity dispersion and metallicity of a much brighter galaxy. The solid line shows the fit $\log M_\bullet/M_{\text{bulge}} = -1.79 - 0.04 \log M_{\text{bulge}}$ (from Magorrian et al. 1998).

What is the origin of the very high specific frequencies of globular clusters in some of the brighter ellipticals? Schweizer (1987) suggested that globular clusters may form from the colliding gas of disk galaxies that are merging to form the ellipticals. Ashman & Zepf (1992) further developed this picture, which was supported by the discovery of large numbers of bright young star clusters in some nearby merging systems. The antennae system NGC 4038/9 is a spectacular example. It shows several thousand young star clusters luminous enough to be identified as young globular clusters (Whitmore et al. 1999), plus a small number of old globular clusters which presumably came into the merger from the two parent spirals. The luminosity function of these young clusters is different from that of the old clusters seen now in elliptical galaxies. While the old clusters have a gaussian-like luminosity function, the luminosity function for the young clusters in the merger systems are typically power laws, with increasing numbers of fainter young clusters. NGC 3256 is another example of a merging system, with about 1000 young star clusters, again with a power law luminosity function (Zepf et al. 1999). This difference between the luminosity functions of the young star clusters in mergers and the old star clusters in ellipticals is probably not a serious objection to Schweizer's suggestion, because it is very likely that a large fraction of the fainter young clusters will disrupt on timescales shorter than 10^8 years through the effects of stellar mass loss and the galactic tidal fields. It seems clear, however, that if the high specific frequencies of globular clusters in ellipticals are due to to cluster formation in the merging of fully formed spirals, then most of the clusters should have the metallicities of the interstellar gas of the merging spirals, which we would expect to be near solar.

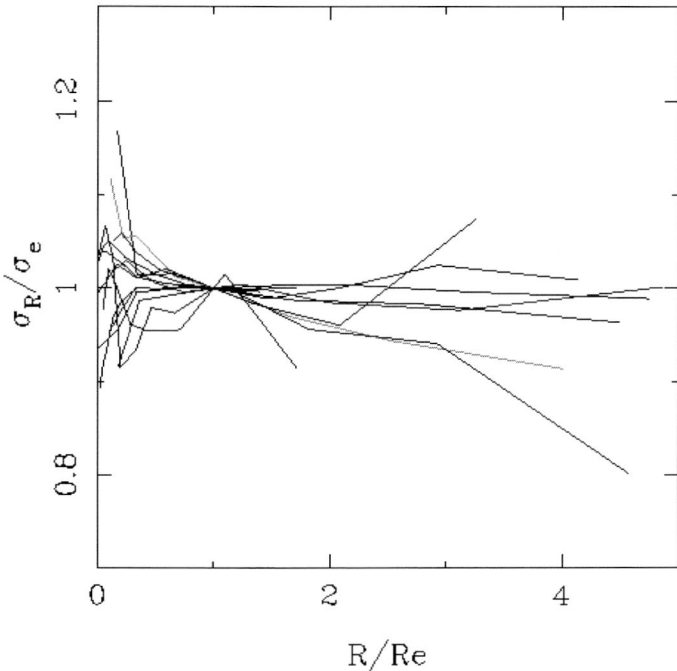

FIGURE 17. The luminosity weighted line of sight velocity dispersion within an aperture of radius R, normalised to the effective radius: each curve represents an early-type galaxy.

Are the excess clusters in high-S_N ellipticals mainly metal-poor or metal-rich ? Forbes et al. (1997) find that galaxies with large S_N have predominantly metal-poor clusters. This argues against the formation of the excess clusters in the merging of evolved spirals. It may make more sense to consider the formation of clusters during the merging of the dark matter and baryon lumps early in the hierarchical formation of large galaxies. We could then associate the very populous globular cluster systems with the vigorous merger activity that took place at high redshift while the galaxy was still metal-poor. Taniguchi et al. (1999) have developed a picture in which globular clusters form in molecular clouds associated with supernova events shortly after the first burst of massive star formation in the young galaxies. In this context, we note the molecular detections of some ultra-luminous infrared galaxies at redshifts $z > 4$ (e.g. Ohta et al. 1996). Rapid chemical enrichment must have occurred in some of these massive forming galaxies, and it seems entirely possible that the metal-rich mode of globular clusters in ellipticals also formed during the hierarchical merging at this time.

The specific frequency of globular clusters is a measure of the number of globular clusters per unit stellar luminosity of the parent galaxy; it shows a systematic variation with galaxy luminosity, and within some galaxies the local value of S_N (i.e. the ratio of the globular cluster density to the surface brightness) increases with radius. The specific frequency is sometimes regarded as a measure of the globular cluster formation efficiency (i.e. the ratio of the mass of clusters to the initial mass of gas). With the usual definition of S_N there is then an indication that the cluster formation efficiency varies from galaxy to galaxy and even within individual galaxies. McLaughlin (1999) has argued that it makes more sense to include the hot X-ray emitting gas plus the stars in this kind of estimate of the cluster formation efficiency. With this definition, there is a remarkable uniformity of the cluster formation efficiency, both within individual galaxies and from galaxy to galaxy. This universal cluster formation efficiency is about 0.0025 by mass. This value refers to old clusters which have survived disruption to the present time.

8. The Inner Regions of Ellipticals

8.1. Surface Photometry

The structure of the innermost regions of ellipticals is closely related to their largescale structure and dynamics. HST surface photometry of the inner few hundred parsecs of a large sample of early-type galaxies (Faber et al. 1997) showed that the inner surface brightness distribution follows a power law $I(r) \propto r^{-\gamma}$ where γ lies in the range 0 to 1. It turns out that γ has a bimodal distribution within this range, with γ depending on the luminosity of the galaxy. Figure 14 shows the inner surface brightness profiles for this sample: the bimodal distribution of the slope in the central regions is clear.

For the brightest galaxies, with $M_V < -22$, the surface brightness distribution changes slope at the break radius r_b; within r_b they have flat inner cores with $\gamma < 0.3$. One the other hand, the fainter galaxies, with $M_V > -20.5$, have steep inner cusps with $\gamma = 0.8 \pm 0.3$. Few galaxies have γ values between these two modes. As mentioned already in section 3, most of the more luminous galaxies (low γ) have boxy isophotal shapes, and most of the less luminous galaxies (steep γ) are disky. Remarkably, in the region of intermediate luminosity ($-22 < M_V < -20.5$) where galaxies are found in both of the γ modes, the galaxies with low γ are again boxy systems and the galaxies with steep γ are disky: see Figure 3. Why should there be such a close relationship between the morphology of the innermost 100 pc of early type galaxies and their global boxiness/diskiness which is defined on much larger scales ? Before considering this question, we should discuss (1) the scaling laws for the parameters of the inner regions of early type galaxies, and (2) the central massive dark objects that are detected in most early-type galaxies.

8.2. Scaling Laws for the Inner Regions

The scaling laws provide further connections between the inner and outer regions. Faber et al. fit the so-called Nuker profile to the surface brightness profiles:

$$I(r) = I_b 2^{(\beta-\gamma)/\alpha}(r_b/r)^{\gamma}[1 + (r/r_b)^{\alpha}]^{(\gamma-\beta)/\alpha}$$

so the asymptotic logarithmic slope for small r is $-\gamma$, the break radius r_b is the point of maximum curvature of the profile in log-log coordinates, and I_b is the surface brightness at r_b. Figure 15 shows how r_b and I_b scale with the absolute magnitude of the galaxy. The scaling of the global parameters r_e and central velocity dispersion σ are also shown.

8.3. Central Massive Dark Objects in Early Type Galaxies

Most early-type galaxies appear to have massive dark objects(MDOs) at their centers: is there any connection between these dark objects and the structural properties of the inner parts of these galaxies ? Magorrian et al. (1998) made dynamical models for a sample of 36 galaxies with HST photometry and groundbased kinematics. They used axisymmetric two-integral models with constant M/L ratio plus the central MDO and found good fits to the structure and kinematics for 32 of their galaxies. The dependence of the derived M/L ratios for the underlying galaxies on the galaxy luminosity is similar to that found from fundamental plane arguments (section 4): $M/L \propto L^{0.2}$. For most of their sample, the data are consistent with a gaussian distribution of $\log M_\bullet/M_{\rm bulge}$ (where $M_\bullet$ is the mass of the MDO), with a mean of -2.28 and a standard deviation of 0.51: *i.e.* with a typical value of $M_\bullet/M_{\rm bulge} \simeq 0.005$. See Figure 16.

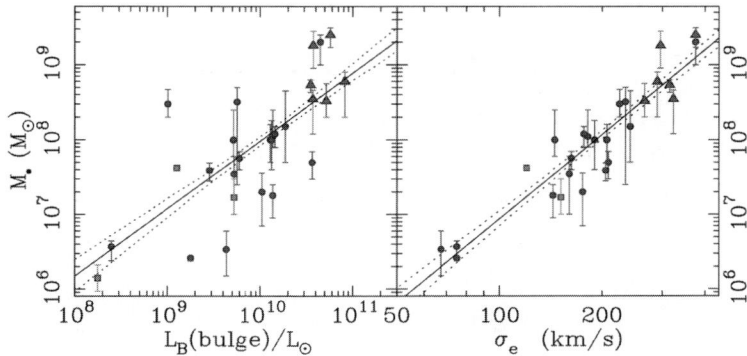

FIGURE 18. The *left* panel shows the black hole mass against the luminosity of the bulge for 26 early-type galaxies. The *right* panel shows the black hole mass against the luminosity weighted velocity dispersion within an effective radius. The solid line is the best fit correlation and the dotted lines are the 68% confidence limits (From Gebhardt *et al.* 2000).

The correlation of $M_\bullet$ with $M_{\rm bulge}$ is not particularly tight. In a striking development, it was found that the black hole mass $M_\bullet$ correlates much more tightly with the velocity dispersion of the bulge $\sigma(r_e)$ as measured at the effective radius r_e of the bulge. (Gebhardt *et al.* 2000; Ferrarese & Merritt 2000). Figure 17 shows the typical run of velocity dispersion within an aperture of radius r against radius: the dispersion within r_e is a good measure of the typical dispersion of the bulge itself (*i.e.* in the absence of the black hole). Figure 18 compares the two correlations. The tight $M_\bullet - \sigma$ relation relates the mass of the black hole to the depth of the potential well of the bulge itself. The implication is still obscure. It may just reflect a tight correlation between the black hole mass and the bulge mass, which was obscured by inaccurate estimates of the bulge mass: *i.e.* the velocity dispersion alone may be more accurate estimator of the bulge mass. This does not seem very plausible, because we have seen already in §4 that ellipticals are two-parameter systems.

Alternatively the depth of the potential well of the bulge may somehow limit the mass of the black hole that can grow through accretion processes. See Silk & Rees (1998) for discussion of how the effect of quasar outflows on the accretion of gas can limit the growth of black holes.

8.4. *Black Holes and the Inner Structure of Early Type Galaxies*

The presence of a steep inner density cusp in an early type galaxy is not necessarily associated with a central MDO or black hole. For example the singular isothermal sphere is a simple equilibrium system with a density distribution $\rho(r) \propto r^{-2}$ and no central point mass. Also, simulations of hierarchical galaxy formation consistently produce dark halos with central cusps at least as steep as $\rho(r) \propto r^{-1}$ (e.g. Navarro *et al.* 1996). Nevertheless, there is the possibility that the density cusps and the central black holes are associated.

Several authors have discussed the formation of stellar cusps by the slow growth of a

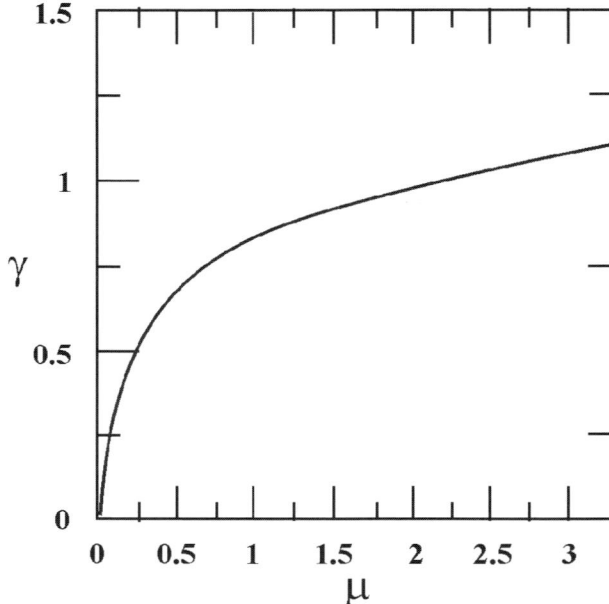

FIGURE 19. The dependence of the cusp slope γ on the dimensionless mass of the central black hole, for models of the adiabatic growth of a central black hole (from van der Marel 1999).

black hole in a pre-existing core. Most recently, van der Marel (1999) investigated the effects of growing a black hole adiabatically in an initially non-singular isothermal core. The initial cores have a central density $\rho_\circ$ and a core radius $r_\circ$, so the dimensionless black hole mass can be defined as $\mu = M_\bullet/M_{\rm core}$ where $M_{\rm core} = \frac{4}{3}\pi\rho_\circ r_\circ^3$. Fitting a Nuker law to the systems after the adiabatic growth of the black hole shows how the Nuker parameters depend on μ. Figure 18 shows the inner slope γ against μ: γ increases rapidly to about 0.7 as μ rises from 0 to 0.5 and then more slowly to $\gamma \simeq 1$ at $\mu \simeq 3$.

Using the observed scaling laws for the Nuker parameters, the observed scaling of $M_\bullet$ and M/L with luminosity L, and the dependence of $r_b/r_\circ$ on μ from the adiabatic growth models, it follows that the observations require that

$$\log \mu = 4.03 - 0.48 \log L$$

So the dimensionless black hole mass μ decreases with increasing L and the cusp slope γ is then flatter for the more luminous galaxies. van der Marel suggests that the bimodality in the distribution of the slope γ may be associated with a bimodality in the distribution of $M_\bullet/L$ which in turn goes with the boxy/disky dichotomy in global structure. Dissipation during the formation of the disky systems may contribute to the higher values of $M_\bullet/L$ which are needed to give their higher γ values.

Merritt (1999) argues that it is probably not realistic to consider the growth of a black hole in an initial isothermal sphere with a large core radius, because simulations of violent relaxation tend to produce systems that are already cusped. The growth of a black hole in such a cusped system would give a yet steeper cusp. The shallow cusps of the bright ellipticals are then more difficult to understand in this adiabatic model. An alternative picture for the central regions of these bright ellipticals is through the coalescence of two black holes following a merger of two galaxies, which would transfer energy from the binary black holes to the stars in the inner regions (see for example Makino & Ebisuzaki 1996). This process creates a shallow core with mass comparable to the mass of the black holes. The break radius r_b is then the scale of the region affected by the binary black hole.

REFERENCES

AGUILAR, L. & WHITE, S. 1986, ApJ, 307, 97.
ARNABOLDI, M. et al. 1994, ESO Messenger, 76, 40.
ARNABOLDI, M. et al. 1998, ApJ, 307, 759.
ASHMAN, K. & ZEPF, S. 1992, ApJ, 384, 50.
BARNES, J. 1988, ApJ, 331, 699.
BINNEY, J. & TREMAINE, S. 1987, Galactic Dynamics (Princeton:PUP).
BOWER, R., LUCEY, J., ELLIS, R. 1992, MNRAS, 254, 601.
BUSARELLO, G., CAPACCIOLI, M., CAPOZZIELLO, S., LONGO, G., PUDDU, E. 1997, A&A, 320, 415.
CAON, N., CAPACCIOLI, M., D'ONOFRIO, M. 1993, MNRAS, 265,1013.
CAPACCIOLI, M., HELD, E., LORENZ, H., VIETRI, M. 1990, AJ, 99, 1813.
COHEN, J., RYZHOV, A. 1997, ApJ, 486, 230.
DAVIES, R. et al. 1983, ApJ, 266, 41.
DE VAUCOULEURS, G. 1948, Ann. Astrophys., 11, 247.
DJORGOVSKI, S. 1995, ApJ, 438, L29.
DRESSLER, A. et al. 1997, ApJ, 490, 577.
FABER, S. et al. 1997, AJ, 114, 1771.
FABER, S. & JACKSON, R. 1976, ApJ, 204, 668.
FALL, S.M. 1983, *"Internal Kinematics and Dynamics of Galaxies", (IAU Symposium 100) (Dordrecht: Reidel)*, p 391.
FERGUSON, H. & SANDAGE, A. 1988, AJ, 96, 1520.
FERGUSON, H. & SANDAGE, A. 1989, ApJ, 346, L53.
FERRARESE, L. & MERRITT, D. 2000, astro-ph/0006053.
FISHER, D., FRANX, M., ILLINGWORTH, G. 1995, ApJ, 448, 119.
FORBES, D., BRODIE, J., GRILLMAIR, C. 1997, AJ, 113, 1652.
GEBHARDT, K. et al. 2000, astro-ph/0006289.
GRAHAM, A., & COLLESS, M. 1997, MNRAS, 287, 221.
GRAHAM, A., LAUER, T. R., COLLESS, M. & POSTMAN, M. 1996, ApJ, 465, 534.
GREBEL, E. 1999, *"The Stellar Content of Local Group Galaxies", (ASP)*, p 17.
HUI, X., FORD, H., FREEMAN, K., DOPITA, M. 1995, ApJ, 449, 592.
JAFFE, W. 1983, MNRAS, 202, 995.
JERJEN, H., BINGGELI, B., FREEMAN, K. 1999, AJ, in press.
JERJEN, H. & TAMMANN, G. 1997, A&A, 321, 713.
KAUFFMANN, G. 1996, MNRAS, 281, 487.
KODAMA, T., ARIMOTO, N., BARGER, A., ARAGON-SALAMANCA, A. 1998, A&A, 334, 99.
KUNTSCHNER, H. & DAVIES, R. 1998, MNRAS, 295, L29.
MAKINO, J. & EBISUZAKI, T. 1996, ApJ, 465, 527.
MALIN, D., QUINN, P., GRAHAM, J. 1983, ApJ, 272, L5.
MCLAUGHLIN, D. 1999, AJ, 117, 2398.
MERRITT, D. 1999, astro-ph/9906047.
MOBASHER, B., GUZMAN, R., ARAGON-SALAMANCA, A., ZEPF, S. 1999, MNRAS, 304, 225.
MOORE, B., LAKE, G., KATZ, N. 1998, ApJ, 495, 139.

Moore, B, Lake, G., Quinn, T., Stadel, J. 1999, MNRAS, 304, 465.
Naab, T., Burkert, A., Hernquist, L. 1999, astro-ph/9908129.
Navarro, J., Frenk, C., White, S. 1996, ApJ, 462, 563.
Ohta, K. et al. 1996, Nature, 382, 426.
Pahre, M., de Carvalho, R., Djorgovski, S. 1998, AJ, 116, 1606.
Porter, A., Schneider, D., Hoessel, J. 1991, AJ, 101, 1561.
Quinn, P. 1984, ApJ, 279, 596.
Sandage, A., Binggeli, B., Tammann, G. 1985, AJ, 90, 1759.
Schweizer, F. 1987, "Nearly Normal Galaxies" (Springer), p 18.
Sersic, J. 1968, Atlas de Galaxias Australes. Observatorio Astronómico, Cordoba.
Taniguchi, Y., Trentham, N., Ikeuchi, S. 1999, astro-ph/9909170.
Toomre, A. 1977, "The Evolution of Galaxies and Stellar Populations" (Yale University Press), p 401.
Toomre, A. & Toomre, J. 1972, ApJ, 178, 623.
van Albada, T. 1982, MNRAS, 201, 939.
van der Marel, R. 1999, AJ, 117, 744.
van Dokkum, P., Franx, M., Kelson, D., Illingworth, G. 1998, ApJ, 504, L17.
van Dokkum, P., Franx, M., Fabricant, D., Kelson, D., Illingworth, G. 1999, ApJ, 520, L95.
Weil, M., Bland-Hawthorn, J., Malin, D. 1997, ApJ, 490, 664.
Whitmore, B. et al. 1999, astro-ph/9907430.
Zepf, S. et al. 1999, astro-ph/9904247.
Zurek, W. H., Quinn, P. J., Salmon, J. K. 1988, ApJ, 330, 519.

Disk Galaxies

By KENNETH C. FREEMAN

Mt. Stromlo Observatory, Canberra, Australia

In these lecture, we discuss the disk galaxies. Dynamically, disks are very simple. Their equilibrium is primarily between gravity and rotation, so it is possible to study their gravitational potential and dark matter content with confidence. On the other hand, disks are highly dissipated structures, so some of the information about the early dynamical history of their baryons has been lost. In this sense, they are probably less useful than ellipticals as probes of events at high redshift. Another complication is that most disks are still forming stars, so their evolution (dynamical, structural and chemical) continues.

1. The Structure of Disk Galaxies

Some disk galaxies have very substantial bulges and others do not have bulges at all. See Figure 1) for examples of galaxies with extreme bulge-to-disk ratios. The relative strength of the bulge and the disk is a major element of the Hubble classification, with the pure disk galaxies like IC 5249 (Figure 1) classified late in the Hubble sequence. It is clear that bulges are not an essential element of the formation of disk galaxies. The bulge-to-disk ratio is an important factor in determining the morphology of disk galaxies, not only in its own right as a classification criterion but also through its effect on the shape of the rotation curve and so on the optical morphology of the star-forming disk.

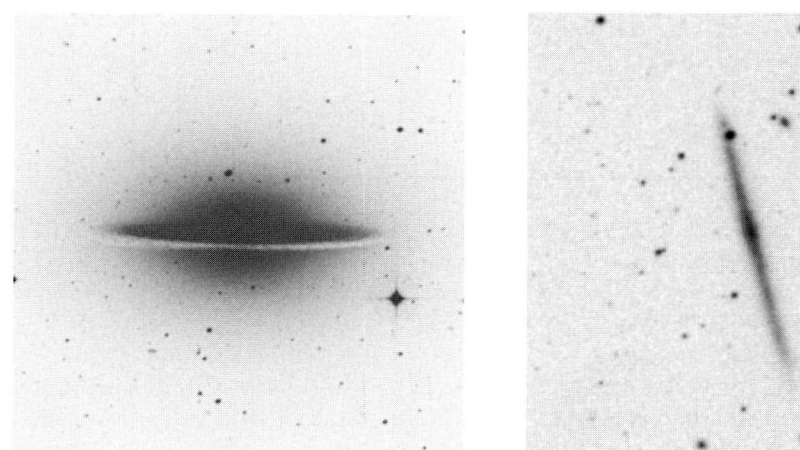

FIGURE 1. Examples of galaxies with a very large bulge (left: the Sombrero galaxy NGC 4594) and a very small bulge (right: IC 5249). Images from the DSS.

The relationship of the bulge-to-disk ratio to the Hubble type was clearly quantified by Simien & de Vaucouleurs (1986) who made bulge-disk decompositions of the surface brightness distributions of 98 galaxies. Figure 2 shows the relative brightness of the bulge and disk against the morphological type parameter T (T = −6 to −4 for ellipticals, −2 to −1 for S0 galaxies, and then increases from 0 at S0/a through 5 at Sc to 9 at Sm).

Most disks have a simple exponential surface brightness distribution of the form

$I(r) = I_\circ e^{-r/h}$ where $I_\circ$ is the central surface brightness and h is the radial scalelength (e.g. Freeman 1970). This distribution typically extends out to about 5 radial scalelengths, beyond which the disks are often truncated (e.g. van der Kruit 1988). What are the typical central surface brightnesses and scale lengths of disk galaxies ? This is an important issue, because these parameters for equilibrium disks reflect the disk's angular momentum and mass. Angular momentum transport between baryonic and dark matter during hierarchical galaxy formation affects the angular momentum of the equilibrium disks, and theories of galaxy formation need to reproduce their typical surface brightnesses and sizes (which is currently a troublesome issue). Estimating the distribution of these parameters is not easy, because there are strong biases from the selection effects that go into defining samples of disk galaxies. For example, samples usually have some kind of selection limit on the apparent diameter, and explicit or implicit limits on the isophotal surface brightness. Samples are dominated by large galaxies of higher surface brightness, because such galaxies are included from a larger volume of space. It is necessary to correct the observed distributions of disk parameters for this visibility volume effect.

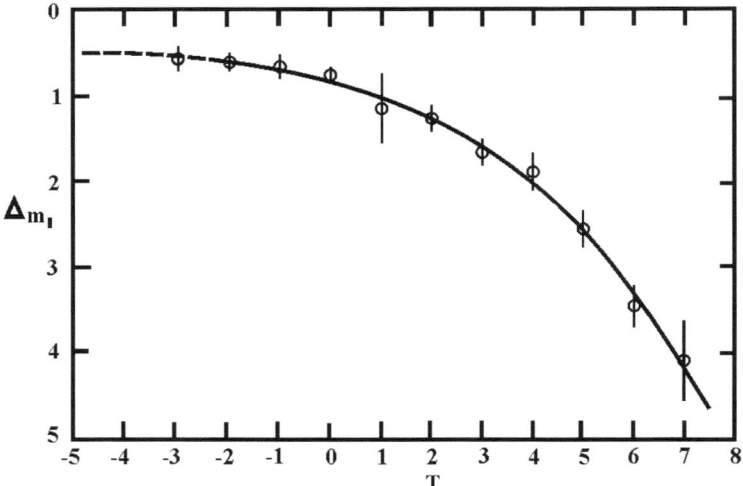

FIGURE 2. Mean values of the relative brightness in magnitudes between bulge and disk *vs.* the morphological type parameter T. Note the continuity between the lenticular (T < 0) and the spiral sequences. (From Simien & de Vaucouleurs 1986).

de Jong & Lacey (1999) have recently derived the bivariate distribution of absolute magnitude, surface brightness and scale length for disks, using a large diameter-limited sample of Sa to Sdm spirals with I-band surface photometry. The surface brightness profiles were decomposed into bulge + disk components, and the volume density of galaxies per Mpc^3 in $(M_I, h, I_\circ)$ was calculated, with correction for the visibility volume. Figures 3 and 4 show the distribution of galaxies over central surface brightness and scalelength, in bins of absolute magnitude. Figure 5 shows a grey-scale representation of the density distribution of galaxies over effective radius r_e and effective surface brightness μ_e: the left panel shows the unweighted distribution and the right panel shows the luminosity-weighted distribution. We live in the kind of galaxy that contributes most to the luminosity of the local universe.

The possible existence of giant galaxies with low surface brightness as a major compo-

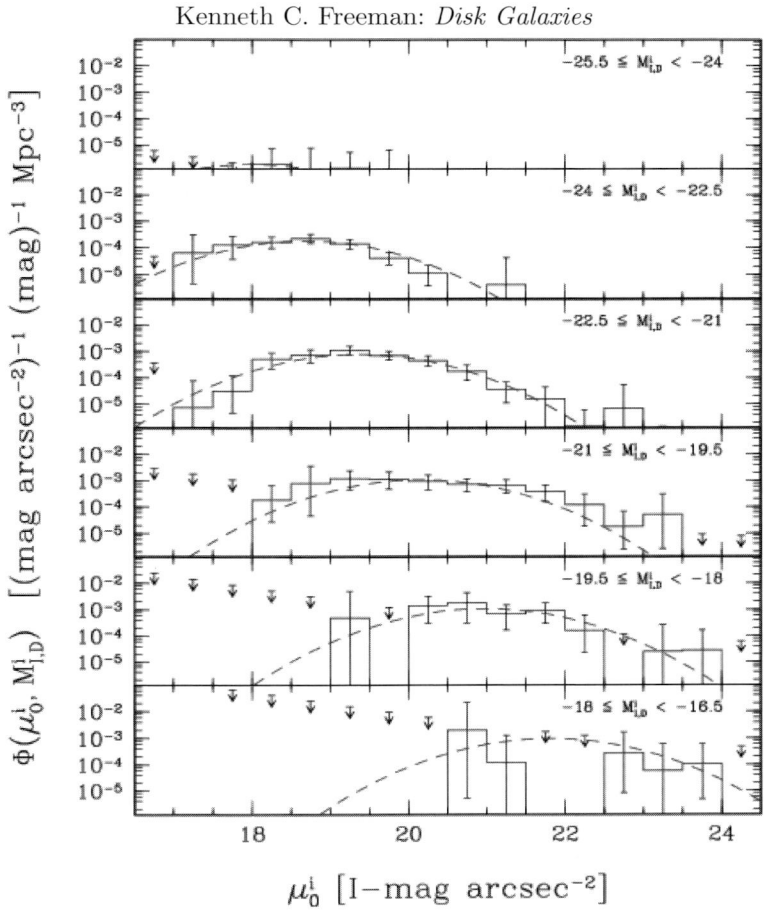

FIGURE 3. The density distribution of Sa-Sm galaxies as a function of disk central surface brightness μ_o^i in bins of absolute magnitude (from de Jong & Lacey 1999). The dashed curves show a model based on simple collapse arguments and the Tully-Fisher law.

nent of the galaxy population has been an issue for many years (see for example Impey & Bothun 1997). Although a few such galaxies are known, it seems from the work of de Jong & Lacey (1999) that such galaxies are relatively rare. If significant numbers of gas rich low surface brightness giants do exist, they will soon be detected in the Parkes HIPASS survey.

In simulations of the growth of density fluctuations in the expanding universe, blobs of matter acquire angular momentum from each other by tidal torques. The dimensionless parameter $\lambda = J|E|^{\frac{1}{2}}G^{-1}M^{-\frac{5}{2}}$ is a measure of the ratio (rotational velocity)/(virial velocity) for a blob. (Here J, E and M are the angular momentum, binding energy and mass of the system). For disks in centrifugal equilibrium, $\lambda \simeq 0.45$. Simulations of hierachical galaxy formation show that the typical value of λ is 0.05 ± 0.03 (e.g. Zurek et al. 1988). In the absence of significant transport of angular momentum from the disk to the dark halo as the galaxy is forming, λ is a useful measure of the collapse factor for the baryons, and hence for the surface brightness of the ensuing disk. Simple arguments (Fall & Efstathiou 1980), assuming that the specific angular momentum of the dark matter and baryons are similar, show that the collapse factor for the baryons is $r_t/h = \sqrt{2}/\lambda \simeq 30$ where r_t is the truncation radius of the halo. For our Galaxy, $h \simeq 4$

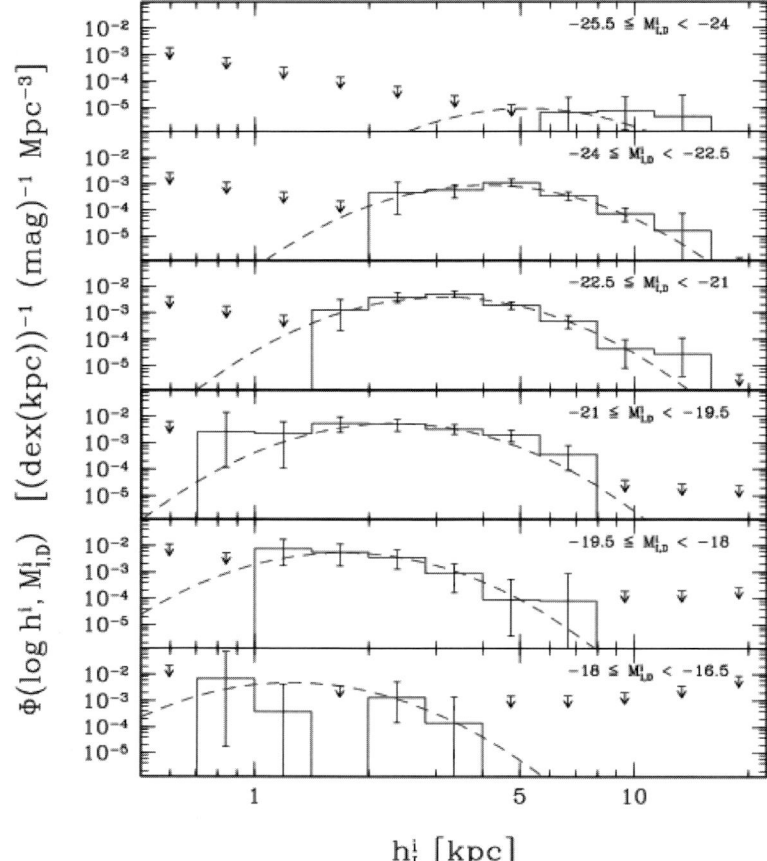

FIGURE 4. The density distribution of Sa-Sm galaxies as a function of disk scale length h in bins of absolute magnitude (from de Jong & Lacey 1999). Dashed curves as in Figure 3.

kpc so the truncation radius of the dark halo should be about 120 kpc. This is consistent with the observed extent of the galactic dark halo (e.g. Freeman 1996). If the baryons and dark matter maintain the same specific angular momentum throughout the collapse of the disk, disks with higher λ-values are initially closer to centrifugal equilibrium, have a smaller collapse factor and lead to disks of lower surface density. The distributions of surface brightness and scale length from de Jong & Lacey (1999) indicate that the initial spread in λ may be smaller than given by the cosmological simulations (if the assumptions about conservation of specific angular momentum are correct).

2. Star Formation Law in Disks

Theories of galaxy formation need a prescription for the star formation rate as a function of gas density and physical conditions. At this time, star formation is not well understood, and empirical prescriptions are needed. Kennicutt's law (1989) is widely used for the star formation in galactic disks. Kennicutt showed that the local rate of massive star formation, as measured by the Hα surface brightness, follows the HI surface density Σ_{HI} with a relation of the Schmidt law form SFR $\propto \Sigma_{HI}^{1.3\pm0.3}$ in dense regions. This law breaks down at lower gas densities; below a critical threshold density (which is

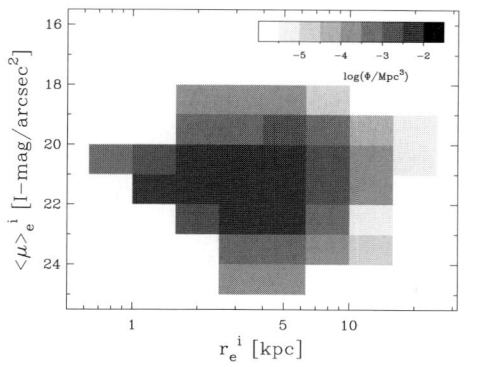 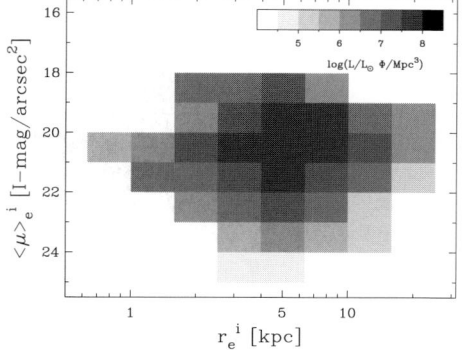

FIGURE 5. The density distribution of Sa-Sm galaxies over effective radius r_e and effective surface brightness μ_e: the left panel shows the unweighted dstribution and the right panel shows the luminosity-weighted distribution (from de Jong & Lacey 1999).

typically about a few $M_\odot$ pc^{-2} and varies from galaxy to galaxy), massive star formation is completely suppressed. Figure 6 gives examples of the functional dependence of star formation rate against the HI surface density for several galaxies, showing the variable threshold. A more recent study (Kennicutt 1998) of the disk-averaged star formation rate and gas densities gives a law SFR $\propto \Sigma_{\rm HI}^{1.4\pm0.15}$.

Kennicutt argued that this threshold is associated with the onset of gravitational instability of the disk, as given by Toomre's criterion $\Sigma_{\rm HI} \simeq \frac{\kappa c}{3.36}$, where κ and c are the epicyclic frequency and the velocity dispersion of the gas. Figure 7 shows the ratio $\Sigma_{\rm HI}/\Sigma_{\rm crit}$ against radius normalised to the radius of the HII region disk for a sample of galaxies; here $\Sigma_{\rm crit}$ is the critical surface density given by the Toomre criterion. The values of $\Sigma_{\rm HI}/\Sigma_{\rm crit}$ at the edge of the HII region disk all lie in a narrow range of values, even though the absolute values of the threshold densities vary by more than a factor 10 from galaxy to galaxy.

3. Star Formation History of Disks

As the gas dissipated to form the disk, it seems likely that star formation started in the inner disk where the surface density was highest, and then propagated outwards as disk settling continued. Many disk galaxies still have extended envelopes of HI, in which star formation has not yet begun. Bell & de Jong (1999) have attempted to measure the radial age gradient in the disks of spirals. Measuring the ages of stellar populations from their integrated properties is difficult, because changes in age and in metallicity have similar effects on the spectra of composite stellar populations. This age-metallicity degeneracy can be broken by using particular combinations of spectral lines or some combinations of broad band colors. Bell & de Jong used optical and near-IR colors and stellar population synthesis models to partially break the degeneracy and estimate the radial variation of age within a sample of disk galaxies. Some assumptions are needed: they adopt a maximum age of 12 Gyr for their populations, assume an exponentially declining star formation rate, and then derive the average age of the stellar population as a function of radius.

They find that most spiral galaxies have stellar population gradients, in the sense that their inner regions are older and more metal rich. The star formation history of the stellar population appear to correlate most strongly with the K-band central surface brightness: galaxies of higher surface brightness have weaker radial age gradients, and their mean age at the disk half-light radius is older. While galaxies with a central surface brightness $\mu_K(0) = 16$ K mag arcsec^{-2} have a mean age at $1r_e$ of about 10 Gyr, the corresponding age for galaxies with $\mu_K(0) = 20$ is only 6 Gyr. The mean radial age gradients range from near zero at $\mu_K(0) = 16$ to -1.2 Gyr per disk scale length at $\mu_K(0) = 20$. The gas fraction for the high surface brightness disks is near zero, and rises rapidly to about 60% at $\mu_K(0) = 20$, which is consistent with the age gradient pattern. The luminosity of the galaxy appears to have less effect on its star formation history of its disk, although the mean metallicity of the disk does show the usual metallicity-luminosity relation. The dependence of mean age, gas fraction and metallicity on the K-band surface brightness are shown in Figure 8.

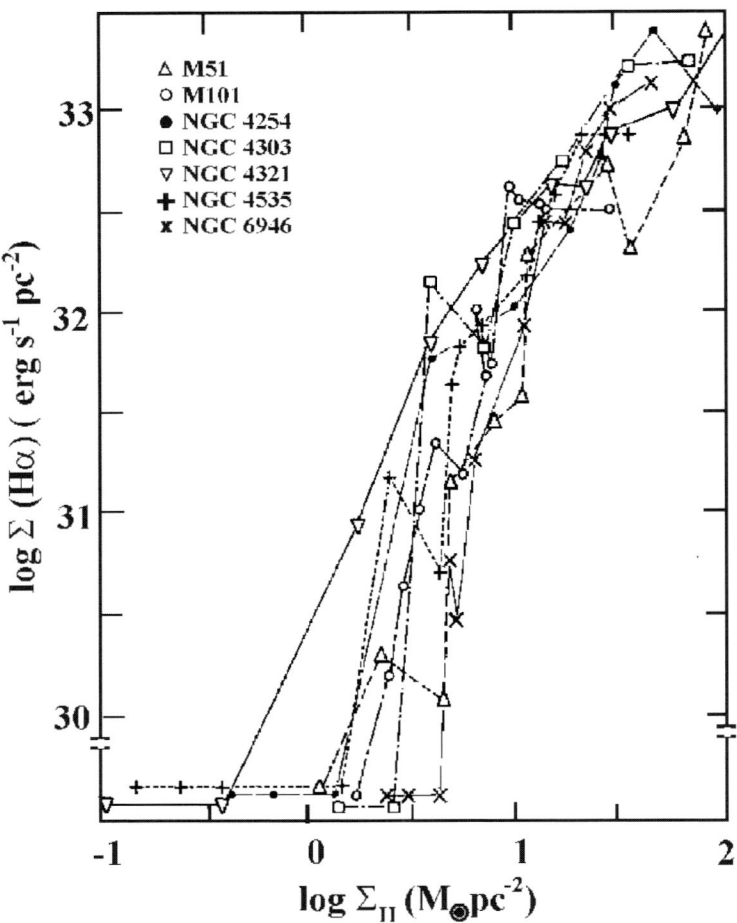

FIGURE 6. Examples of Hα surface brightness against HI surface density within several galaxies. The points at the bottom show regions below the star formation threshold, where no Hα emission was detected (from Kennicutt 1989).

A related optical and near-IR study of a sample of 26 LSB disk galaxies (Bell et al. 1999) came up with a rather similar conclusion. These LSB galaxies cover a wide range of abundance, from about -2 to solar. LSB galaxies with lower metallicities have lower surface brightness, lower luminosities and higher gas fractions. Most of the sample show color gradients consistent with a mean stellar age gradient, in the sense that the outer regions of the disks are younger in the mean. Again, the mean age correlates well with $\mu_K(0)$ and also with gas fraction.

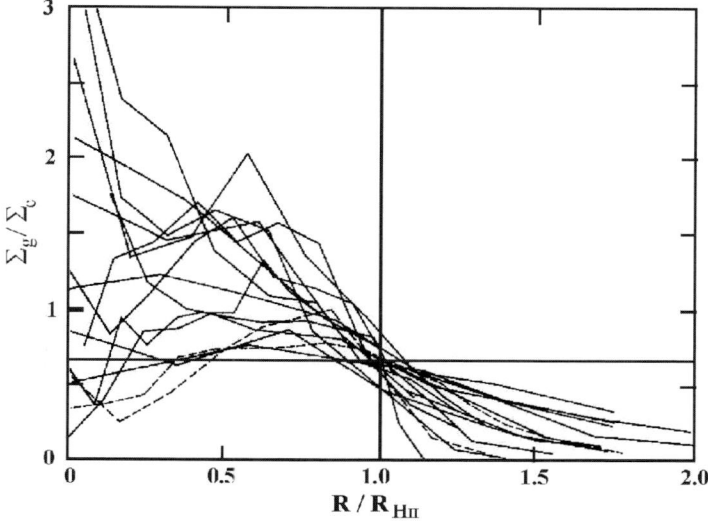

FIGURE 7. Radial dependence of the ratio of HI surface density to the critical density for gravitational stability. Each curve represents a galaxy with an actively star-forming disk. The radius is normalised to the outer radius of the HII region disk (from Kennicutt 1989).

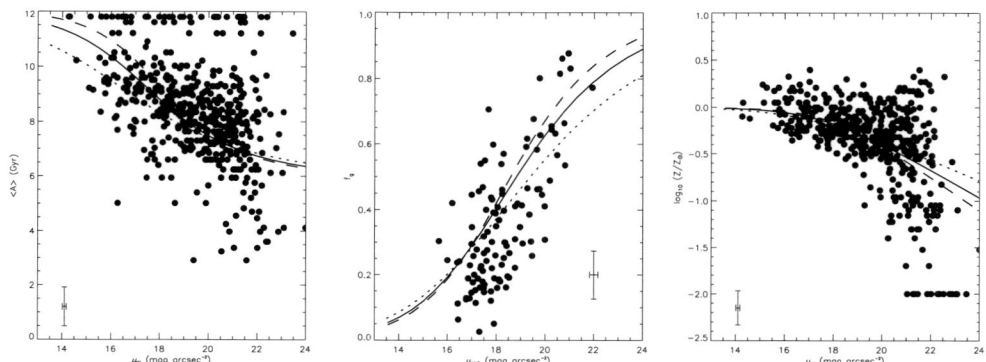

FIGURE 8. Local average age (left), gas fraction (middle) and metallicity (right) against K-band surface brightness. The curves show some simple star formation history models: see Bell & de Jong (1999).

4. Dark Matter in Disk Galaxies

We will cover this subject in more detail later but some brief comments are needed here in preparation for the next section. The dynamics of the disks of disk galaxies are relatively simple, particularly for the HI component which is close to a state of circular rotation. So it is relatively straightforward to estimate the radial distribution of the radial component of the gravitational field in the plane of the disk. The process is complicated by many HI disks showing warps out of the plane, which are usually handled with tilted ring models. Also, some disks have oval distortions or lop-sided asymmetries which can affect the velocity field.

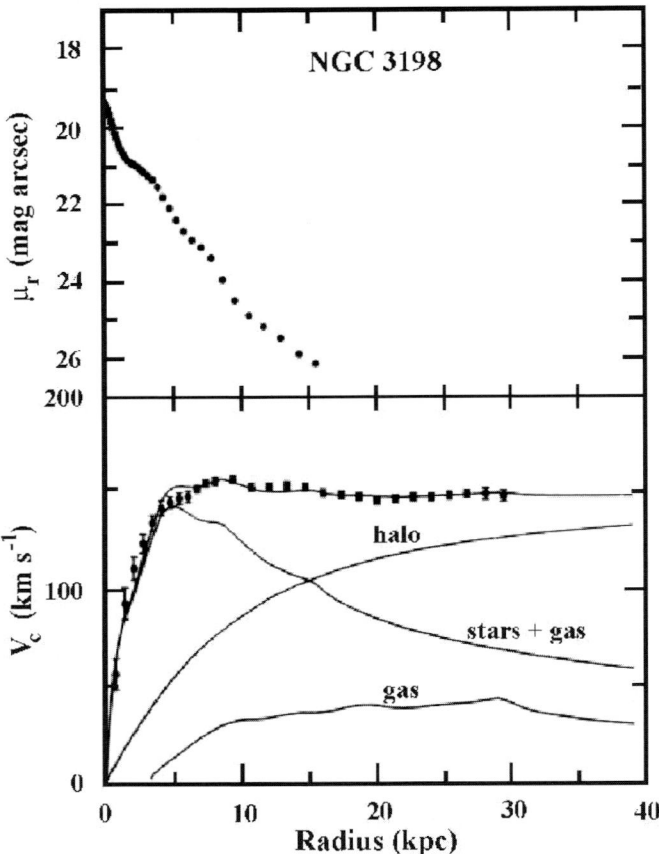

FIGURE 9. Surface brightness distribution (upper) and HI rotation curve (lower) for NGC 3198. The points show the observed rotation curve. The curve labelled "gas" is the rotation contribution of the gas. The other curve includes also the predicted rotation contribution from the stars, calculated from the surface brightness distribution with an adopted M/L_B ratio of 3.8. Note the large discrepancy beyond 7 kpc (from Begeman 1989).

Rotation curves show a wide range of morphologies, from almost solid body rotation at one extreme to almost constant (flat) rotation curves at the other. Can these rotation curves be understood from the gravitational field of the luminous component of the galaxies ? In the inner regions of the galaxies (within say 3 scale lengths) the answer appears to be yes, at least for galaxies of normal surface brightness. In almost all cases, the observed rotation curve is well fit with a rotation curve calculated by assuming that the surface density follows the surface brightness distribution. There is one free parameter in this fit, the M/L ratio, which affects only the amplitude of the rotation curve, not the shape. The mean I-band M/L ratio derived in this way is around 5: the spread in M/L can be accounted for almost entirely by the uncertainties in the distances of the galaxies (see Freeman 1992).

Many galaxies have HI rotation curves that extend to many scalengths. Usually these rotation curves are approximately flat at large radius and they cannot be explained by the luminous component alone. Figure 9 shows the surface brightness profile and the extended HI rotation curve of NGC 1398 (Begeman, 1989). This rotation curve has been decomposed into the contributions from the luminous disk, the gas and the dark halo which is needed to maintain the shape of the rotation curve at large radius. The decomposition is a "maximum disk" decomposition, such that the M/L ratio for the disk is made as large as possible, so that the disk accounts for as much of the rotation as possible in the inner regions. As mentioned above, this procedure works for almost all disk galaxies of normal surface brightness: the luminous component can account for the shape of the rotation curve in the inner few scalelengths. This does not necessarily mean that the maximum disk procedure is correct; we will return to this contentious maximum disk notion later.

In Figure 9, the peak of the rotation contribution from the disk is similar to the amplitude for the dark halo's contribution to the rotation curve. This similarity is known as the conspiracy, and it holds for most disk galaxies of normal surface brightness. Only a few galaxies are known for which the conspiracy is broken by having a disk rotation curve that rises significantly above the peak of the halo rotation curve. This conspiracy need not have happened, because the rotation amplitude for the exponential disk alone just depends on its surface density and scalelength, which in turn depend on its mass and angular momentum. The conspiracy means that the disk dynamics have been tuned to the properties of the dark halo during the process of galaxy formation. (If the maximum disk assumption were wrong, then the contribution of the disk to the rotation curve would of course be less).

5. The Tully-Fisher Law

The Tully-Fisher law for disk galaxies is the analog of the Faber-Jackson law for ellipticals. Tully & Fisher (1977) discovered a correlation between the global HI profile width and the absolute magnitude of disk galaxies. This law is now widely used as a distance indicator, with the HI profile width or the amplitude of the optical rotation curve as the luminosity estimator. Both work well, as one might expect from the typical shape of rotation curves (see Figure 9).

What kind of luminosity - velocity relation would we expect ? For isolated exponential disks, simple equilibrium arguments lead to

$$L \propto V^4/[I_\circ (M/L)^2]$$

where $I_\circ$ is the central surface brightness of the disk and M/L its mass to light ratio, and we have seen that both are roughly constant from galaxy to galaxy. This is close to the observed Tully-Fisher law, at least for the I-band and the near-IR. Sakai et al. (1999) have made a recent calibration of the Tully-Fisher law in the BVRIH bands. They find that the slope of the relationship increases from about 3.2 ± 0.3 in B to about 4.5 ± 0.3 in H. See Figure 10.

These small departures from the expected slope probably reflect a weak dependence of $I_\circ$ and M/L on the luminosity, in the same way that M/L can produce the tilt of the fundamental plane for ellipticals. For galaxy formation theory, the zero point of the Tully-Fisher law is also important. For example, Sakai et al. give the I-band Tully-Fisher

law in the form
$$M_I = (-10.00 \pm 0.08)(\log W_{50} - 2.5) - 21.32$$
where W_{50} is the width of the HI profile at half peak height corrected for inclination. For the exponential disk alone, the zero point depends on $\log[I_\circ (M/L)^2]$. The M/L ratio is just a measure of the stellar population in the disk, but the central surface density $\Sigma_\circ = I_\circ M/L$ depends on the total mass M and angular momentum J: $\Sigma_\circ \propto M^7/J^4$. So, as in the discussion of the conspiracy above, the $J(M)$ relation for the disk is defined by the dynamics of galaxy formation and determines the zero point of the Tully-Fisher law. Because of the conspiracy for galaxies of normal surface brightness, this argument is not greatly modified by the presence of the dark halo.

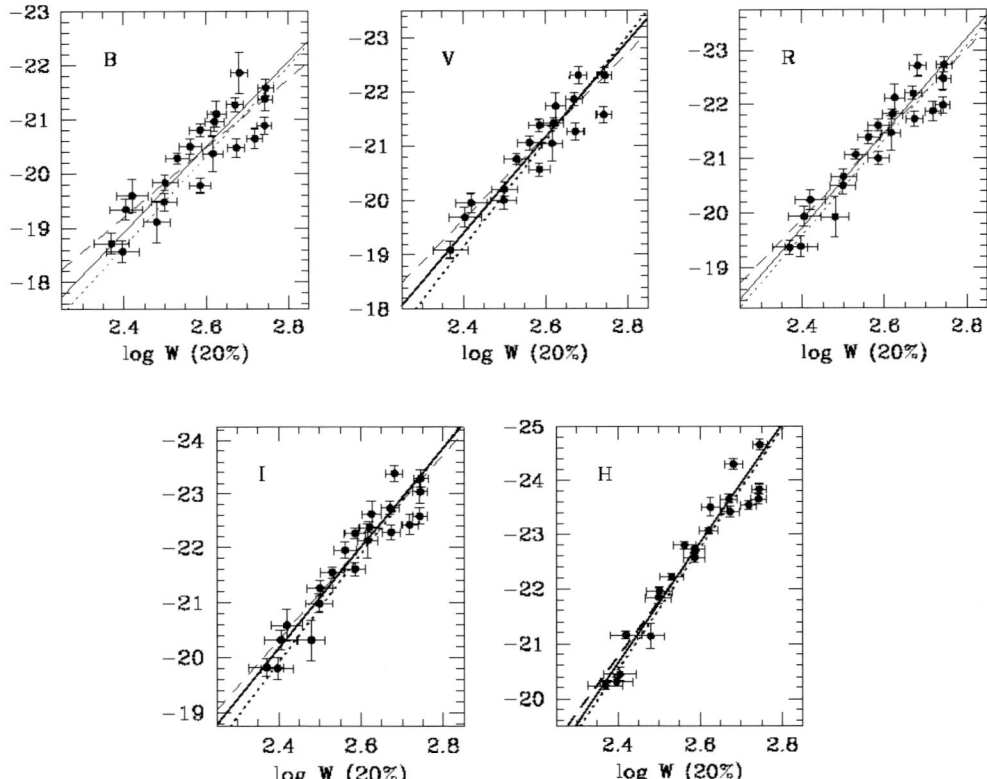

FIGURE 10. BVRIH Tully-Fisher relations for spiral galaxies with cepheid distances, using 20% linewidths. The solid lines show the bivariate fits (from Sakai et al. 1999).

Willick (1999) made an optical R-band Tully-Fisher study of a sample of galaxies in clusters, in which he explicitly included the dependence on surface brightness that one might expect from the discussion above. He recast the Tully-Fisher relation into the usual fundamental plane form, and found the relation $r_e \propto V^{1.79} I_e^{-0.75}$. This can be compared with the R-band fundamental plane for elliptical galaxies, $r_e \propto \sigma^{1.38} I_e^{-0.82}$. The surface brightness dependence is quite similar.

Now we look at the Tully-Fisher law for low surface brightness galaxies. In these galaxies, the gravitational field is dominated everywhere by the dark halo. Zwaan et al.

(1995) showed that the Tully-Fisher law is very similar, in slope and zero point, to the law for the galaxies of high surface brightness. See Figure 11. These LSB galaxies have low surface brightness because they have significantly longer scalelengths than the HSB at the same W_{50}: see Figure 12.

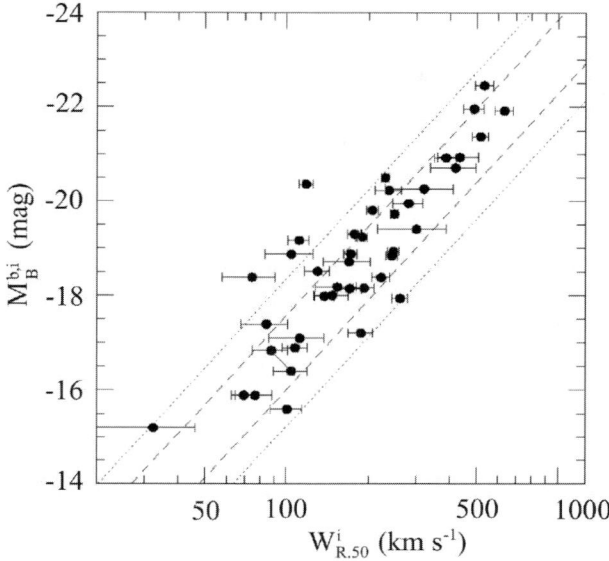

FIGURE 11. Tully-Fisher relation for a sample of LSB galaxies. The long and short dashed lines show the 1σ and 2σ range for a sample of normal galaxies (from Zwaan et al. 1995).

For the LSB galaxies, the dark halo determines the W_{50} value, almost independent of the dynamics of the baryons, and the baryonic matter determines the luminosity as before. Although the baryons can be regarded as a tracer population which hardly affects the potential, it turns out that the baryon mass is related to the dynamics of the dark halo. It is not clear why this should be so. One possibility is that the formation of dark halos produces a Faber-Jackson relation between halo mass and halo velocity dispersion of the form $M_{dark} \propto \sigma^4$. Indeed, Navarro et al. (1997) find tight power law relationships between M and maximum circular velocity for their simulated halos; the power law slopes depend on the cosmology and range from about 3.3 to > 5. Their power law relationship reflects the higher characteristic density of the lower-mass halos, which form at higher redshift. If the ratio of baryonic to dark mass is similar from galaxy to galaxy, then this would lead directly to a Tully-Fisher relation between baryonic mass and rotational velocity.

A rotation curve study of a small sample of LSB galaxies by Swaters (1999) raises an interesting possibility. He found that the rotation curves calculated from the light distribution gave a good fit to the observed rotation curve in the inner few scale lengths, as they do for HSB galaxies. But, because these disk have LSB, the M/L ratios required were in some cases very high, up to 17 in the R-band. This is well outside the range predicted by current population synthesis models. Nevertheless, the similarity between the shapes of the observed calculated rotation curves suggests that the total mass density and the luminous mass density are closely coupled over the inner few scale lengths of the disk, despite the very large M/L ratio that this implies. It makes one wonder if the high M/L ratio of these LSB galaxies could be associated with a disk of baryonic dark matter, perhaps in the form of cold gas or white dwarfs, that follows the surface density distribution of the LSB luminous disk. If this is a common feature of disk galaxies, then the disk would need to be relatively thick in order not to violate recent low estimates of the matter density near the sun (e.g. Creze *et al.* 1998).

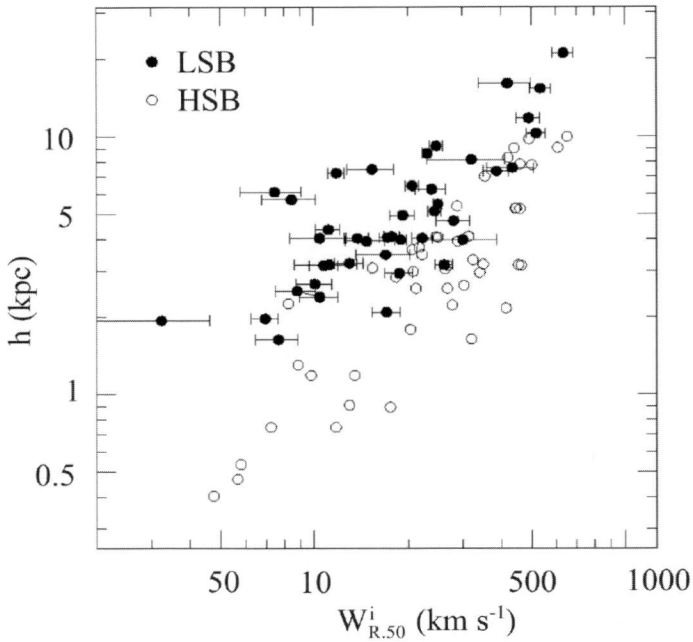

FIGURE 12. Scale length against velocity width for LSB and HSB galaxies (from Zwaan *et al.* 1995).

Although most disk galaxies do lie on the Tully-Fisher relation, within the usual scatter of about 0.2 to 0.3 mag, some very gas-rich galaxies are found to be significantly underluminous for their HI line width. Figure 13 adapted from Meurer *et al.* (1996), shows the Tully-Fisher relation for a sample of normal disk galaxies, and the two dwarf galaxies NGC 2915 and DDO 154. These two galaxies are unusually gas-rich, with gas masses that are an order of magnitude larger than their stellar masses. They lie 2 to 3 magnitudes below the Tully-Fisher relation. This appears to be due to the large fraction of their baryons that have not formed into stars. If the gas of these two galaxies is notionally converted into stars with an M/L ratio of 1, then they rise to the usual Tully-Fisher relation, as shown by the vertical bars in Figure 13. Again, this is an indication that the TF law is about the relationship of total baryon content of disk galaxies to the circular velocity of their dark halos.

6. Bars

The morphological classification of galaxies recognizes two parallel sequences, normal (SA) and barred (SB). Some galaxies are transition systems with weak central bars and are classified SAB. It turns out that many of the galaxies that appear normal in optical images do show bars in their near-IR images which are less affected by dust and star-

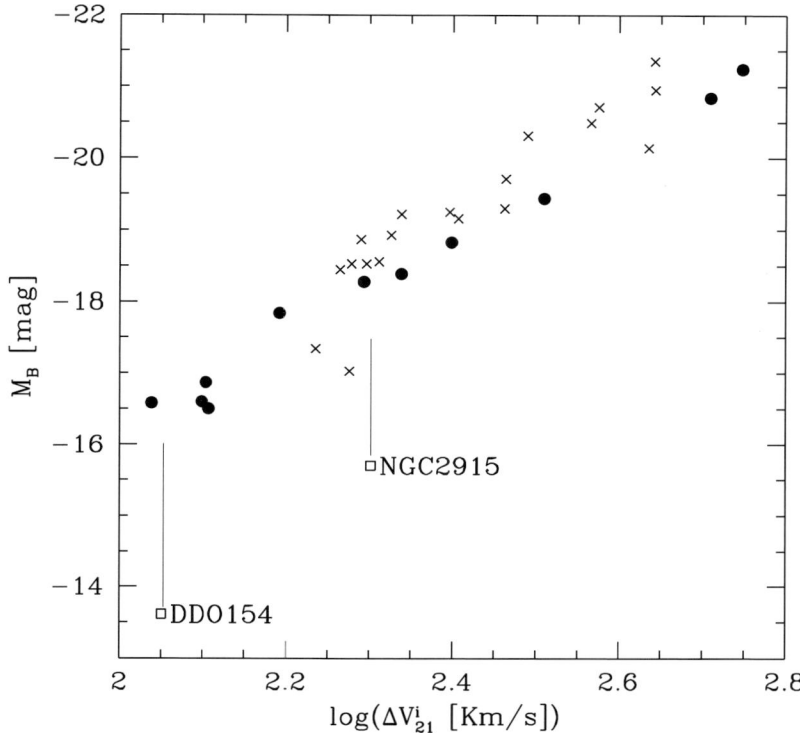

FIGURE 13. The Tully-Fisher relationship for a sample of nearby disk galaxies compared with the two very gas-rich galaxies NGC 2915 and DDO 154 (open squares). The upper ends of the vertical bars show where they would lie if their gas were turned into stars with $M/L = 1$ (adapted from Meurer et al. 1996).

forming regions. About 72% of spirals appear barred in the H-band, and the fraction of bars does not vary much from Sa to Sd (Eskridge et al. 1999).

The origin of bars is still not well understood. Stellar disks can be unstable to bar-forming modes, and the bars that form in this way thicken through vertical instabilities and appear like boxy or peanut-shaped bulges when seen edge-on (we will discuss bulges in more detail later). Sellwood (1999) has recently cast doubt on disk instability as the main mechanism that leads to bar formation. He points out that the high central densities of most disk galaxies should have inhibited bar-forming instabilities.

Another interesting argument against bar formation by instability comes from observations of disk galaxies at intermediate redshift. If bars do form through the instability of stellar disks, then they should form early in a galaxy's life. But this does not appear to be so: Abraham et al. (1999) studied the incidence of bars among galaxies in the Hubble Deep Field North & South, and found that bars are relatively rare for redshifts $z > 0.7$. This suggests that bars form long after the disks of these galaxies have assembled. Early disks are apparently stable against bar-forming instabilities, so some kind of secular bar-forming mechanism may be involved. This observation by Abraham et al. seems important for understanding disk formation and stability, but I don't believe we fully understand the implications yet.

7. Bulges

We have already seen that some disk galaxies have very substantial bulges and others do not have bulges at all (see Figure 1 for examples). Bulges show a wide variety of form, from the dominant bulge of the Sombrero galaxy at one extreme to the small bulges like the bulge of the Milky Way at the other. The goal of the following subsections is to discuss the possible origin of the different kinds of bulges. At this time, it seems likely that the small bulges and large bulges are formed through different processes.

7.1. *Large Bulges*

Large bulges, as in M31 and M104 (the Sombrero galaxy), follow the $r^{1/4}$ surface brightness distribution ($\log I \propto -r^{1/4}$). For example, the light distribution in the bulge of M31 follows the $r^{1/4}$ law from a radius of 200 pc out to 20 kpc (Pritchet & van den Bergh 1994). An $r^{1/4}$ light distribution is usually associated with the dynamical relaxation from a fairly violent merger or aggregation history (e.g. van Albada & van Gorkom 1977; Barnes 1988). Chemically, the bulges of spirals show an Mg/Fe *vs.* absolute magnitude relation in the same sense as for ellipticals: the brighter bulges show a more marked overabundance of Mg relative to Fe (Jablonka *et al.* 1996). The usual interpretation of this effect is that, after the first major burst of star formation, supernova-driven winds in the more luminous systems remove the remaining gas quickly, and so reduce the subsequent Fe-enrichment by the slower type I SNe.

For the bulge of M31, photometric studies of fields in the outer bulge show that the bulge stars have a wide range of chemical abundance, from about [Fe/H] = -2 to -0.2, with little or no abundance gradient out to a radius of 40 kpc (e.g. Durrell *et al.* 1994; Couture *et al.* 1995; Rich *et al.* 1999; Holland *et al.* 1996). All of this indicates that the formation of large bulges occurred early and quickly, with a high degree of dynamical mixing and relaxation, much as for giant ellipticals. The present view about the formation of ellipticals through mergers would then suggest that large bulges form through major or minor mergers of gas-rich galaxies; the gas is needed for the subsequent formation of the disk of the merged system.

There are differences between giant ellipticals and large bulges, most notably in their rotational behaviour. Bulges lie close to the oblate isotropic rotator curve in the ($V/\sigma - \epsilon$) plane while, from the kinematics of their *inner* regions ($r \leq r_e$, where r_e is the half-light radius), the giant ellipticals mostly lie well below the oblate curve. We now know that, in at least some of the giant ellipticals, a large fraction of their angular momentum resides in their outermost regions (e.g. Arnaboldi *et al.* 1998), probably as a result of angular momentum transport while the system was out of equilibrium. The total specific angular momenta of large spirals and large ellipticals appear to be fairly similar, although the internal distribution of the angular momentum is different. It is not yet clear why the inner regions of large bulges rotate more rapidly than for large ellipticals. One possibility is that bulge formation went on in the presence of a substantial envelope of high angular momentum gas, which later dissipated to form the disk. Some of this gas may have funnelled into the inner bulge, through the torques that redistribute angular momentum, and so produced a more rapidly rotating inner bulge. Alternatively, Naab *et al.* (1999) have shown that mergers of equal mass disks produce boxy merger products with slow rotation and flattened by anisotropic velocity dispersion, while the mergers of disks with a 3:1 mass ratio produce disky systems that are rotationally supported. This suggests that the rotationally supported large bulges may be preferentially the products of unequal mass mergers. (This may also be true for the lower luminosity elliptical galaxies, which are mostly rotationally supported and disky systems: see for example Rix *et al.* 1999.)

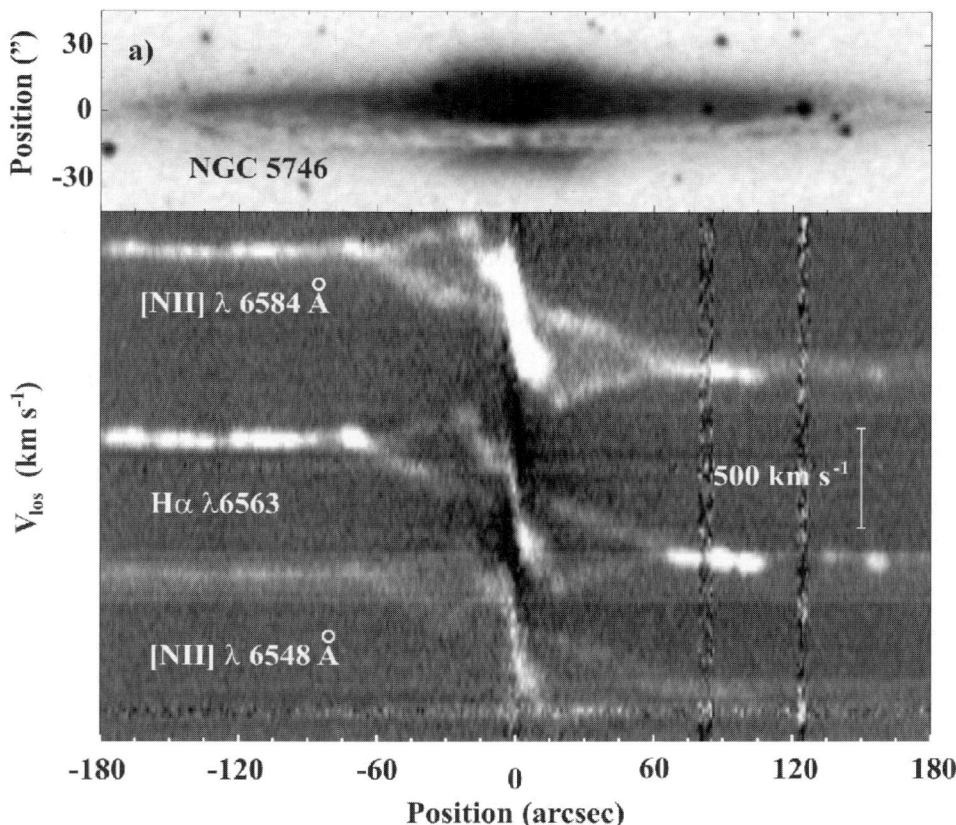

FIGURE 14. The upper panel is an image of the nearly edge-on galaxy NGC 5746, showing its boxy bulge. The lower panel shows the emission lines of [NII] and Hα, from a long slit located along the major axis of the galaxy. The spatial scales of the upper and lower panels are the same. The figure-of-eight structure seen in the emission lines from the region of the boxy bulge is believed to result from the orbit properties in the gravitational field of a rotating bar (from Bureau & Freeman 1999).

7.2. Small Bulges

A large fraction of the small bulges seen in later-type disk galaxies have a boxy or peanut-like structure when seen edge-on. Their photometric structure is usually exponential rather than $r^{1/4}$ (Courteau et al. 1996). What is the origin of these smaller bulges ? N-body simulations of self-gravitating disks strongly suggest that most of the small boxy/peanut bulges, like the bulge of the Milky Way, are bars arising from planar and vertical instabilities of disks (e.g. Combes et al. 1990; Pfenniger & Friedli 1991). Some of these structures may themselves be triggered by interactions (e.g. Noguchi 1987).

Until recently, observational verification of the bar-like nature of boxy bulges has been difficult, because boxy bulges are seen most clearly in edge-on spirals, and then it is not so clear whether they are barred. Kuijken & Merrifield (1995) devised a kinematical test of the bar-like nature of near-edge-on boxy/peanut bulges. This test, which depends on the properties of the two principal orbit families in the gravitational field of a rotating bar, is particularly effective and direct for galaxies with extended emission lines in the region of the bulge. An example of this test (from Bureau & Freeman 1999) is shown in Figure 14 for the boxy/peanut bulge of NGC 5746. The upper panel shows an image of the galaxy, and the lower panel shows the [NII] and Hα emission lines from a long slit placed along the major axis of the galaxy. The spatial scales of the image and the spectrum are the same, so position along the emission lines corresponds to position along the major axis of the image. The striking figure-of-eight shape of the emission lines arises from the orbit properties in a bar: it is an excellent diagnostic of the presence of the multiple orbit families within a bar, and hence of the presence of a bar. Bureau & Freeman (1999) applied the Kuijken-Merrifield test to 15 edge-on boxy/peanut bulges with extended emission lines, and a non-boxy control sample. Of these 15, 11 show the effect clearly, 3 are very dusty so the effect may be masked, and one galaxy is disturbed by interaction. None of the 7 galaxies in the non-boxy control sample shows the effect. One can conclude from his work that most boxy/peanut bulges are indeed bar-like. Merrifield & Kuijken (1999) present further observational results and insight into this diagnostic.

We can infer from these observations that a large fraction of the small bulges seen in later-type edge-on disk galaxies are in fact bars, which probably arose from instabilities of the disk before or after the disk reached equilibrium. This view is supported by the work of Courteau *et al.* (1996), who investigated the bulge/disk structure of a large sample of disk galaxies. They found that

• most of their galaxies with Hubble types Sb and later are better fitted by an exponential disk plus an exponential bulge rather than with an $r^{1/4}$ law bulge.

• the ratio of the scale lengths of the bulge to the disk has a restricted range, 0.09±0.04.

This again favors a picture in which the properties of the bulge (for these later-type galaxies) is tightly linked to the properties of the disk. Courteau *et al.* propose that their results are best understood in a model where the disk forms first and the bulge emerges from the disk. Kormendy (1993) has argued further, that some of the apparent bulges seen in more face-on systems are not really bulges at all, but rather are flat substructures within the parent disk. This is not in contradiction with the double exponential light distribution seen by Courteau *et al.* nor with the view that the apparent "bulges" of these later-type systems are closely related to their disks.

Why do some disk galaxies appear to have no bulge ? If the above discussion is correct, we would infer that these pure disk galaxies have managed to avoid the bar-forming events that lead to small bulges (and the mergers that lead to large bulges). These pure disk galaxies include systems like M33, and are mostly of lower luminosity. We will see later that these fainter galaxies have denser dark halos, and it may be that the dark halos of these pure disk systems are more effective in suppressing the bar-forming instabilities that would lead to the emergence of a small bulge. For example, compare two disks, one with $M_B = -22.5$ and the other with $M_B = -18.5$, corresponding to a massive spiral and to a galaxy like M33 respectively. Assume that each has the same baryonic disk surface density and disk velocity dispersion, and that the dark halos follow the observed scaling relations which we will discuss later. Then we would expect the Toomre Q-parameter to be about twice as large for the disk of the fainter galaxy. Similarly the stability index of Efstathiou *et al.* (1982) would indicate that the fainter disk is more stable against bar forming modes.

At this stage, the indications are that there is more than one way to form bulges. Small exponential bulges apparently form from their parent disks, while the large $r^{1/4}$ bulges probably form via mergers of gas-rich systems of unequal mass. If this is correct, then the early merger history of individual disk galaxies and the stabilising effects of

7.3. Globular Clusters in Disk Galaxies

The specific frequencies of globular clusters in disk galaxies do appear to change along the Hubble sequence. Early type galaxies with large bulges, like NGC 7814 and the Sombrero galaxy in Figure 1, have specific frequencies that are similar to those for the normal ellipticals (2 ± 1 and 3.5 ± 1 for the two examples cited). On the other hand, the later type disk galaxies have specific frequencies that are typically about 0.5. Kissler-Patig et al. (1999) studied the globular cluster populations of two later-type systems, NGC 4565 and NGC 5907, and compared them with the globular cluster system of the Galaxy. These three galaxies have similar absolute magnitude ($M_V \simeq -21.3$) and specific frequencies ($S_N \simeq 0.55$) but their bulges and disks are different. The Galaxy and NGC 4565 have boxy bulges and thick disks that contribute several percent of the thin disk light, while NGC 5907 has an insignificant bulge and a thick disk that is less than about 0.5% of the thin disk. So it seems that the processes that build boxy bulges and thick disks are not very relevant to the building of globular cluster systems.

The formation of globular clusters in merging galaxies indicates that globular cluster formation requires fairly violent conditions. From the discussion of small bulges above, it seems likely that small bulges are produced primarily from the disk, through disk instabilities. We will argue later that thick disks are probably produced through the heating of the early stellar thin disk in minor mergers and accretion events. So the formation of small bulges and thick disks may be a fairly quiescent process, and so may have little to contribute to forming globular clusters. On the other hand, the large bulges probably formed through early mergers of gas-rich objects, so the higher specific frequency of globular clusters in the early-type disk galaxies makes sense in this picture.

8. Pure Disk Galaxies

These systems (like the example of IC 5249 in Figure 1) lie late in the Hubble sequence and are mostly systems of relatively low mass and luminosity. Many authors have discussed the fragility of disks and have shown how the existence of a thin disk limits the amount of mass such galaxies can have accreted after the stars of the disk had formed. This limit is only a few percent of the mass of the stellar disk (e.g. Toth & Ostriker 1992). So the thinness of the disks of these pure disk galaxies indicates a relatively undisturbed history.

Some galaxies that appear to have a pure disk structure show a weak underlying stellar halo or thick disk from very deep surface photometry (see Figure 15). NGC 5907 is an example (Morrison et al. 1994; Sackett et al. 1994; Lequeux et al. 1998). This galaxy is unusually luminous for an Sc system ($M_B \simeq -20.5$), and its faint extended structure may well be the result of a very minor merger or interaction at some point in the galaxy's history. The apparently metal-rich nature of this halo or thick disk (from its relatively red color) argues against the possibility that it came from a weak episode of star formation before the disk had settled.

Abe et al. (1999) made deep surface photometry of the less luminous edge-on galaxy IC 5249 (see Figure 1) and again find evidence for a very faint thick disk in this system. But not all pure disk galaxies show a stellar halo or thick disk. Very deep surface photometry of the edge-on galaxy NGC 4244 by Fry et al. (1999) shows no evidence for a halo or thick disk. This galaxy is relatively faint ($M_B \simeq -18.4$) and its surface brightness distribution shows only a single exponential disk with a radial cutoff at about 5 radial scalelengths.

NGC 4244 appears to be a true undisturbed pure disk system. The existence of such systems is interesting because it demonstrates that in at least some of the late-type disk galaxies

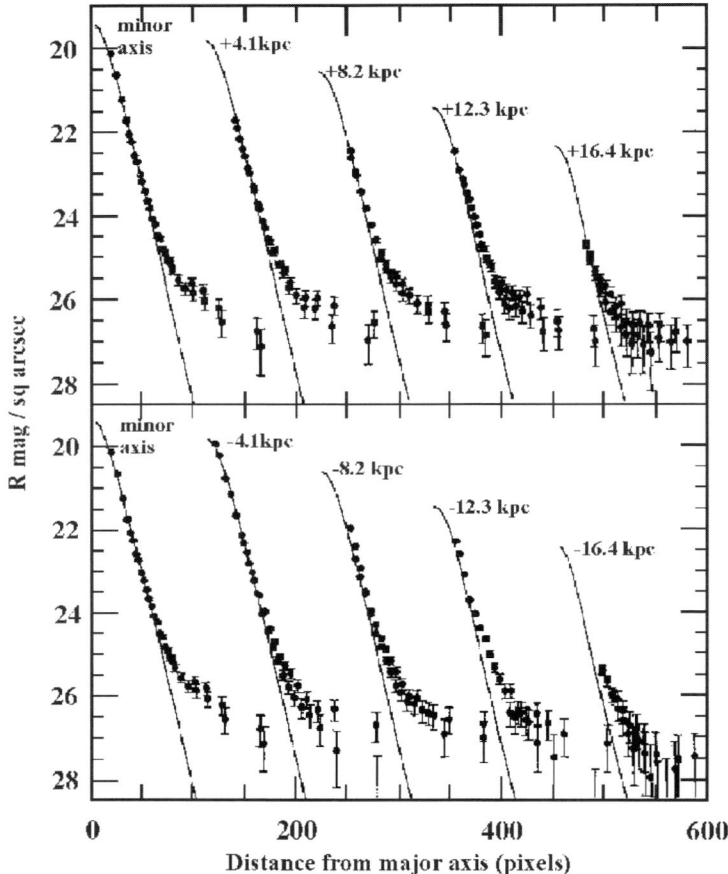

FIGURE 15. R-band surface brightness profiles parallel to the minor axis of NGC 5907. The distance from the minor axis is shown for each profile. The curves show the expected surface brightness distribution for a pure disk that is exponential in radius and height above the plane, with scaleheight 430 pc and scalelength 4.8 kpc. The faint excess light associated with the stellar halo or thick disk is evident below a surface brightness of about 25 mag arcsec^{-2}. (From Morrison et al. 1994)

- star formation did not begin until the gas of the disk had settled to the plane, and
- since the onset of star formation in the disk, the galaxy has suffered no significant dynamical disturbance from internal or external sources, *i.e.* no mergers or accretion of small satellites or dark matter lumps, and no growth of a bar/bulge by disk instabilities.

9. The Velocity Dispersion of the Gas Disks

A striking feature of the gas disks of disk galaxies is that the velocity dispersion of the gas is typically 8 to 10 km s^{-1}. This is seen for galaxies over a wide range of absolute magnitudes, from large spirals to small dIrr galaxies. These gas disks provide a reservoir of gas for future star formation, and the velocity dispersion is a regulating factor in the stability of the gas layer.

What is the energy input that maintains this velocity dispersion of the gas ? In some galaxies at least, it is likely to be the star formation in the disk. This is nicely seen in the detailed HI study of the LMC, by Kim *et al.* (1998). Here the spatial resolution is about 15 pc and the HI study reveals the small scale structure that contributes to the observed velocity dispersion. They find that the most massive HI clouds have the smallest dispersion, indicating a hierarchy of cloud-cloud collisions and cloud merging. Kim *et al.* fitted a rotation curve and then derived the peculiar motions of each pixel of their HI image from this rotation field. The distribution of the peculiar velocities of the brightest HI at each pixel is very close to exponential (Figure 16). This suggests to the authors that the HI clouds derive their velocity dispersion from a stochastic process in velocity space, through a balance between momentum input from stellar winds and momentum "dissipation" by cloud-cloud collisions.

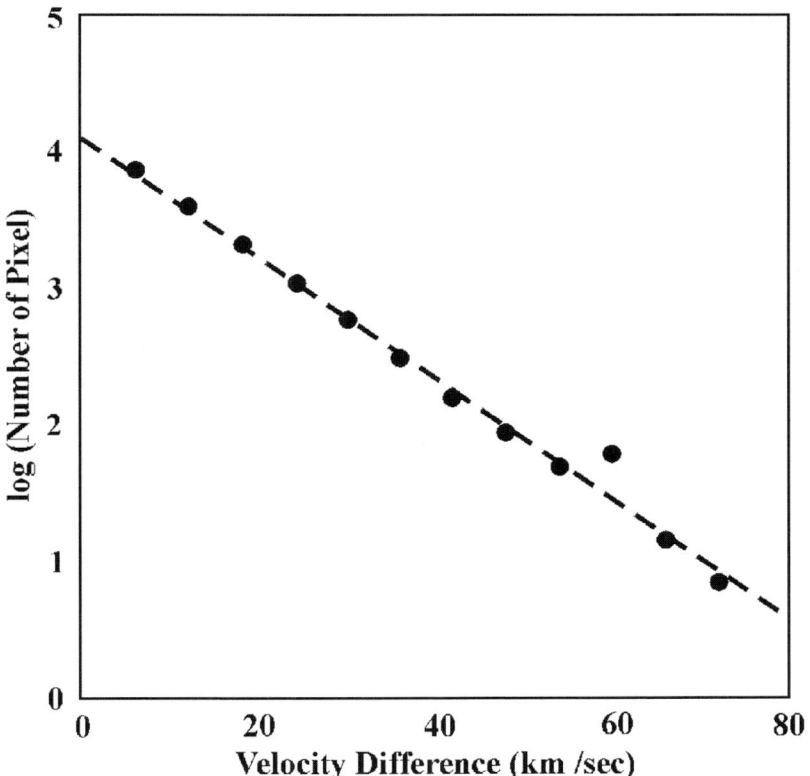

FIGURE 16. The histogram of the modulus of the velocity difference for individual HI clouds from the mean velocity field of the LMC. The distribution is close to an exponential across the full range of peculiar velocities. This indicates the stochastic nature of the turbulent HI motions in the LMC (from Kim *et al.* 1998).

In some other galaxies, however, star formation is unlikely to be the source of the energy input. An example is the disk galaxy NGC 2915. This galaxy has a very extended HI distribution, reaching out to about 22 optical scalelengths, with a well-defined rotation field and a velocity dispersion at the usual level of about 8 km s^{-1} (Meurer *et al.* 1996).

The HI disk shows spiral structure far beyond the limits of the optical galaxy, but there is no visible indication of star formation in this HI disk, and a very deep Hα study of the HI disk by Meurer et al. (unpublished) with the TAURUS tuneable filter on the AAT shows only a very few tiny HII regions powered by single early B stars. So what is the energy input that maintains the velocity dispersion of the the HI gas in this galaxy ? Bureau et al. (1999) suggested that the HI spiral structure in this galaxy is driven by the torque of an underlying rotating triaxial dark halo. If that is correct, then the dark halo could provide a mechanical source of energy input to the HI. It would be interesting to know how common this is. Other potential sources of energy for the HI gas include the ambient UV radiation field, and turbulence that feeds on the rotation field of the galaxy.

10. S0 Galaxies

The S0 galaxies are disk systems without significant spiral structure. Figure 17 shows an example.

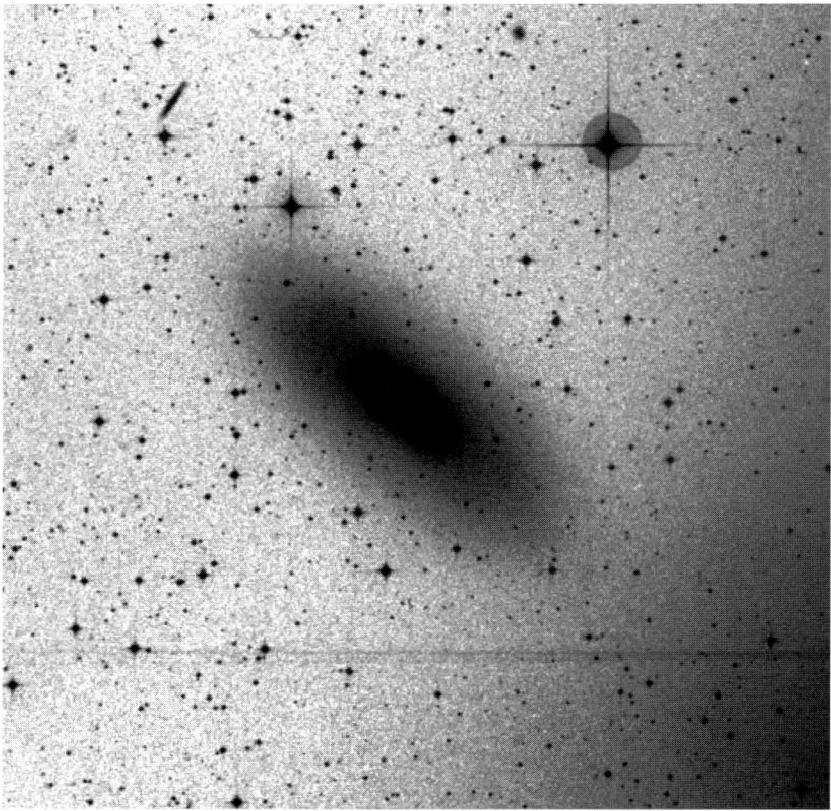

FIGURE 17. NGC 5102: an S0 galaxies in the nearby Centaurus A group. Note the relatively structureless disk. Image from the DSS.

What is their origin ? Are they simply spirals that have used up most of their interstellar gas through star formation ? S0 galaxies are found predominantly in denser galactic environments, and opinion at this time is that S0 galaxies have lost their gas through interaction with other galaxies or with their environment. For example, Bekki (1998) suggests that S0 galaxies form through unequal-mass mergers; the enhanced star formation rate exhausts the interstellar medium, leaving a single gas poor remnant S0. Jones *et al.* (1999) studied the stellar content of E and S0 galaxies in several clusters at intermediate redshift and find no significant difference between the luminosity-weighted ages of the E and S0 galaxies in these clusters. They conclude that the progenitors of S0 galaxies in rich clusters are mostly early-type galaxies which have had their star formation truncated through interaction with the cluster environment. The fate of later-type lower-surface-brightness spirals in clusters is different: Moore *et al.* (1999) showed that these less dense systems are likely to end up as dwarf spheroidal or dwarf elliptical galaxies through the effects of the cluster harassment process. Ram pressure stripping in clusters is also likely to be a significant (though not on its own sufficient) element in the production of S0 galaxies (Abadi *et al.* (1999).

Some S0 galaxies do have a significant interstellar medium and a low level of ongoing massive star formation (e.g. Pogge & Eskridge 1998). A nearby example is the S0 galaxy NGC 5102 in the Centaurus A group: this galaxy has a substantial HI content (van Woerden *et al.* 1993) and shows a sprinkling of OB stars and small HII region over part of its disk (Pritchet 1979).

REFERENCES

ABADI, M., MOORE, B., BOWER, R. 1999, astro-ph/9903436.
ABE, F. *et al.* 1999, AJ, 118, 261.
ABRAHAM, R. *et al.* 1999, MNRAS, 308, 596.
ARNABOLDI, M., FREEMAN, K., GERHARD, O., MATTHIAS, M., KUDRITZKI, R., MENDEZ, R, CAPACCIOLI, M, FORD, H. 1998, ApJ, 507, 759.
BARNES, J. 1988, ApJ, 331, 699.
BEGEMAN, K. 1989, A&A, 223, 47.
BEKKI, K. 1998, astro-ph/9806106.
BELL, E. & DE JONG, R. 1999, astro-ph/9909402.
BELL, E. *et al.* 1999, astro-ph/9909401.
BUREAU, M. & FREEMAN, K. 1999, astro-ph/9904015.
BUREAU, M., FREEMAN, K., PFITZNER, D., MEURER, G. 1999, astro-ph/9906498.
COMBES, F., DEBBASCH, F., FRIEDLI, D., PFENNIGER, D. 1990, A&A, 233, 82.
COURTEAU, S., DE JONG, R., BROEILS, A. 1996, ApJ, 457, 73.
COUTURE, J., RACINE, R., HARRIS, W., HOLLAND, S. 1995, AJ, 109, 2050.
CREZE, M., CHEREUL, E., BIENAYME, O., PICHON, C. 1998, A&A, 329, 920.
DE JONG, R. & LACEY, C. 1999, astro-ph/9910066.
DURRELL, P., HARRIS, W., PRITCHET, C. 1994, AJ, 108, 2114.
EFSTATHIOU, G., LAKE, G., NEGROPONTE, J. 1982, MNRAS, 199, 1069.
ESKRIDGE, P. *et al.* 1999, atro-ph/9910479.
FALL, S.M. & EFSTATHIOU G. 1980, MNRAS, 193, 189.
FREEMAN, K. 1970, ApJ, 160, 811.
FREEMAN, K. 1992, "Physics of Nearby Galaxies: Nature or Nurture" (Editions Frontieres), p 201.
FREEMAN, K. 1996, "Unsolved Problems of the Milky Way" (Kluwer), p 645.
FRY, A., MORRISON, H., HARDING, P., BOROSON, T. 1999, astro-ph/9906019.
HOLLAND, S., FAHLMAN, G., RICHER, H. 1996, AJ, 112, 1035.

Impey, C. & Bothun, G. 1997, ARA&A, 35, 267.
Jablonka, P., Martin, P., Arimoto, N. 1996, AJ, 112, 1415.
Jones, L, Smail, I., Couch, W. 1999, astro-ph/9907423.
Kennicutt, R. 1989, ApJ, 344, 685.
Kennicutt, R. 1998, ApJ, 498, 541.
Kim, S., Staveley-Smith, L., Dopita, M., Freeman, K., Sault, R., Kesteven, M., McConnell, D. 1998, ApJ, 503, 674.
Kissler-Patig, M., Ashman, K., Zepf, S., Freeman, K. 1999, AJ, 118, 197.
Kormendy, J. 1993, "Galactic Bulges", IAU Symposium 153. Ed. H. Dejonge & H. Habing (Kluwer:Dordrecht), p 209.
Kuijken, K. & Merrifield, M. 1995, ApJ, 443, L13.
Lequeux, J., Combes, F., Dantel-Fort, M., Cuillandre, J-C., Fort, B., Mellier, Y. 1998, A&A, 334, L9.
Merrifield, M. & Kuijken, J. 1999, A&A, 345, L47.
Meurer, G., Carignan, C., Beaulieu, S., Freeman, K. 1996, AJ, 111, 1551.
Moore, B, Lake, G., Quinn, T., Stadel, J. 1999, MNRAS, 304, 465.
Morrison, H., Boroson, T., Harding, P. 1994, AJ, 108, 1191.
Naab, T., Burkert, A., Hernquist, L. 1999, astro-ph/9908129.
Navarro, J., Frenk, C., White, S. 1997, ApJ, 490, 493.
Noguchi, M. 1987, MNRAS, 228, 635.
Pfenniger, D. & Friedli, D. 1991, A&A, 252, 75.
Pogge, R. & Eskridge, P. 1998, astro-ph/9808136.
Pritchet, C. 1979, ApJ, 231, 354.
Pritchet, C. & van den Bergh, S. 1994, AJ, 107, 1730.
Rich, M., Mighell, K., Neill, J. 1999, "Formation of the Galactic Halo . . . Inside and Out", ed H. Morrison and A. Sarajedini (San Francisco: ASP), p 544.
Rix, H-W., Carollo, M., Freeman, K. 1999, ApJ, 513, L25.
Sackett, P, Morrison, H., Harding, P., Boroson, T. 1994, Nature, 370, 441.
Sakai, S. et al. 1999, astro-ph/9909269.
Sellwood, J. 1999, astro-ph 9909489.
Simien, F. & de Vaucouleurs, G. 1986, ApJ, 302, 564.
Swaters, R. 1999, PhD thesis, Rijkuniversiteit te Groningen.
Toth, X. & Ostriker, J. 1992, ApJ, 389, 5.
Tully, R.B. & Fisher, J.R. 1977, A&A, 54, 661.
van Albada, T. & van Gorkom, J. 1977, A&A, 54, 121.
van der Kruit, P. 1988, A&A, 192, 117.
van Woerden, H., van Driel, W., Braun, R., Rots, A. 1993, A&A, 269, 15.
Willick, R. 1999, ApJ, 516, 47.
Zurek, W., Quinn, P., Salmon, J. 1988, ApJ, 330, 519.
Zwaan, M., van der Hulst, J.M., de Blok, W.J.G., McGaugh, S. 1995, MNRAS, 273, L35.

Dark Matter in Disk Galaxies

By KENNETH C. FREEMAN

Mt. Stromlo Observatory, Canberra, Australia

In our current picture of galaxy formation, we believe that galaxies formed through a hierarchy of merging. The merging elements were a mixture of baryonic and dark matter. The dark matter settled into a partially mixed dark halo while, in disk galaxies, the baryons settled into a rotating disk and bulge. What can we learn about the properties of the dark halos ? Do the properties of the dark halos predicted by simulations correspond to what is inferred from observational studies ? There is no clear answer yet.

This lecture will consider only the dark matter properties of disk galaxies. There is abundant evidence for dark matter in the brightest and the faintest elliptical galaxies, but ellipticals of intermediate mass are more difficult to study. The techniques for studying the dark matter content of ellipticals are different than for the disk galaxies, and would be the subject of another lecture.

1. Dark Halos in Spiral Galaxies

Spiral galaxies are probably the most straightforward systems in which to study the distribution of dark matter. The equibilibrum of the gas in the disk is primarily a balance between the radial potential gradient and the acceleration of circular motion. For a rotational velocity $V(R)$ where R is the radius, the equilibrium is given by

$$\frac{V^2}{R} = -\frac{\partial \Phi}{\partial R} \approx \frac{GM(R)}{R^2}$$

where $M(R)$ is the mass enclosed within radius R. The shape of $V(R)$ can be anything from solid body to $V \approx$ constant (flat). For larger spirals like the Galaxy, $V(R)$ is usually fairly flat, so the enclosed mass $M(R) \propto R$.

Is this evidence for a dark halo ? Not necessarily: it depends on how far the rotation curve extends. Most spirals have a roughly exponential surface brightness distribution $I(R) = I_\circ \exp(-R/h)$: for a large galaxy like the Milky Way, the scale length $h \approx 4$ kpc. Optical rotation curves, measured from the spectra of ionized gas, extend to $R \approx 3h$. Now assume that the surface *density* distribution follows the optical surface *brightness* distribution. Can this observed surface density distribution with its gravitational potential $\Phi(R)$ explain the observed rotation curve $V(r)$? The answer to this question is

- yes, for optical rotational curves which extend out to about 3 disk scale lengths: Figure 1 shows examples of very different kinds of optical rotation curves which are reproduced very well from the *luminous* surface density alone. There is only one free parameter in the fits of the calculated rotation curve to the observed rotation curve. This is the M/L ratio which, in the I-band, turns out to be about 4.2 in the mean: the rms scatter of about 1.3 is not much larger than the scatter expected from the distance uncertainties alone. The mean value of M/L_I is consistent with the expectation from stellar population synthesis models.
- no, for galaxies with 21 cm HI rotation curves that extend far out, to $R \gg 3h$. Figure 2 shows the HI rotation curve for the spiral NGC 3198, extending out to $11h$. Here the expected $V(R)$ from the stars and the gas falls far below the observed rotation

curve in the outer regions of the galaxy. This is seen for almost all spirals with rotation curves that extend out to many scale lengths.

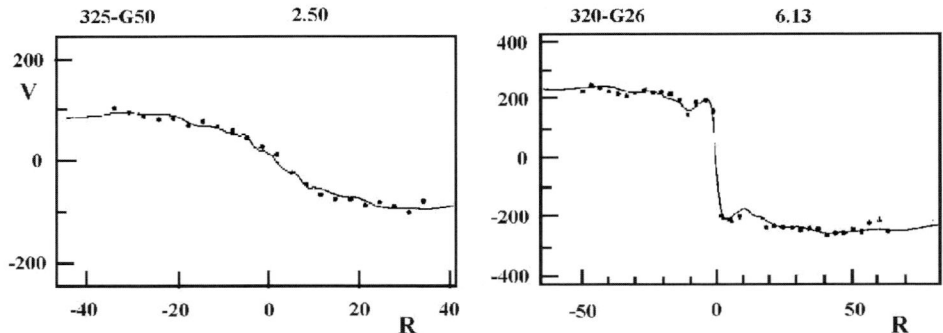

FIGURE 1. Two examples of Hα rotation curves with very different morphology for galaxies from Buchhorn's (1991) sample. The points represent the data, and the curve shows the rotation curve calculated from the I-band surface brightness distribution. The M/L_I ratio adopted for the fit is shown at the top right of each panel. The units of velocity and radius are km s^{-1} and arcsec.

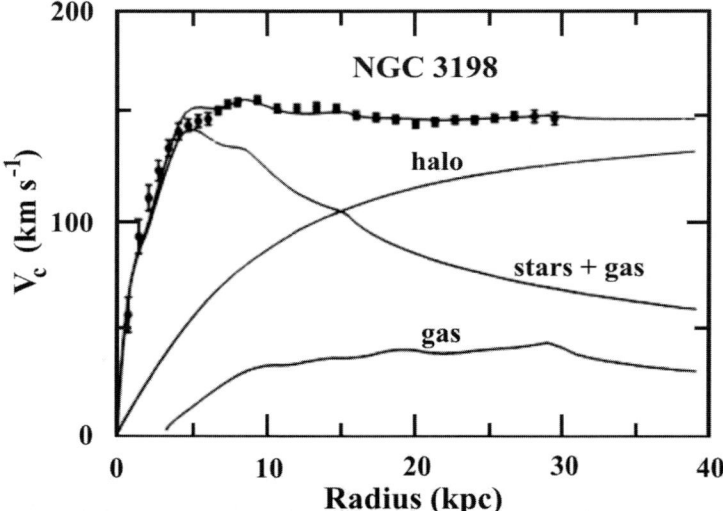

FIGURE 2. HI rotation curve for NGC 3198, from Begeman (1987). The adopted contributions to the rotation curve from the disk, the gas and the dark halo are labelled. The sum of these contributions is shown by the curve that passes through the observed rotation curve.

It is interesting that the optical rotation curves end (*i.e.* the HII regions and diffuse HII emission end) just at the point where the gravitational field of the (stars + gas) becomes visibly inadequate. We conclude that the luminous matter dominates the radial potential gradient $\partial \Phi/\partial R$ for $R \lesssim 3h$, but beyond this radius the dark halo becomes progressively more dominant. Typically, out to the radius where the HI data ends, the ratio of dark to luminous mass is 3 to 5: values up to about 10 are found in a few examples.

For the decomposition of the rotation curve shown in Figure 2, the stellar M/L ratio is taken to be as large as possible without leading to a hollow dark halo. Many galaxies have been analysed in this way: the decomposition often works out like the one in Figure 2, with comparable peak circular velocity contributions from the disk and the dark halo. The inferred stellar M/L ratios are again usually consistent with those expected from synthetic stellar populations, at least for the brighter galaxies (see section 4 for the situation with dwarf galaxies). Nevertheless, many people do not believe that this maximum disk, minimum halo approach is correct. One reason is that the estimated surface density of the galactic disk near the sun appears to be low (e.g. Crézé et al. 1998) and may be inconsistent with a maximum disk.

The maximum disk question is important for us here because inferences about the structure of dark halos from rotation curves depend so much on the correctness of the maximum disk interpretation. For example, if the maximum disk decompositions are correct, then the inferred dark halos have approximately uniform cores with a relatively large ratio of (core radius)/(disk scale length), as in the example of Figure 2. The halos that form in cosmological simulations, on the other hand, have steeply cusped inner halos.

Optical rotation curves provide support for the maximum disk interpretation. In the inner regions of the disks of larger spirals (radius $R \gtrsim 3h$ where h is the disk scalelength), the rotation curves are well fit by assuming that mass follows light. Buchhorn (1991) analysed 552 galaxies with optical rotation curves and I-band surface photometry, and this process worked well for about 97% of his sample. The mean M/L_I ratio is $4.2 \pm 1.3(\sigma)$: the scatter in the M/L_I values is not much larger than the scatter expected from the Tully-Fisher distance uncertainties. This sample included galaxies with the whole range of rotation curve morphologies, from nearly solid body to nearly flat. Figure 1 shows two extreme examples. The implication of the good fits over the whole range of rotation curve morphology is that *either* the stellar disk dominates the radial potential gradient within $R \gtrsim 3h$ *or* the radial potential gradient of the halo faithfully mimics the potential gradient of the disk in almost every example: in terms of surface desity distributions, the halo mass distribution must actually over-mimic or exaggerate the mass distribution of the disk, because of the relative weakening of the radial gravitational field of the halo by its presumably thicker mass distribution.

Athanassoula et al. (1987) used spiral structure theory to give a dynamical constraint on the M/L ratio of the stellar disk. From the number of spiral arms observed in a sample of spiral galaxies with well-measured rotation curves, they argued that most of the disks must indeed be close to maximum.

Debattista & Sellwood (1998) give an interesting argument from the dynamics of barred spirals that favors the maximum disk picture. In a dense halo, dynamical friction slows down the rotation of the bar. In a low density halo (like in a maximum disk environment), the bar rotation rate stays high. Evidence from gas flows in barred galaxies, and from direct measurement of bar rotation rates, indicates that bars do rotate rapidly, with their Lagrange points typically at 1.2 to 1.4× the bar semi-length.

So I conclude that the maximum disk picture is probably correct, at least for galaxies of normal surface brightness (as in Buchhorn's sample). For a contrary view, see Courteau & Rix (1999).

2. The Shapes of Dark Halos

Cosmological simulations of dissipationless galaxy formation produce halos that are typically weakly prolate. For example, Warren et al. 1992) found mean axial ratios $\langle b/a \rangle \simeq 0.7$, $\langle c/a \rangle \simeq 0.5$ where a, b, c are the major, intermediate and minor axes of the halos. With baryonic infall (e.g. Katz & Gunn 1991), the halos come out more spherical, with $\langle b/a \rangle \simeq 0.9$, $\langle c/a \rangle \simeq 0.65$. So it would be interesting to know if real halos conform to these expectations. Measuring the shapes of dark halos has proved difficult in practice.

Polar ring galaxies have matter rotating in two more or less orthogonal planes, so in principle the shape of the density distribution can be derived from the estimates of the potential gradient in these two planes. This kind of study has been done in most detail for the polar ring galaxy NGC 4650A (Sackett et al. 1994; Combes & Arnaboldi 1995). Indications are again that the halo is flattened (E6) but there is still some ambiguity about to which plane it is flattened. The halos of polar ring galaxies may not be typical in shape: the presence of a well developed polar ring may *require* a flattened and triaxial halo.

The thickness and flaring of the HI layer in spirals gives some measure of the shape of the halo: a more spherical halo permits more flaring at large radius, where the vertical gravitational field is reduced. Olling (1995) finds indications that the halos of some galaxies are relatively flat, with axial ratios as low as 0.2. Olling & Merrifield (2000) argue that the dark halo of our Galaxy probably has an axial ratio that is closer to 0.8.

For our Galaxy, the good agreement between the dynamical estimate of the total density of matter near the sun and the density derived from a census of stars and gas (e.g. Crézé et al. 1998) leaves little room for a highly flattened dark matter distribution. The scaleheight of the dark halo is probably larger than about 2 kpc.

At this stage, it is not really clear what the observations are indicating about the shapes of dark halos. This problem remains an interesting observational challenge.

3. Dark Matter in Dwarf Irregular Galaxies

For the brighter gas-rich dwarfs, with circular velocities $V_c > 30$ km s^{-1}, we can use rotation curve techniques as for the larger spirals. For the smallest dIrr galaxies, however, the rotational velocities are small and the gas is supported primarily by its velocity dispersion. The mass can be derived from the HI velocity distribution by assuming hydrostatic equilibrium: this gives high M/L ratios (> 12) for several faint dwarfs with $M_B > -10.5$ (Lo et al. 1993). The HI velocity dispersions are usually close to 10 km s^{-1}, similar to those in much larger galaxies. The origin of this 10 km s^{-1} velocity dispersion is not really understood; it does not appear to be due to energy and momentum input from star formation because similar values for the velocity dispersion are seen in the outer parts of larger spirals where the star formation is negligible. With this uncertainty about the origin of the velocity dispersion, we must ask whether the assumption of hydrostatic equilibrium is correct in these very small dIrr systems. (I hope it is correct, because there is no other way to estimate their dark matter content). These faint dwarfs are interesting systems: the example of GR8 ($M_B = -10.6$, Carignan et al. 1990) indicates that the ratio of (total mass)/(gas + stellar mass) out to the edge of the HI distribution is about 8. For this galaxy, hydrostatic equilibrium leads to the conclusion that the luminous matter lies entirely within the approximately uniform density core of its dark halo. This provides some evidence for a core in a dark halo, independent of the rotation curve analyses. The density of this core is about $0.07 M_\odot$ pc^{-3}, which is much higher than the typical inferred core densities for spirals ($< 0.01 M_\odot$ pc^{-3}) but typical of the high core densities for very small dwarfs.

Swaters (1999) studied the rotation curves of 35 dwarf galaxies with peak circular velocities $\lesssim 100$ km s^{-1}. He decomposed the rotation curves into three components

$$V_{rot}^2 = (M/L)_* V_{d,1}^2 + \eta V_{HI}^2 + V_{halo}^2$$

where $V_{d,1}$ is the rotation curve calculated from the surface photometry with $(M/L) = 1$, V_{HI} is the rotation curve calculated from the HI surface density distribution, and V_{halo} is the rotation curve for a pseudo-isothermal halo model. The η factor in the second term is there to test the idea (e.g. Bosma 1978) that the surface density of dark halos sometimes appears to track the surface density of the HI. Regarding the degeneracy and relative contributions of the three terms, Swaters found that

- for 30 of the 35 galaxies, the stellar disk can be scaled to reproduce most of the inner observed rotation curve, though some high M/L values (~ 20) were needed
- a combination of stellar disk and scaled HI (no pseudo-isothermal halo) also works well, with η-values between 3 and 12 to represent the dark matter
- a zero disk (*i.e.* HI and pseudo-isothermal halo only) also works well.

The most striking result here is that mass appears to follow light in the inner parts of these late-type dwarfs, and also for a sample of LSB galaxies studied by Swaters. Although mass follows light, a wide range of M/L ratios is indicated, some much higher that plausible from stellar population models.

What does this mean? We recall the ideas of Pfenniger *et al.* (1994) who suggested that the dark matter may be in the form of cold gas, and also the recent work on old white dwarfs in the Galaxy. The MACHO project (Alcock *et al.* 2000) detected ~ 15 microlensing events toward the LMC, far more than expected from known stellar populations, but representing less than 20 % of the dark halo mass if they lie in the halo. If these microlenses are galactic, their typical masses are of order $0.5 M_\odot$, which suggests that these microlenses could possibly be old white dwarfs. To produce old white dwarfs in such numbers raises many problems with chemical enrichment and the baryon budget. Nevertheless, these old white dwarfs may have been detected: see Ibata *et al.* (1999, 2000) and Méndez & Minniti (2000). To explain the MACHO detections, Gates & Gyuk (1999) proposed a new component of our Galaxy: a thick disk of white dwarfs with mass comparable to that of the thin disk. This could be dynamically attractive for reconciling the "maximum disk" rotation curves with the low density of the stellar disk in Swaters' dwarfs and also in our Galaxy. A truncated stellar IMF, with stars between about 1 and $8 M_\odot$ would be needed to avoid a surviving luminous counterpart.

4. Dark Matter in Our Galaxy

For most spirals, it is possible to derive the mass only out to the last measured point of the HI rotation curve. But because we are located within our Galaxy, it is possible to derive the distribution of mass out a radius of more than 100 kpc, starting with the HI rotation curve which can be measured out to about 20 kpc. At larger radii, other dynamical tracers can be used, such as halo stars and halo satellites. The escape velocity of halo stars in the solar neighborhood is also a useful mass indicator. The Local Group timing arguments, first proposed by Kahn & Woltjer (1959), give total mass estimates for M31 and our Galaxy. See Kochanek (1996) for more details. It appears that the dark halo of the Galaxy reaches out well beyond 100 kpc. The dark mass of the Galaxy is more than 90% of the total mass, which is a much higher fraction than can be measured for most disk galaxies. Only for a few other galaxies in which the HI extends out to very large radii is such a large dark matter fraction found: NGC 2915 (Meurer *et al.*1996) is an example.

5. The Tully-Fisher Law

This law relates the rotation velocity V_{rot} and the luminosity of disk galaxies. The luminosity comes from the baryons. It is not so clear which component is determining V_{rot}. For galaxies of normal surface brightness, like NGC 3198 shown in Figure 2, the peak rotational velocity contributions from the dark halo and from the luminous disk are similar. If V_{rot} reflected mainly the gravitational field of the baryons, then the Tully-Fisher law would simply be a statement about the equilibrium of the baryonic component of disk galaxies and would not be so interesting. In this case, the observed slope of the TF law would be consistent with disks having a similar surface density and similar stellar M/L ratio. An important paper by Zwaan et al. (1995) shows that there is more to the TF law than this. They found that the slope and zero point of the TF laws for disks of high and low surface brightness are indistinguishable. In the low surface brightness disks, the rotational velocity (and therefore the width of the integrated HI profile) is believed to be determined primarily by the dark halo. The TF law is then a relationship between the baryon luminosity and the depth of the potential well for the dark halo. Going one step further, some of the fainter disk galaxies have a large fraction of their baryon mass in the form of HI (e.g. DDO154 and NGC 2915: see Freeman 1999). These gas-rich galaxies are typically one or two magnitudes fainter than expected from their HI velocity widths and the TF law. However, if we notionally convert the gas in these systems into stars with a M/L ratio ~ 1, then their total luminosity falls right on the TF law. We can interpret the TF law as a relation between the total mass of baryons and the binding energy of the halo, which is presumably established very early in the galaxy formation process. See McGaugh et al. (2000) for more on this subject.

6. Dark Matter in High Velocity Clouds

At least some of the HVCs may be intergalactic objects falling into the Local Group (Blitz et al. 1999, Braun & Burton 1999). Dynamical simulations of hierarchical galaxy formation predict far more satellites than observed for large disk galaxies like our Galaxy (e.g. Moore et al. 1999). Do the HVCs represent the missing satellites of the Milky Way ? It would be very interesting to know if the HVCs are also dark matter dominated. Estimates of their HI mass and their virial mass both depend (differently) on distance, so this issue will remain unresolved while the distances to HVCs remain mostly unknown. This is a long-standing problem.

7. Cusped Halos

Since about 1985, observers have used dark halo models with constant density cores to interpret rotation curves. Commonly used models include the non-singular isothermal sphere, which has a well defined core radius and central density, and the pseudo-isothermal sphere with density distribution

$$\rho = \rho_\circ \left[1 + (r/r_c)^2\right]^{-1}$$

which again has a well-defined core radius r_c and central density $\rho_\circ$. Why were these models used ? I think it was because
• rotation curves of spirals appeared to have an inner solid-body component which indicated a core of roughly constant density
• hot stellar systems like globular clusters and elliptical galaxies were successfully modelled by King models at the time: these models are modified isothermal system with cores.

On the other hand, we now know that CDM simulations consistently produce halos that are cusped at the center. For example, the NFW halo distribution (Navarro et al. 1996) has the form

$$\rho = \rho_s (r/r_s)^{-1} \left[1 + (r/r_s)\right]^{-2}$$

This apparent difference between the apparently observed cores of galaxies and the predictions of CDM remains a problem.

Kravtsov et al. (1998) compiled rotation curves for a sample of dwarf and LSB galaxies. In these galaxies, the effect of the baryonic component on the rotation curve is believed to be minimal, so the rotation curve reflects more or less the gravitational field of the halo. Recall that a solid body inner rotation curve corresponds to a mass distribution with a constant density core. Kravtsov et al. found that $\rho(r) \sim r^{-0.2}$ to $r^{-0.4}$ for their galaxies, *i.e.* a distribution with a shallow central cusp. They made some ΛCDM simulations which gave halos with $\rho(r) \sim r^{-0.0}$ to $r^{-0.4}$ at small r but with significant scatter. Other high resolution simulations (e.g. Moore et al. 1999) find inner halo slopes that are even steeper than the NFW distribution, with $\rho(r) \sim r^{-1.5}$.

Rotation curves and simulations appear to disagree. The simulations mostly give a steep central $\rho(r)$ for the halos, while the observed rotation curves indicate a flat central $\rho(r)$. What is wrong: observations or theory ? Does it matter ? Yes: the density distribution of the dark halos provides a critical test of the nature of dark matter and of galaxy formation theory. For example, the proven presence of cusped halos can exclude some dark matter particles (Gondolo 2000). The halo density profiles and other halo relations can be used to derive the fluctuation spectrum (Ma & Fry 2000).

Maybe CDM is wrong. For example, self-interacting dark matter (e.g. matter with elastic scattering, annihilation) can give a flat central $\rho(r)$ via heat transfer into the cold inner regions. But further evolution can lead to core collapse and even steeper r^{-2} cusps: see for example Burkert (2000), Dalcanton & Hogan (2000), Kochanek & White (2000).

How about problems with the data ? There is a growing view that the observed HI rotation curves may be misleading. In particular, there is evidence that HI beam smearing, which can seriously affect the inferred halo structure, has not always been properly corrected. New high resolution optical rotation curves of some of the prototype flat-cored dwarfs rise more steeply with radius than do their HI rotation curves, and they fit better to the more cuspy halos like the NFW model. For example, see Swaters (2000) for an optical rotation curve of the prototype system DDO154. Blais-Ouellette et al. (1999) show examples of some galaxies with both optical and HI rotation curves. For the prototype spiral NGC 3198 (see Figure 2), they find that the optical rotation curve and the beam-smearing-corrected HI rotation agree well. On the other hand, for the dwarf system NGC 5585 they find that the optical rotation curve rises more steeply.

Although at first sight this steeper optical rotation curve appears to favor a cusped halo, I think the opposite conclusion may be correct. NGC 5585 is a well-known example of a system in which the shape of the (beam-smeared) HI rotation curve required a very low M/L ratio for the stellar disk: *i.e.* the contribution of the stellar disk to the rotation curve was constrained to be very small. With the higher resolution optical rotation curve, the derived M/L ratio for the stellar disk is more realistic (relative to stellar population models), and the revised halo has a *lower* $\rho_\circ$ and a *longer* core radius r_c (if the maximum disk decomposition is correct). Although the new rotation curve rises more steeply near the center, this actually exacerbates the discrepancy between the shallow cores derived from rotation curves and the steeper cusps derived from simulations (because the stellar disk can now contribute more to the inner rotation curve, so the remaining dark halo is less concentrated). See also van den Bosch et al. (2000) for further discussion of this beam-smearing problem. Optical and CO observations with higher spatial resolution are clearly valuable to constrain the inner structure of disk galaxies, but they will not necessarily determine the structure of the inner dark halos.

8. Scaling Laws for Dark Halos

How do the properties of dark halos scale with the luminosity of mass of the parent galaxy ? For example, maximum disk decompositions of rotation curves using nonsingular isothermal sphere models for the dark halo give three parameters for the halo: the central density $\rho_\circ$, the core radius r_c and the (constant) velocity dispersion σ. These three parameters are related through the usual King model convention: $4\pi G \rho_\circ r_c^2 = 9\sigma^2$.

Kormendy (1990, see also Kormendy & Freeman 2001) has compiled parameters for the halos of a sample of spiral and dwarf irregular galaxies with well-observed rotation curves. The sample also includes some of the nearby dwarf spheroidal galaxies, for which the halo parameters are derived from the stellar kinematics and a King model analysis. The halo parameters are strongly correlated with the absolute magnitude of the parent galaxy. The brighter galaxies have larger values of the core radius r_c and the velocity dispersion σ. The most striking result is that the *faintest galaxies* have the *densest* dark halos; over the 15 magnitude range of galaxies represented in the sample, the halo density $\rho_\circ$ changes by more than 2 dex, with values of $\rho_\circ$ approaching $1 M_\odot$ pc^{-3} for the faintest galaxies. This is consistent with the view that the smallest halos collapse first. Their high density reflects the density of the universe at the time they collapsed. In principle, one could estimate the redshift of their formation from the density of their dark halos. With their very dense dark halos, we would expect these low-mass galaxies to be very robust in mergers, despite their apparently fragile optical appearance.

9. Summary

• HI rotation curves are the major source of what we know about the distribution of dark matter in gas-rich galaxies
• the dark matter fraction of the total mass of disk galaxies can be as high as about 90% in systems for which the kinematical observations extend to large enough radii
• the inner structure of dark halos remains a problem. Simulations predict central cusps in the density distribution: the observations probably favor flat cores (despite recent claims to the contrary)
• the contribution of the luminous disk to the rotation curve (the maximum disk problem) remains controversial. The core/cusp problem will be difficult to resolve until this issue is sorted out
• even for dwarf and LSB galaxies, mass appears to follow light in the inner regions, but with a large range of M/L values, some higher than plausible from stellar population models. What does this mean ?

REFERENCES

ALCOCK, A. *et al.* 2000, ApJ, astro-ph/0001272.
BEGEMAN, K. 1987, Thesis, Rijkuniversiteit te Groningen.
BLAIS-OUELLETTE, S. *et al.* 1999, AJ, 118, 2123.
BLITZ, L. *et al.* 1999, ApJ, 514, 818.
BOSMA, A. 1978, Thesis, Rijkuniversiteit te Groningen.

BRAUN, R. & BURTON, W.B. 1999, A&A, 341, 437.
BUCHHORN, M. 1991, personal communication.
BURKERT, A. 2000, ApJ, 534, 143.
CARIGNAN, C., BEAULIEU, S., FREEMAN, K. 1990, AJ, 99, 178.
COURTEAU, S. & RIX, H-W. 1999, ApJ, 513, 561.
CRÉZÉ, M. et al. 1998, A&A, 329, 920.
DALCANTON, J. & HOGAN, C. 2000, astro-ph/0004381.
DEBATTISTA, V. & SELLWOOD, J. 1998, ApJ, 513, 107.
FREEMAN, K. 1999, "Low Surface Brightness Universe" (ASP Conference Series 170), (ASP, San Francisco), ed J. Davies, C. IMpey & S. Phillipps, p 3.
GATES, E. & GYUK, G. 1999, astro-ph/9911149.
GONDOLO, P. 2000, astro-ph/0002226.
IBATA, R. et al. 1999, ApJ, 524, 95.
IBATA, R. et al. 2000, ApJ, 532, L41.
KOCHANEK, C. 1996, ApJ, 457, 228.
KOCHANEK, C. & WHITE, M. 2000, astro-ph/0003483.
KORMENDY, J. 1990, ASP Conf. Ser. 10: Evolution of the Universe of Galaxies, 33.
KORMENDY, J. & FREEMAN, K. 2001, in preparation.
KRAVTSOV, A. et al. 1998, ApJ, 502, 48.
LO, K.Y. et al. 1993, AJ, 106, 507.
MA, C. & FRY 2000, astro-ph/0003343.
MCGAUGH, S. et al. 2000, ApJ, 533, 99.
MÉNDEZ, R. & MINNITI, D. 2000, Ap,J 529, 911.
MOORE, B. et al. 1999, ApJ, 524, L19.
NAVARRO, J., FRENK, C., WHITE, S. 1996, ApJ, 462, 563.
OLLING, R. P. 1995, Ph.D. Thesis, U. of Columbia.
OLLING, R. P. & MERRIFIELD, M. 2000, MNRAS, 311, 361.
PFENNIGER, D., COMBES, F. & MARTINET, L. 1994, A&A, 285, 79.
SWATERS, R. 1999, Ph.D. Thesis, Rijkuniversiteit te Groningen.
SWATERS, R. 2000, poster paper at this conference.
VAN DEN BOSCH, F. et al. 2000, AJ, 119, 1579.
ZWAAN. M., VAN DER HULST, J.M., DE BLOK, W.J.G., MCGAUGH, S. 1995, MNRAS, 273, L35.